Fractional Calculus: Theory and Applications

Fractional Calculus: Theory and Applications

Special Issue Editor

Francesco Mainardi

MDPI • Basel • Beijing • Wuhan • Barcelona • Belgrade

Special Issue Editor
Francesco Mainardi
University of Bologna
Italy

Editorial Office
MDPI
St. Alban-Anlage 66
Basel, Switzerland

This is a reprint of articles from the Special Issue published online in the open access journal *Mathematics* (ISSN 2227-7390) from 2017 to 2018 (available at: http://www.mdpi.com/journal/mathematics/special_issues/Fractional_Calculus_Theory_Applications)

For citation purposes, cite each article independently as indicated on the article page online and as indicated below:

LastName, A.A.; LastName, B.B.; LastName, C.C. Article Title. *Journal Name* **Year**, *Article Number*, Page Range.

ISBN 978-3-03897-206-8 (Pbk)
ISBN 978-3-03897-207-5 (PDF)

Contents

About the Special Issue Editor

Francesco Mainardi Is a retired professor of Mathematical Physics from the University of Bologna where he taught for 40 years (he retired in November, 2013 at the age of 70). Even in his retirement, he continues to carry out teaching and research activities. His fields of research concern several topics of applied mathematics, including diffusion and wave problems, asymptotic methods, integral transforms, special functions, fractional calculus and non-Gaussian stochastic processes. At present his h-index is more than 50. He has published more than 150 refereed papers and some books as an author or editor. His homepage is http://www.fracalmo.org/mainardi.

mathematics

MDPI

Editorial

Fractional Calculus: Theory and Applications

Francesco Mainardi

Department of Physics and Astronomy, University of Bologna, and The National Institute of Nuclear Physics (INFN), Via Irnerio, 46, I-40126 Bologna, Italy; francesco.mainardi@bo.infn.it; Tel.: +39-051-209-1068

Received: 22 July 2018; Accepted: 17 August 2018; Published: 21 August 2018

Fractional calculus is allowing integrals and derivatives of any positive order (the term fractional is kept only for historical reasons). It can be considered a branch of mathematical physics that deals with integro-differential equations, where integrals are of convolution type and exhibit mainly singular kernels of power law or logarithm type.

It is a subject that has gained considerably popularity and importance in the past few decades in diverse fields of science and engineering. Efficient analytical and numerical methods have been developed but still need particular attention.

The purpose of this Special Issue is to establish a collection of articles that reflect the latest mathematical and conceptual developments in the field of fractional calculus and explore the scope for applications in applied sciences.

The papers in this Special Issue can be divided according to the following scheme considering their main purposes:

(1) Analytical Theory
(2) Numerical Methods
(3) Applications

1. Analytical Theory

We start with a brief note by the Guest Editor Francesco Mainardi [1]: A Note on the Equivalence of Fractional Relaxation Equations to Differential Equations with Varying Coefficients. This equivalence is indeed shown for the simple fractional relaxation equation for which the solution in terms of the Mittag–Leffler function is known. This simple argument may lead to the equivalence of more general processes governed by evolution equations of fractional order with constant coefficients to processes governed by differential equations of integer order but with varying coefficients. Our main motivation is to solicit researchers to extend this approach to other areas of applied science to have a deeper knowledge of certain phenomena, both deterministic and stochastic ones, investigated nowadays with the techniques of fractional calculus.

Then, we consider two notes about the fractional Marchaud derivative from different perspectives that surely constitute a novelty in the actual literature of fractional calculus.

In the paper by Fausto Ferrari [2]: Weyl and Marchaud Derivatives: A Forgotten History, the author recalls the contribution given by Hermann Weyl and André Marchaud to the notion of fractional derivative. In addition, he discusses some relationships between the fractional Laplace operator and Marchaud derivative in the perspective to generalize these objects to different fields of the mathematics.

The aim of the paper by Sergei Rogosin and Maryna Dubatovskaya [3]: Letnikov vs. Marchaud: A Survey on Two Proinent Constructions of Fractional Derivatives, is to present the essence of two important approaches in Fractional Calculus, namely, those developed by Letnikov (or by Grünwald and Letnikov) and by Marchaud. The authors collect here the most important results for the corresponding fractional derivatives, compare these constructions and highlight their role in Fractional Calculus and its applications.

In the paper by Trifce Sandev [4]: Generalized Langevin Equation and the Prabhakar Derivative, the generalized Langevin equation is considered with regularized Prabhakar derivative operator. The author analyzes the mean square displacement, time-dependent diffusion coefficient and velocity autocorrelation function. Further, he introduces the so-called tempered regularized Prabhakar derivative and analyzes the corresponding generalized Langevin equation with friction term represented through the tempered derivative.

In the paper by Roberto Garra, Enzo Orsingher and Federico Polito [5]: A Note on Hadamard Fractional Differential Equations with Varying Coefficients and Their Applications in Probability, the authors establish a connection between some generalizations of the COM–Poisson distributions and integro-differential equations with time-varying coefficients involving Hadamard integrals or derivatives. Moreover, they suggest a new interesting application in probability of a recently introduced generalized Le Roy function (see [6]).

In the paper by Yuri Luchko [7]: On Some New Properties of the Fundamental Solution to the Multi-Dimensional Space- and Time-Fractional Diffusion-Wave Equation, the Mellin–Barnes integrals technique is employed to deduce some new analytical properties of solutions to the multi-dimensional space- and time-fractional diffusion-wave equation. Indeed, some new closed-form formulas for particular cases of the fundamental solution are derived. In particular, the author solves the open problem of the representation of the fundamental solution to the two-dimensional neutral-fractional diffusion-wave equation in terms of the known special functions.

In the paper by Khadidja and Lamine Nisse [8]: An Iterative Method for Solving a Class of Fractional Functional Differential Equations with "Maxima", the authors deal with nonlinear fractional differential equations with "maxima" and deviating arguments. The nonlinear part of the problem under consideration depends on the maximum values of the unknown function taken in time-dependent intervals. Proceeding by an iterative approach, they obtain the existence and uniqueness of the solution, in a context that does not fit within the framework of fixed-point theory methods for the self-mappings, frequently used in the study of such problems. An example illustrating their main result is also given.

2. Numerical Methods

The paper "Numerical Solution of Fractional Differential Equations: A Survey and a Software Tutorial" by Roberto Garrappa [9] aims to provide a tutorial for the numerical solution of fractional differential equations (FDEs). In particular, numerical methods for solving systems of FDEs, as well as of multi-order type (i.e., in which each equation has a different order), and multi-term FDEs (i.e., equations in which derivatives of different order appears in the same equation), are presented. Some aspects related to the efficient implementation of the methods are discussed and the corresponding MATLAB routines are made freely available.

The paper "Numerical Solution of Multiterm Fractional Differential Equations Using the Matrix Mittag–Leffler Functions" by Marina Popolizio [10] focuses on a numerical approach to solve Multiterm Fractional Differential Equations (MTFDEs), that is, equations involving derivatives of different orders. They are very common to model many important processes, particularly for multi-rate systems. The analyzed approach is based on the possibility to equivalently write MTFDEs in terms of a linear system of Fractional Differential Equations of the same order; the,n the solution is computed by means of the Mittag-Leffler function evaluated in the coefficient matrix by means of very recent tools [11]. This matrix approach turns out to be very accurate and fast, also in comparison with other numerical methods, as shown by several numerical tests presented in the paper.

The paper by Vladimir D. Zakharchenko and Ilya G. Kovalenko [12]: Best Approximation of the Fractional Semi-Derivative Operator by Exponential Series, considers the implementation of a fractional-differentiating filter of the order of $1/2$ by a set of automation astatic transfer elements, which greatly simplifies practical implementation. Real technical devices have the ultimate time delay, albeit small in comparison with the duration of the signal. As a result, the real filter will process the signal with some error. In accordance with this, this paper introduces and uses the concept of a "pre-derivative" of 1/2 of magnitude. An optimal algorithm for realizing the structure of the filter is proposed based on the criterion of minimum mean square error. Relations are obtained for the quadrature coefficients that determine the structure of the filter. This technique is shown to be useful for a significant reduction in the time required to obtain an estimate of the mean frequency of the spectrum of Doppler signals when seeking to measure the instantaneous velocity of dangerous near-Earth cosmic objects.

3. Applications

Among the many interesting applications of fractional calculus to physical systems, in this Special Issue, we find the paper devoted to the fractional viscoelasticity.

In the paper "Storage and Dissipation of Energy in Prabhakar Viscoelasticity" by Ivano Colombaro, Andrea Giusti and Silvia Vitali [13], the authors clarify some aspects of the attenuation processes emerging in a Fractional Maxwell model of viscoelasticity involving Prabhakar derivatives. On this topic we refer the reader to A. Giusti and I. Colombaro [14].

A further application related to fractional calculus is devoted to the free electron laser (FEL) and carried out by a group led by a well-known specialist on this topic (Prof. Dattoli). In the paper "Fractional Derivatives, Memory Kernels and Solution of a Free Electron Laser Volterra Type Equation" by Marcello Artioli, Giuseppe Dattoli, Silvia Licciardi and Simonetta Pagnutti [15], the authors recall that the high gain FEL equation is a Volterra type integro-differential equation amenable for analytical solutions in a limited number of cases. In this note, a novel technique, based on an expansion employing a family of two variable Hermite polynomials, is shown to provide straightforward analytical solutions for cases hardly solvable with conventional means. The possibility of extending the method by the use of expansion using different polynomials (such as two variable Legendre) expansion is also discussed.

The paper "Application of Tempered-Stable Time Fractional-Derivative Model to Upscale Subdiffusion for Pollutant Transport in Field-Scale Discrete Fracture Networks" by Bingqing Lu, Yong Zhang, Donald M. Reeves, HongGuang Sun and Chunmiao Zheng [16] aims to explore the relationship between real-world aquifer properties and non-Fickian transport dynamics. According to the authors, the fractional partial differential equations built upon fractional calculus can be reliably applied with appropriate hydro-geologic interpretations. They use the Monte Carlo approach to generate field-scale multiple discrete fracture network (DFN) flow and transport scenarios where the fracture properties change systematically, and then to simulate groundwater flow and pollutant transport through the complex DFNs. For a point source located initially in the mobile phase or fracture, the late-time behavior for the pollutant breakthrough curves (BTCs) simulated by the Monte Carlo approach is then explained by the tempered–stable time fractional advection–dispersion equation. The relationship between medium heterogeneity and transport dynamics through the combination of numerical experiments and stochastic analysis is built.

In the paper by Guoxing Lin [17]: Analysis of PFG Anomalous Diffusion via Real-Space and Phase-Space Approaches, two significantly different methods are proposed to analyze the pulsed-field gradient (PFG) anomalous diffusion: the effective phase-shift diffusion equation (EPSDE) method and a method based on observing the signal intensity at the origin. The EPSDE method describes the phase evolution in virtual phase space, while the method to observe the signal intensity at the origin describes the magnetization evolution in real space. However, these two approaches give the same general PFG signal attenuation including the finite gradient pulse width (FGPW) effect, which can be numerically evaluated by a direct integration method. The direct integration method is fast and without overflow. It is a convenient numerical evaluation method for Mittag–Leffler function-type PFG signal attenuation. The methods here provide a clear view of spin evolution under a field gradient, and their results will help the analysis of PFG anomalous diffusion.

Funding: This research received no external funding.

Conflicts of Interest: The author declares no conflict of interest.

References

1. Mainardi, F. A Note on the Equivalence of Fractional Relaxation Equations to Differential Equations with Varying Coefficients. *Mathematics* **2018**, *6*, 8. [CrossRef]
2. Ferrari, F. Weyl and Marchaud Derivatives: A Forgotten History. *Mathematics* **2018**, *6*, 6. [CrossRef]
3. Rogosin, S.; Dubatovskaya, M. Letnikov vs. Marchaud: A Survey on Two Prominent Constructions of Fractional Derivatives. *Mathematics* **2018**, *6*, 3. [CrossRef]
4. Sandev, T. Generalized Langevin Equation and the Prabhakar Derivative. *Mathematics* **2017**, *5*, 66. [CrossRef]
5. Garra, R.; Orsingher, E.; Polito, F. A Note on Hadamard Fractional Differential Equations with Varying Coefficients and Their Applications in Probability. *Mathematics* **2018**, *6*, 4. [CrossRef]
6. Garrappa, R.; Rogosin, S.; Mainardi, F. On a Generalized Three-Parameter Wright Function of Le Roy Type. *Fract. Calc. Appl. Anal.* **2017**, *20*, 1196–1215. [CrossRef]
7. Luchko, Y. On Some New Properties of the Fundamental Solution to the Multi-Dimensional Space- and Time-Fractional Diffusion-Wave Equation. *Mathematics* **2017**, *5*, 76. [CrossRef]
8. Nisse, K.; Nisse, L. An Iterative Method for Solving a Class of Fractional Functional Differential Equations with "Maxima". *Mathematics* **2017**, *6*, 2. [CrossRef]
9. Garrappa, R. Numerical Solution of Fractional Differential Equations: A Survey and a Software Tutorial. *Mathematics* **2018**, *6*, 16. [CrossRef]
10. Popolizio, M. Numerical Solution of Multiterm Fractional Differential Equations Using the Matrix Mittag–Leffler Functions. *Mathematics* **2018**, *6*, 7. [CrossRef]
11. Garrappa, R.; Popolizio, M. Computing the Matrix Mittag-Leffler Function with Applications to Fractional Calculus. *J. Sci. Comput.* **2018**. [CrossRef]
12. Zakharchenko, V.; Kovalenko, I. Best Approximation of the Fractional Semi-Derivative Operator by Exponential Series. *Mathematics* **2018**, *6*, 12. [CrossRef]
13. Colombaro, I.; Giusti, A.; Vitali, S. Storage and Dissipation of Energy in Prabhakar Viscoelasticity. *Mathematics* **2018**, *6*, 15. [CrossRef]
14. Giusti, A.; Colombaro, I. Prabhakar-like fractional viscoelasticity. *Commun. Nonlinear Sci. Numer. Simul.* **2018**, *56*, 138–143. [CrossRef]
15. Artioli, M.; Dattoli, G.; Licciardi, S.; Pagnutti, S. Fractional Derivatives, Memory Kernels and Solution of a Free Electron Laser Volterra Type Equation. *Mathematics* **2017**, *5*, 73. [CrossRef]
16. Lu, B.; Zhang, Y.; Reeves, D.; Sun, H.; Zheng, C. Application of Tempered-Stable Time Fractional-Derivative Model to Upscale Subdiffusion for Pollutant Transport in Field-Scale Discrete Fracture Networks. *Mathematics* **2018**, *6*, 5. [CrossRef]
17. Lin, G. Analysis of PFG Anomalous Diffusion via Real-Space and Phase-Space Approaches. *Mathematics* **2018**, *6*, 17. [CrossRef]

mathematics

MDPI

Article

A Note on the Equivalence of Fractional Relaxation Equations to Differential Equations with Varying Coefficients

Francesco Mainardi

Department of Physics and Astronomy, University of Bologna, and the National Institute of Nuclear Physics (INFN), Via Irnerio, 46, I-40126 Bologna, Italy; francesco.mainardi@bo.infn.it; Tel.: +39-0512091068

Received: 14 December 2017; Accepted: 5 January 2018; Published: 9 January 2018

Abstract: In this note, we show how an initial value problem for a relaxation process governed by a differential equation of a non-integer order with a constant coefficient may be equivalent to that of a differential equation of the first order with a varying coefficient. This equivalence is shown for the simple fractional relaxation equation that points out the relevance of the Mittag–Leffler function in fractional calculus. This simple argument may lead to the equivalence of more general processes governed by evolution equations of fractional order with constant coefficients to processes governed by differential equations of integer order but with varying coefficients. Our main motivation is to solicit the researchers to extend this approach to other areas of applied science in order to have a deeper knowledge of certain phenomena, both deterministic and stochastic ones, investigated nowadays with the techniques of the fractional calculus.

Keywords: Caputo fractional derivatives; Mittag–Leffler functions; anomalous relaxation

MSC: 26A33; 33E12; 34A08; 34C26

1. Introduction

Let us consider the following relaxation equation

$$\frac{d\Psi}{dt} = -r(t)\,\Psi(t)\,, \quad t \geq 0\,, \tag{1}$$

subjected to the initial condition, for the sake of simplicity,

$$\Psi(0^+) = 1\,, \tag{2}$$

where $\Psi(t)$ and $r(t)$ are positive functions, sufficiently well-behaved for $t \geq 0$. In Equation (1), $\Psi(t)$ denotes a non-dimensional field variable and $r(t)$ the varying relaxation coefficient.

The solution of the above initial value problem reads

$$\Psi(t) = \exp[-R(t)]\,, \quad R(t) = \int_0^t r(t')\,dt' > 0\,. \tag{3}$$

It is easy to recognize from re-arranging Equation (1) that, for $t \geq 0$,

$$r(t) = -\frac{\Phi(t)}{\Psi(t)}\,, \quad \Phi(t) = \Psi^{(1)}(t) = \frac{d\Psi}{dt}(t)\,. \tag{4}$$

The solution (3) can be derived by solving the initial value problem by separation of variables

$$\int_1^{\Psi(t)} \frac{d\Psi(t')}{\Psi(t')} = \int_0^t \frac{\Phi(t')}{\Psi(t')}\, dt' = -\int_0^t r(t')\, dt' = R(t). \tag{5}$$

>From Equation (3), we also note that

$$R(t) = -\log[\Psi(t)]\,. \tag{6}$$

As a matter of fact, we have shown well-known results that will be relevant for the next sections.

2. Mittag–Leffler Function as a Solution of the Fractional Relaxation Process

Let us now consider the following initial value problem for the so-called fractional relaxation process

$$\begin{cases} {}_*D_t^\alpha \Psi_\alpha((t) = -\Psi_\alpha(t), & t \geq 0, \\ \Psi_\alpha(0^+) = 1, \end{cases} \tag{7}$$

with $\alpha \in (0,1]$. Above, we have labeled the field variable with Ψ_α to point out its dependence on α and considered the *Caputo fractional derivative*, defined as:

$$ {}_*D_t^\alpha \Psi_\alpha(t) = \begin{cases} \dfrac{1}{\Gamma(1-\alpha)} \displaystyle\int_0^t \frac{\Psi_\alpha^{(1)}(t')}{(t-t')^\alpha}\, dt', & 0 < \alpha < 1, \\ \dfrac{d}{dt}\Psi_\alpha(t)\,, & \alpha = 1. \end{cases} \tag{8}$$

As found in many treatises of fractional calculus, and, in particular, in the 2007 survey paper by Mainardi and Gorenflo [1] to which the interested reader is referred for details and additional references, the solution of the fractional relaxation problem (7) can be obtained by using the technique of the Laplace transform in terms of the Mittag–Leffler function. Indeed, we get in an obvious notation by applying the Laplace transform to Equation (7)

$$s^\alpha \widetilde{\Psi}_\alpha(s) - s^{\alpha-1} = -\widetilde{\Psi}_\alpha(s), \quad \text{hence} \quad \widetilde{\Psi}_\alpha(s) = \frac{s^{\alpha-1}}{s^\alpha + 1}, \tag{9}$$

so that

$$\Psi_\alpha(t) = E_\alpha(-\,t^\alpha) = \sum_{n=0}^{\infty} (-1)^n \frac{t^{\alpha n}}{\Gamma(\alpha n + 1)}\,. \tag{10}$$

For more details on the Mittag–Leffler function, we refer to the recent treatise by Gorenflo et al. [2]. In Figure 1, for readers' convenience, we report the plots of the solution (10) for some values of the parameter $\alpha \in (0,1]$.

It can be noticed that, for $\alpha \to 1^-$, the solution of the initial value problem reduces to the exponential function $\exp(-t)$ with a singular limit for $t \to \infty$ because of the asymptotic representation for $\alpha \in (0,1)$,

$$E_\alpha(-\,t^\alpha) \sim \frac{t^{-\alpha}}{\Gamma(-\alpha+1)}, \quad t \to \infty\,. \tag{11}$$

Now, it is time to carry out the comparison between the two initial value problems described by Equations (1) and (7) with their corresponding solutions (3), (10). It is clear that we must consider the derivative of the Mittag–Leffler function in (10), namely

$$\Phi_\alpha(t) = \frac{d}{dt}\Psi_\alpha(t) = \frac{d}{dt}E_\alpha(-t^\alpha) = -t^{\alpha-1}\, E_{\alpha,\alpha}(-t^\alpha)\,. \tag{12}$$

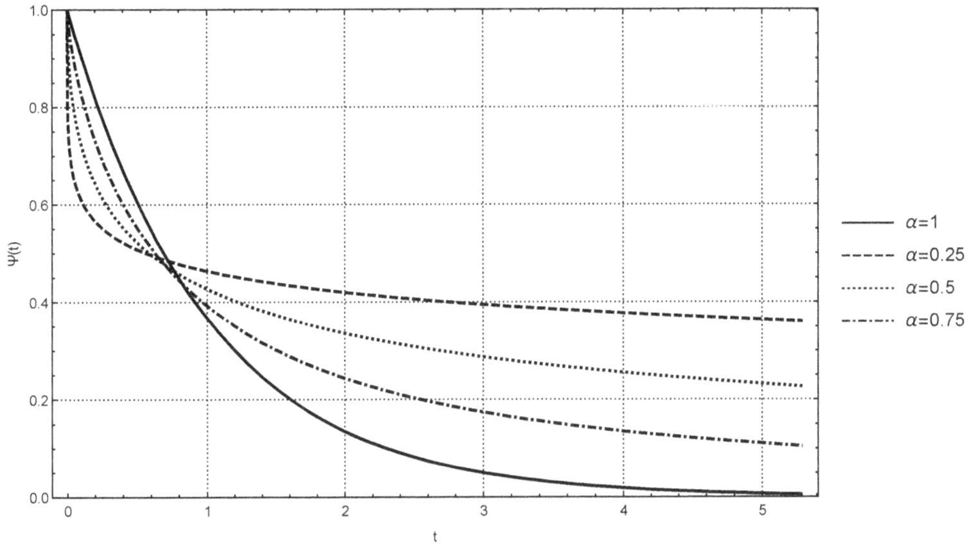

Figure 1. Plots of the Mittag–Leffler function $\Psi_\alpha(t)$ for $\alpha = 0.25, 0.50, 0.75, 1$ versus $t \in [0, 5]$.

In Figure 2, we show the plots of positive function $-\Phi_\alpha(t)$ for some values of $\alpha \in (0, 1]$.

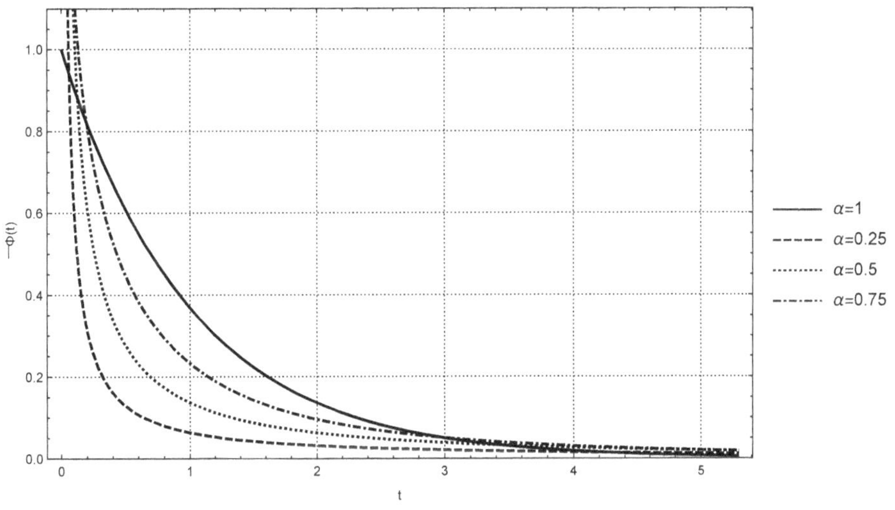

Figure 2. Plots of the positive function $-\Phi_\alpha(t)$ for $\alpha = 0.25, 0.50, 0.75, 1$ versus $t \in [0, 5]$.

The above discussion leads to the varying relaxation coefficient of the equivalent ordinary relaxation process:

$$r_\alpha(t) = -\frac{\Phi_\alpha(t)}{\Psi_\alpha(t)} \Longrightarrow r_\alpha(t) = \frac{t^{\alpha-1} E_{\alpha,\alpha}(-t^\alpha)}{E_\alpha(-t^\alpha)} \sim \begin{cases} \frac{t^{\alpha-1}}{\Gamma(\alpha)}, & t \to 0^+, \\ \alpha t^{-1}, & t \to +\infty. \end{cases} \tag{13}$$

Figure 3 depicts the plots of $r_\alpha(t)$ for some rational values of α, including the standard case $\alpha = 1$, in which the ratio reduces to the constant 1.

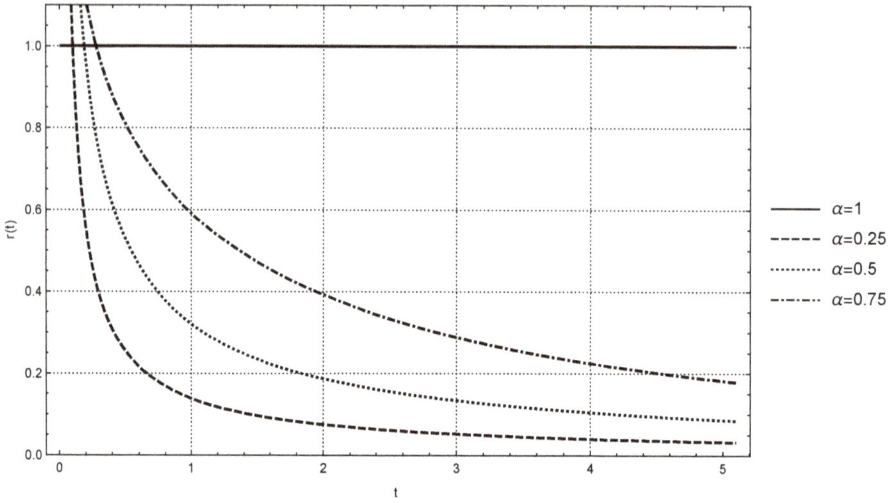

Figure 3. Plots of the ratio $r_\alpha(t)$ for $\alpha = 0.25, 0.50, 0.75, 1$ versus $t \in [0, 5]$.

We conclude by plotting in Figure 4 the function $R_\alpha(t) = -\log[\Psi_\alpha(t)]$ for some values of the parameter $\alpha \in (0, 1]$.

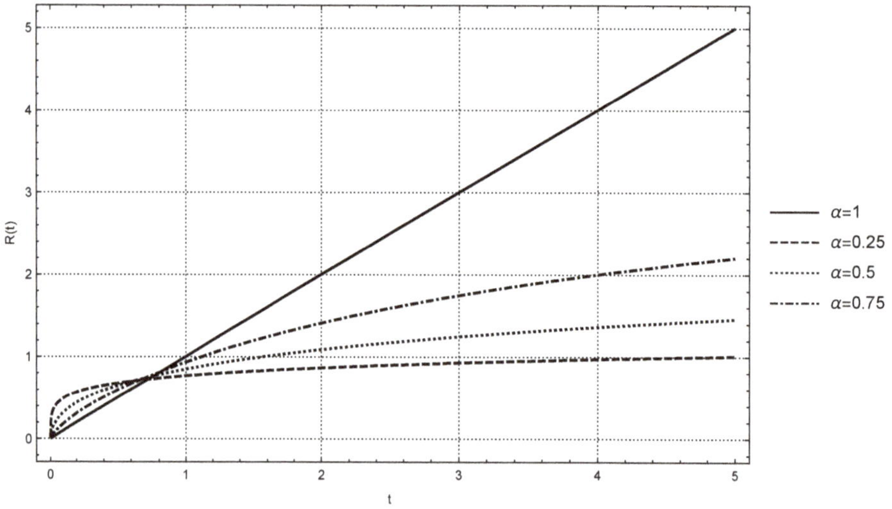

Figure 4. Plots of $R_\alpha(t) = -\log[\Psi_\alpha(t)]$ for $\alpha = 0.25, 0.5, 0.75, 1$ versus $t \in [0, 5]$.

3. Conclusions

In this note, we have shown how the fractional relaxation process governed by a fractional differential equation with a constant coefficient is equivalent to a relaxation process governed by an ordinary differential equation with a varying coefficient. These considerations provide a different look at this fractional process over all for experimentalists who can measure the varying relaxation coefficient versus time. Indeed, if this coefficient is found to fit the analytical or asymptotical expressions in Label (13), the researcher cannot distinguish if the governing equation is fractional or simply ordinary. To make the difference, we thus need other experimental results.

We are convinced that it is possible to adapt the above reasoning to other fractional processes, including anomalous relaxation in viscoelastic and dielectric media and anomalous diffusion in complex systems. This extension is left to perceptive readers who can explore these possibilities.

Last but not least, we do not claim to be original in using the above analogy in view of the great simplicity of the argument: for example, a similar procedure has recently been used by Sandev et al. [3] in dealing with the fractional Schrödinger equation.

Acknowledgments: The author is very grateful to Leonardo Benini, a Master's student in Physics (University of Bologna), for his valuable help in plotting. As a matter of fact, he has used the MATLAB routine for the Mittag–Leffler function appointed by Roberto Garrappa (see https://it.mathworks.com/matlabcentral/fileexchange/48154-the-mittag-leffler-function). The author would like to devote this note to the memory of the late Rudolf Gorenflo (1930–2017), with whom for 20 years he had published joint papers. The author presumes that this note is written in the spirit of Gorenflo, being based on the simpler considerations. This work has been carried out in the framework of the activities of the National Group of Mathematical Physics (GNFM-INdAM). Furthermore, the author has appreciated constructive remarks and suggestions of the anonymous referees that helped to improve this note.

Conflicts of Interest: The author declares no conflict of interest.

References

1. Mainardi, F.; Gorenflo, R. Time-fractional derivatives in relaxation processes: A tutorial survey. *Fract. Calc. Appl. Anal.* **2007**, *10*, 269–308.
2. Gorenflo, R.; Kilbas, A.A.; Mainardi, F.; Rogosin, S. *Mittag–Leffler Functions, Related Topics and Applications*; Springer: Berlin/Heidelberg, Germany, 2014.
3. Sandev, T.; Petreska, I.; Lenzi, E.K. Effective potential from the generalized time-dependent Schrödinger equation. *Mathematics* **2016**, *4*, 59–68.

mathematics

MDPI

Article

Weyl and Marchaud Derivatives: A Forgotten History †

Fausto Ferrari

Dipartimento di Matematica, Università di Bologna, Piazza di Porta S. Donato 5, 40126 Bologna, Italy; fausto.ferrari@unibo.it; Tel.: +39-051-2094471

† Dedicated to Sandro Salsa for his 68th birthday.

Received: 22 November 2017; Accepted: 29 December 2017; Published: 3 January 2018

Abstract: In this paper, we recall the contribution given by Hermann Weyl and André Marchaud to the notion of fractional derivative. In addition, we discuss some relationships between the fractional Laplace operator and Marchaud derivative in the perspective to generalize these objects to different fields of the mathematics.

Keywords: fractional derivatives; Grünwald–Letnikov derivative; Weyl derivative; Marchaud derivative; fractional Laplace operator; extension operator

1. Introduction

Exactly one century ago, while we are writing, in 1917, a paper by Hermann Weyl, *Bemerkungen zum Begriff des Differentialquotienten gebrochener Ordnung*, appeared, [1]. It dealt with the definition of a fractional derivative in a weaker sense with respect to the approach classically known at that time with the name of Riemann–Liouville derivative.

Ten years later, in 1927, the thesis of a misunderstood French mathematician, Adré Paul Weyl, was published, who discussed at the age of forty his PhD work entitled *Sur les dérivées et sur les différences des fonctions de variables réelles*, [2].

In [3], the names Weyl and Marchaud appear associated with the notion of fractional derivative more than two hundred times. Nevertheless, in my opinion, the name Marchaud is not so popular even among the mathematicians dealing with fractional calculus, in particular among scientists coming from Western countries. Due to the huge quantity of papers dealing with fractional subjects, my previous statement could appear debatable. In any case, this opinion can be tested just consulting, for instance a database. We tried, for instance, with the American Mathematical Society database MathShiNet. In fact, inserting the keyword "Marchaud" anywhere, we obtain around two hundred files. Among these two hundred files, improving the request by also searching the word Marchaud in the titles of the papers, we find around fifty files. In addition, by reading these titles, covering for example the last twenty years, we realize at a first glance that the frequency of mathematicians from Eastern countries is prevalent. Indeed, on the contrary of what we stated about Western mathematicians, Marchaud's name is recurrent in fractional calculus literature and among mathematicians coming from Eastern Europe, let us recall one more time the number of citations that appear in [3].

Concerning Hermann Weyl, of course, we are considering a very popular mathematician for many other mathematical reasons. Nevertheless, we have to say that also in this case Weyl's name is not usually associated with the fractional calculus even if the specialists in the field are aware of the importance of his contribution in fractional calculus. It could be interesting to understand whether many of them know why a fractional derivative is entitled to him, but this is another story.

For different reasons, the authors of the two cited papers will not publish any more results explicitly amenable to the fractional derivative. Thus, accepting the previous interpretation, they appear as isolated points in the *mare magnum* of the fractional calculus, where the more popular names are nowadays others.

In Marchaud's doctoral thesis, see p. 47, Section 27 Formula (23) in [2], or the definition (23) in the published paper [4] at p. 383, he defined the following fractional differentiation for sufficiently regular real functions $f : (0,1) \to \mathbb{R}$ extended with 0 for $x \leq 0$, whenever $\alpha \in (0,1)$:

$$\mathbf{D}^{\alpha} f(x) = \frac{\alpha}{\Gamma(1-\alpha)} \int_0^{+\infty} \frac{f(x) - f(x-t)}{t^{1+\alpha}} dt.$$

This definition can be easily given for a function defined in all of $\mathbb{R}$ and for every $\alpha \in (0,1)$ distinguishing two types of derivative, see [3], respectively from the right and from the left:

$$\mathbf{D}^{\alpha}_{+} f(x) = \frac{\alpha}{\Gamma(1-\alpha)} \int_0^{+\infty} \frac{f(x) - f(x-t)}{t^{1+\alpha}} dt$$

and

$$\mathbf{D}^{\alpha}_{-} f(x) = \frac{\alpha}{\Gamma(1-\alpha)} \int_0^{+\infty} \frac{f(x) - f(x+t)}{t^{1+\alpha}} dt.$$

The construction of these operators will be briefly described in the next section following the original motivation contained in Marchaud's thesis.

The problem of giving a coherent definition of derivative of a function for all positive real numbers has a long history—for instance, see [3,5–8] for some detailed information. In any case, Abel's contribution for solving the tautochrone problem, [9], and the work by Liouville [10] and Riemann [11] in application to geometry are fundamental and well known at the beginning of the fractional calculus. Many other authors have written papers that contributed to improving the knowledge of this subject. Nevertheless, I think that a very special role has to be recognized to Hermann Weyl because, probably following the path traced by Riemann, as Weyl himself writes in [1], he introduced, maybe first, the nonlocal operator that is known as Marchaud derivative, for people who know it, in a significative, even if particular, case. We shall dedicate Section 5 to this aspect.

Our interest to this subject comes out after the celebrated contribution given by the paper [12]. Indeed, the authors developed an idea that was already contained in [13]. In any case, in [12], the authors dedicated their interest to a different type of nonlocal operator with respect to the fractional derivative: the fractional Laplace operator. In particular, in [12], a different perspective in the interpretation of the nonlocal operators was introduced using a method based on an *extension* approach (see also [13]). We do not want to bore the reader too much with this subject. However, some words are in order. Heuristically, following the extension approach, idea it is possible to deduce the properties of a nonlocal operator from the ones of a local operator. In [12], the authors were concerned with the fractional Laplace operator, while the local operator obtained after the extension construction was a degenerate elliptic operator in divergence form. This approach can be developed considering the solution of ad hoc Dirichlet problem formulated in an unbounded set, where an auxiliary variable has been added, and then taking the limit of a weighted normal derivative of the solution of the Dirichlet problem, when this auxiliary variable vanishes. The scientific follow up of [12] produced an enormous amount of papers. Moreover, in [14], such an idea was generalized considering an abstract approach in a very powerful way. Following this stream of ideas, in [15], an intrinsic characterization of the fractional sub-Laplace operators in Carnot groups was obtained. Roughly speaking, the operators considered in this last case are sums of squares of smooth vector fields satisfying the Hörmander condition in a non-commutative structure.

The approach described in [12,14], and then in [15], was also extended to the case of fractional operators in [16] and independently also, as very often it happens when the time is ripe, in [17]. Indeed, with this aim, commenting for instance [17], we faced the problem of defining the Marchaud derivative via an extension approach in order to obtain a Harnack inequality for solutions of homogeneous fractional equations. As a consequence of this research, we realized in particular that Marchaud derivative and Weyl derivative have been, in a sense, perhaps a little put aside in the last time, especially considering the great development and the large popularity that research about nonlocal operators has recently had. This last remark is essentially based on the popularity of other fractional derivatives, for instance the Riemann–Liouville derivative or the Caputo derivative (see [7,18,19]) for a modern approach to these operators. Indeed, see also [20] for a recent example of application involving Caputo derivative.

On the other hand, by reading the monumental opera [3], it is possible to verify, as we pointed out at the beginning of this introduction, that Weyl and Marchaud names are cited many times. Thus, the curiosity of explaining this situation was strong. Why do only few people associate Weyl and Marchaud names to the fractional subject? More precisely, why do only few people utilize these fractional derivatives for applications, simply preferring other definitions, even if it appears natural to use Weyl and Marchaud operators? We do not have any conclusive reply. In any case, it is quite difficult to understand what the true motivations of this apparent amnesia are. Of course, the specialists of the fractional calculus know Marchaud and Weyl derivatives in reading in particular [3] where the right tribute to both of these mathematicians has been given. Perhaps only recently a new awakened interest about these definitions has spread out. For this reason we think, hopefully, that the contribution of this paper might be useful in consolidating this new trend.

Anyhow, a partial reply to the previous questions, partially related with beginning of these events, can be found in the social contour that strongly influenced the life of the two mathematicians. The period during which this research was developed was very uproarious for Europe. Weyl's paper [1] dated back to 1917, Marchaud's thesis [2] was published in 1927 and both the lives of these two people were, for different reasons, affected by the two world wars events (see e.g., [21] for some biography details about Weyl's life and [22,23] for some information about Marchaud).

In this paper, we want to analyze the definition of fractional derivative given by Weyl and Marchaud, concentrating on those aspects that, in perspective, seem to be more flexible for generalizing to other situations the notion of nonlocal operator (see e.g., [24] for facing the case of the semigroup approach in its abstract generality and then for recalling the contribution given in [14], for fractional Laplace operators, and then recalling [25] for several generalizations in an abstract approach, including in principle: Riemannian manifolds, Lie groups, infinite dimensions and non-symmetric operators, see also [15] for the particular case of Carnot groups). We would also like to point out some recent research establishing few relationships between Marchaud derivative and some nonlocal operators in non-commutative structures (see [26]).

With this aim, we summarize the plan of the paper. After this introduction, the reader can find in Section 2 some basic biographic information about Weyl and Marchaud. In Section 3, we discuss briefly how Marchaud came to define his derivative followed by reviewing a part of his Ph.D. thesis. In Section 4, we recall the basic idea already developed earlier by Grünwald and Letnikov that is at the base of the fractional derivative given by Weyl and Marchaud. In Section 5, we recall the seminal Weyl's paper [1] discussing some details about the relationship between his contribution and Marchaud's derivative. In Section 6, we comment the basic ideas of the respective definitions. We face the modern general setting of Marchaud derivative in Section 7 also making some remarks about its properties with respect to partial differential equations. In Section 8, we continue our work by recalling the definition of fractional Laplace operator, while, in Section 9, we deal with the definition of Marchaud derivative via an extension approach and eventually, in Section 10, we conclude our effort making evident the relationship between the fractional Laplace operator and the Marchaud derivative.

In order to outline a few aspects of Weyl's and Marchaud's biographies, we list below only some key facts of the period during which the results about fractional derivative were written. For further curiosities or remarks, we suggest consulting [21–23].

Closing this introduction, we remark that, from a historical point of view, it would be interesting to deepen our knowledge of these two characters of the mathematical world, especially considering the influence and the role of the respective mathematical schools compared to the other mathematicians of their time and their scientific legacy.

2. Short Historical Placement

In this section, we introduce some information about the lives of Weyl and Marchaud, mainly regarding the period of publication of their papers on fractional derivative without pretending to consider this parallel description exhaustive.

Hermann Weyl was born in Germany in 1885. André Paul Marchaud was born in France in 1887.

In 1913, Weyl was professor at the ETH (Swiss Federal Institute of Technology) in Zürich where he interacted also with Einstein. In 1915, Weyl was called up for military service in Germany, but, since 1916, he was exempted from military duties for health reasons. Later on, he came back to Germany as a successor of Hilbert in Göttingen, but, in 1933, he left to go to the *Institute for Advanced Study* in Princeton, escaping from the Nazi regime, where he continued his brilliant career (see e.g., [21] until he died in Zürich in 1955).

In 1913, Marchaud had not gotten his PhD thesis yet, probably because of his health problems. He was professor in a lyceum when he was mobilized by the French army in 1914. In the same year, Marchaud was taken prisoner. He stayed in an *Oflag* (a prison camp for officers only) from 1914 to 1918, at the beginning in Germany and then, thanks to the help of the Red Cross who intervened because he was ill, since 1917 in Switzerland. Marchaud discussed his PhD thesis later on, only in 1927. He continued his career mainly serving as *Rector* (provost) of French universities, even during the Nazi occupation of France in the Second World War, until 1957 (see [22]), when he retired. He died in Paris in 1973.

3. The Marchaud Approach

As we have already announced, in this section, we represent the Marchaud approach following the main steps of a part of his PhD thesis.

The Liouville–Riemann integral of order $\alpha > 0$ of a function $f : [a,b] \to \mathbb{R}$ is defined as:

$$I_a^{(\alpha)} f(x) = \frac{1}{\Gamma(\alpha)} \int_0^{x-a} t^{\alpha-1} f(x-t)dt.$$

In this case, the derivative of order $\alpha < n$ where n is a positive integer is defined by

$$\mathcal{D}_a^\alpha f(x) = D^n I_a^{n-\alpha} f(x).$$

In fact, this definition is well posed because it is independent to n. In particular, it is coherent since the following fundamental identity holds for all the functions in $L^1([a,b])$ that are bounded, for every $\alpha, \alpha' > 0$:

$$I_a^{(\alpha)}[I_a^{(\alpha')} f(x)] = I_a^{(\alpha+\alpha')} f(x).$$

The point of starting by Liouville, as Marchaud observed in [4], is that, if $a = -\infty$, then

$$\mathcal{D}_{-\infty}^\alpha e^{kx} = k^\alpha e^{kx}.$$

Moreover, for $\beta \geq 0$, it results that

$$I_a^\beta(1) = \frac{(x-a)^\beta}{\Gamma(\beta+1)},$$

and in particular $I_a^0(1) = 1$, so that, for every $\beta \geq 0$, and for every $\alpha > 0$, we get

$$\mathcal{D}_a^\alpha I_a^\beta(1) = D^{(n)} I_a^{n-\alpha} I_a^\beta(1) = D^{(n)} I_a^{n-\alpha+\beta} 1 = D^{(n)} \frac{(x-a)^{n-\alpha+\beta}}{\Gamma(n-\alpha+\beta+1)} = \frac{(x-a)^{\beta-\alpha}}{\Gamma(\beta-\alpha+1)}.$$

Hence, for $\beta = 0$,

$$\mathcal{D}_a^\alpha(1) = \mathcal{D}_a^\alpha I_a^0(1) = \frac{(x-a)^{-\alpha}}{\Gamma(-\alpha+1)}.$$

As a consequence, the fractional derivative of order α for α non-integer is in general infinite for $x = a$.

Trying to define the fractional derivative as the fractional integral of negative order α, we obtain a divergent integral. In fact, formally, we should obtain

$$I_a^{(-\alpha)} f(x) = \frac{1}{\Gamma(-\alpha)} \int_0^{x-a} t^{-\alpha-1} f(x-t) dt.$$

Then, Marchaud argues in this way. Taking the integral of order α and assuming to consider the function extended with 0 from $-\infty$ to a, we get that

$$I_{-\infty}^{(\alpha)} f(x) \Gamma(-\alpha) = \int_0^\infty t^{\alpha-1} f(x-t) dt,$$

that is, making clear the definition of Γ,

$$I_{-\infty}^{(\alpha)} f(x) \int_0^\infty t^{\alpha-1} e^{-t} dt = \int_0^\infty t^{\alpha-1} f(x-t) dt.$$

The same formula holds for every positive integer k, so that performing a change of variable like $t = ks$ in both the integrals we get:

$$I_{-\infty}^{(\alpha)} f(x) k \int_0^\infty (ks)^{\alpha-1} e^{-ks} ds = \int_0^\infty (ks)^{\alpha-1} f(x-ks) k ds,$$

which implies

$$I_{-\infty}^{(\alpha)} f(x) \int_0^\infty s^{\alpha-1} e^{-ks} ds = \int_0^\infty s^{\alpha-1} f(x-ks) ds.$$

Then, taking a linear combination od order $p+1$ for a finite sequence of integer positive decreasing number $\{k_i\}_{0 \le i \le p}$, we obtain summing terms by terms

$$I_{-\infty}^{(\alpha)} f(x) \int_0^\infty s^{\alpha-1} \psi(s) ds = \int_0^\infty s^{\alpha-1} \varphi(x,s) ds,$$

where

$$\psi(s) = \sum_{i=0}^p C_i e^{-k_i s}, \quad \varphi(s) = \sum_{i=0}^p C_i f(x-k_i s),$$

and $\{C_i\}_{1 \le 0 \le p} \subset \mathbb{R}$. At this point, Marchaud asks that passing to negative exponent $-\alpha$ the following relation makes sense

$$I_{-\infty}^{(-\alpha)} f(x) \int_0^\infty s^{-\alpha-1} \psi(s) ds = \int_0^\infty s^{-\alpha-1} \varphi(x,s) ds,$$

calling the function $I_{-\infty}^{(-\alpha)} f(x)$ the fractional derivative of order α, that is $\mathcal{D}^\alpha f(x)$ is implicitly defined by

$$\mathcal{D}^\alpha f(x) \int_0^\infty s^{-\alpha-1} \psi(s) ds = \int_0^\infty s^{-\alpha-1} \varphi(x,s) ds,$$

supposing that it is possible to choose ψ in such a way $\gamma(\alpha) := \int_0^\infty s^{-\alpha-1} \psi(s) ds$ does not vanish, and, as a consequence, obtains the expression of φ. Discussing this problem, Marchaud finds that if it is possible to find ψ and φ with previous properties, then

$$\gamma(\alpha)\mathcal{D}^{\alpha}f(x) = \int_0^{\infty} s^{-\alpha-1}\varphi_{\alpha}(x,s)ds,$$

where

$$\varphi_{\alpha}(x,s) = \sum_{i=0}^{p} C_{k_i} f(x - k_i s),$$

with a possible choice for ψ that it is given by

$$\psi(t) = e^{-t}(1-e^{-t})^p = \sum_{j=0}^{p}(-1)^j\binom{p}{j}e^{(-1-j)t}$$

and

$$\varphi_{\alpha}(x,s) = \sum_{i=1}^{p}(-1)^{j-1}\binom{p}{j}f(x-js).$$

After a detailed computation, Marchaud concludes that the existence of the fractional derivative $\mathcal{D}^{\alpha}f(x)$ continuous for continuous functions defined in (a,b) is equivalent to the uniform convergence of the following integral

$$\int_{\epsilon}^{\infty} s^{-1-\alpha}\varphi(x,s)ds$$

in every interval $(a',b) \subset (a,b)$ as $\epsilon \to 0^+$, and it is independent from the choice of the positive numbers $\{k_i\}_{1\leq i\leq p}$. At this point Marchaud defines the fractional derivative of order $\alpha < p$ of a function defined in all of $\mathbb{R}$ implicitly:

$$\mathbf{D}^{\alpha}f(x)\int_0^{\infty} s^{-1-\alpha}e^{-s}(1-e^{-s})^p ds = \int_0^{\infty} s^{-\alpha-1}\sum_{j=1}^{p}(-1)^{j-1}\binom{p}{j}f(x-js)ds,$$

or taking $\psi(t) = (1-e^{-t})^p - (1-e^{-2t})$, it is possible to obtain

$$\mathbf{D}^{\alpha}f(x)\int_0^{\infty} s^{-1-\alpha}\left((1-e^{-s})^p - (1-e^{-2s})^p\right)ds = \int_0^{\infty} s^{-\alpha-1}(\Delta^p_{-s}f(x) - \Delta^p_{-2s}f(x))ds,$$

where

$$\Delta^p_{-s}f(x) = \sum_{j=0}^{p}(-1)^j\binom{p}{j}f(x-(p-j)s).$$

Hence, separating the integral, remarking that

$$\int_0^{\infty} s^{-1-\alpha}(1-e^{-2s})^p ds = 2^{\alpha}\int_0^{\infty} t^{-1-\alpha}(1-e^{-t})^p dt,$$

and

$$\int_0^{\infty} s^{-\alpha-1}\Delta^p_{-2s}f(x)ds = 2^{\alpha}\int_0^{\infty} t^{-\alpha-1}\Delta^p_{-t}f(x)dt,$$

we obtain

$$(1-2^{\alpha})\mathbf{D}^{\alpha}f(x)\int_0^{\infty} s^{-1-\alpha}(1-e^{-s})^p ds = (1-2^{\alpha})\int_0^{\infty} s^{-\alpha-1}\Delta^p_{-s}f(x)ds.$$

As a consequence, we also obtain this representation

$$\mathbf{D}^{\alpha}f(x)\int_0^{\infty} s^{-1-\alpha}(1-e^{-s})^p ds = \int_0^{\infty} s^{-\alpha-1}\Delta^p_{-s}f(x)ds. \tag{1}$$

4. Grünwald–Letnikov Derivative

It is impossible to deal with fractional Marchaud derivative without recalling the contribution of Grünwald and Letnikov (see [27,28]). In fact, for giving a different perspective of the Marchaud derivative, we have to introduce the Grünwald–Letnikov derivative (Indeed, from this point of view, after we had completed this manuscript, Francesco Mainardi pointed out the survey paper [29] dedicated to Marchaud and Grünwald–Letnikov derivatives). To do this, we need some new notation.

We recall that the binomial coefficients can be defined for every $\alpha \in \mathbb{C}$ and $n \in \mathbb{N} \cup \{0\}$ as:

$$\binom{\alpha}{0} = 1, \quad \binom{\alpha}{n} = \frac{\alpha(\alpha-1)\cdots(\alpha-n+1)}{n!} = \frac{(-1)^n(-\alpha)_n}{n!}, \quad n \in \mathbb{N}. \tag{2}$$

It is also true that

$$\binom{\alpha}{n} = \frac{\Gamma(\alpha+1)}{n!\Gamma(\alpha+n-1)}$$

for $\alpha \in \mathbb{C} \setminus -\mathbb{N}$ and $n \in \mathbb{N}$.

We introduce now the following notation concerning the difference of fractional order $\alpha \in \mathbb{R}$ for a function f as follows. Let us denote

$$(\Delta_h^\alpha f)(x) = \sum_{k=0}^{\infty} (-1)^k \binom{\alpha}{k} f(x-kh).$$

We are now in position to define the Grünwald–Letnikov fractional derivative (see [3,27,28]). Let $\alpha \in (0,1)$ be fixed and let $f : \mathbb{R} \to \mathbb{R}$ be a given function. The Grünwald–Letnikov derivative of order α of f is defined, separating the two cases, respectively as:

$$f_+^{\alpha)}(x) = \lim_{h \to 0^+} \frac{(\Delta_h^\alpha f)(x)}{h^\alpha}$$

and

$$f_-^{\alpha)}(x) = \lim_{h \to 0^+} \frac{(\Delta_{-h}^\alpha) f(x)}{h^\alpha},$$

whenever the limit exists.

In order to understand better the reason of this definition, we introduce the following definition.

Definition 1. *We define a non-centered difference of increment h on* $f : \mathbb{R} \to \mathbb{R}$, *as*

$$(I - \tau^{-t}) f(x) = f(x) - f(x-t).$$

Then, we obtain for every $m \in \mathbb{N}$ so that

$$(I - \tau^{-t})^m = \sum_{k=0}^{m} (-1)^k \binom{m}{k} (\tau^{-t})^k$$

and

$$(I - \tau^{-t})^m f(x) = \sum_{k=0}^{m} (-1)^k \binom{m}{k} (\tau^{-t})^k f(x) = \sum_{k=0}^{m} (-1)^k \binom{m}{k} f(x-kt).$$

On the other hand, taking the Taylor expansion of the function $t \to (1+t)^\alpha$ in the center $t_0 = 0$ and $\alpha \in (0,1)$, we get

$$(1+t)^\alpha = \sum_{k=0}^{+\infty} \binom{\alpha}{k} t^k,$$

where

$$\binom{\alpha}{0} = 1, \quad \binom{\alpha}{n} = \frac{\alpha(\alpha-1)\cdots(\alpha-n+1)}{n!} = \frac{(-1)^n(-\alpha)_n}{n!}, \quad n \in \mathbb{N}.$$

Thus, we can extend our definition to the fractional case, and it is possible to define for $\alpha \in (0,1)$

$$(I - \tau^{-t})^{\alpha} f(x) = \sum_{k=0}^{+\infty} (-1)^k \binom{\alpha}{k} (\tau^{-t})^k f(x) = \sum_{k=0}^{\infty} (-1)^k \binom{\alpha}{k} f(x - kt).$$

In this way, we still maintain the semigroup property for the $\Delta_h^{\alpha} = (I - \tau^{-h})^{\alpha}$, because for every $\alpha_1, \alpha_2 \in \mathbb{R}$

$$\Delta_h^{\alpha_1} \Delta_h^{\alpha_2} = (I - \tau^{-h})^{\alpha_1} (I - \tau^{-h})^{\alpha_2} = (I - \tau^{-h})^{\alpha_1 + \alpha_2}$$

and $(I - \tau^{-h})^0 = I$. Here, we simply discuss the case of $\Delta_{+,h}^{\alpha}$ for $\alpha \in (0,1)$, but the results may be generalized to different exponents.

Moreover, the following result holds, see e.g., [7].

Theorem 1. *Let $\alpha, \beta > 0$. Then, for every bounded function:*

$$\Delta_h^{\alpha} \Delta_h^{\beta} f = \Delta_h^{\alpha+\beta} f.$$

In addition, considering one more time [7], and recalling also the contribution given in [30], we have that:

Theorem 2. *Let $\alpha > 0$. Then, for every $f \in L^1(\mathbb{R})$*

$$\mathcal{F}(\Delta_h^{\alpha} f)(x) = (1 - e^{ixh})^{\alpha} \mathcal{F}(f)(x).$$

In particular, it is true that the Grünwald–Letnikov derivative of order $\alpha \in (0,1)$ coincides with the Marchaud derivative of the same order. Indeed, in consideration of the two previous trivial properties, the following result holds.

Theorem 3. *Let $f \in L^p(\mathbb{R})$, $p \geq 1$. Then, for every $q \geq 1$, there exist*

$$f_{\pm}^{\alpha)}(x) = \lim_{h \to 0,\, in\ L^q} \frac{\Delta_{\pm h}^{\alpha} f(x)}{h^{\alpha}}$$

and

$$\mathbf{D}_{\pm}^{\alpha} f(x) = \lim_{\epsilon \to 0,\, in\ L^q} C(\alpha) \int_{\epsilon}^{+\infty} \frac{f(x) - f(x \mp h)}{h^{1+\alpha}} dh.$$

Moreover,

$$f_{\pm}^{\alpha)}(x) = \mathbf{D}_{\pm}^{\alpha} f(x),$$

independently from p and q.

The proof is quite long and can be found in [3], Theorem 20.4. Moreover, about this topic, we recall the very recent contribution [31]. By the way, this last paper can be considered also as a further signal of the renascent interest for a Marchaud derivative. In fact, in that manuscript, it has been recently proved the coincidence of the Marchaud derivative and the Grünwald–Letnikov derivative for functions in Hölder spaces with explicit rates of convergence. Previous results encode many facts. The first concerns the commutativity of the Grünwald–Letnikov derivative as well as the Marchaud derivative, namely $(f^{\alpha)})^{\beta)} = (f^{\beta)})^{\alpha)} = f^{\alpha+\beta)}$ and $\mathbf{D}^{\alpha}\mathbf{D}^{\beta} = \mathbf{D}^{\beta}\mathbf{D}^{\alpha} = \mathbf{D}^{\alpha+\beta}$.

5. Weyl Derivative

Hermann Weyl's name is associated with many important scientific results in physics and mathematics. In particular, concerning fractional derivative, Weyl made an important contribution that is strictly linked to the Marchaud derivative. By the truth Weyl introduced in its paper [1], in p. 302, exactly the definition of the fractional derivative that Marchaud gave in [4]. The paper written by Weyl appeared in 1917, while the Marchaud thesis was published in 1927 (see [2,4]). It is not clear if the two definitions were discovered independently. The cited Weyl paper, whose title is *Bemerkungen zum Begriff des Differentialquotienten gebrochener Ordnung* (remarks on the notion of the differential quotient of a broken order), concerns the notion of a fractional derivative. In the introduction of his paper, Weyl recognizes at first the efforts made by Bernanrd Riemann for obtaining a notion of a derivative for every positive real number. In particular, Weyl cited the contents of the unpublished Riemann notes reported in the XIX paper of the published Riemann opera post. About this fact, Weyl recalls, as the editor of that volume, remarks that Riemann surely did not think that those computations would have been published, at least in that form. In any case, Weyl faces the problem of starting from those notes and having in mind that he wants to obtain a definition that works for periodic functions. In order to avoid the problem of introducing some privileged points, as very often happens in literature concerning fractional derivatives, he assumes that periodic functions have to have a zero mean. We do not enter into the details here (see Section); however, Weyl uses the properties of Fourier series and, on p. 302 [1], the following relationship

$$g(x) = \beta \int_0^\infty \frac{f(x) - f(x-\xi)}{\xi^{1+\beta}} d\xi \tag{3}$$

for functions f Hölder having modulus of continuity with exponent α and $\alpha > \beta$ and knowing that g denotes the fractional derivative of order β. The same argument was reported in [32] on p. 226 (see Formula (3) in IX, 9.81). Nevertheless, the previous formula, apparently, disappeared in the final version of the book [33] published later on, probably because the author was mainly interested in the periodic properties of the functions, but we do not have any proof of this statement.

Anyhow, also in [3], the definition of the Weyl derivative can be found. Starting from Fourier expansion of a periodic function, Weyl defines the kernel

$$\psi^\alpha_\pm(t) = \sum_{k=-\infty, k\neq 0}^{+\infty} \frac{e^{ikt}}{(\pm ik)^\alpha} = 2\sum_{k=1}^{\infty} \frac{\cos(kt \mp \alpha\frac{\pi}{2})}{k^\alpha}.$$

Thus, the so-called Marchaud–Weyl derivative is defined as

$$\mathbf{D}^{(\alpha)}_\pm f(x) = \frac{1}{2\pi}\int_0^{2\pi} (f(x) - f(x-t)) \frac{d}{dt}\psi^{1-\alpha}_\pm(t)dt. \tag{4}$$

If the function f is 2π periodic, then it results that

$$\frac{\alpha}{\Gamma(1-\alpha)}\int_0^{+\infty} \frac{f(x) - f(x-t)}{t^{1+\alpha}}dt = \frac{1}{2\pi}\int_0^{2\pi} (f(x) - f(x-t)) \frac{d}{dt}\psi^{1-\alpha}_\pm(t)dt.$$

The previous relationship has to be correctly expressed in the following sense (see Lemma 19.4 in [3]).

Proposition 1 (Lemma 19.4, [3])**.** *For every $f \in L^p(0, 2\pi)$, $1 \geq p < +\infty$, the following limits converge for almost every $x \in (0, 2\pi)$ simultaneously:*

$$\lim_{\epsilon \to 0^+} \frac{1}{2\pi}\int_\epsilon^{2\pi} (f(x) - f(x-t)) \frac{d}{dt}\psi^{1-\alpha}_\pm(t)dt,$$

$$\lim_{\epsilon\to 0^+}\frac{\alpha}{\Gamma(1-\alpha)}\int_{\epsilon}^{+\infty}\frac{f(x)-f(x-t)}{t^{1+\alpha}}dt,$$

and

$$\begin{aligned}\mathbf{D}_{+}^{(\alpha)}f(x) &= \lim_{\epsilon\to 0^+}\frac{1}{2\pi}\int_{\epsilon}^{2\pi}(f(x)-f(x-t))\frac{d}{dt}\psi_{\pm}^{1-\alpha}(t)dt\\ &= \lim_{\epsilon\to 0^+}\frac{\alpha}{\Gamma(1-\alpha)}\int_{\epsilon}^{+\infty}\frac{f(x)-f(x-t)}{t^{1+\alpha}}dt = \mathbf{D}_{+}^{\alpha}f(x).\end{aligned}$$

By the way, concerning the parallel situation for the fractional Laplace operator on the torus, we point out [34], where similar results to the Proposition 1 have been proved.

We do not enter in the details concerning the question whether the definition of this type of fractional derivative has been invented by Weyl or Marchaud, or maybe by Riemann himself indeed as Weyl seems to suggest in the introduction of his paper [1]. Nevertheless the Formula (3) appeared in [1], as already written, ten years before the Marchaud thesis. In any case, Marchaud correctly cites the Weyl's paper [1]. More precisely Marchaud at p. 50 of his thesis acknowledges to Weyl to have obtained the result in the case of dimension $n = 1$, by referring to the representation (6). In addition Marchaud also admits that Weyl's approach was more powerful with respect to the one established by Montel, see [35]. Montel approach used polynomial approximation, as Marchaud stated. On the contrary, Marchaud remarked, that Weyl's approach is more direct. In [3], XXXIII, the authors faced indirectly that question in the note dedicated to the historical outline of the subject. There, they explained that Formula (6) appeared earlier in [1] by accident. Nevertheless, they concluded that Weyl did not develop his idea, as on the contrary Marchaud did in [2,4]. It would be interesting to know if any interaction between Weyl and Marchaud happened. In any case the Weyl's paper [1] is not one of the most cited among all the important results obtained by Weyl during his fruitful career.

6. Basic Ideas

If we compare the Marchaud derivative with respect to the Riemann–Liouville one, we immediately realize that, in the latter one, the classical derivative operator appears, while, in the first one, it does not. This is one of the key points that Marchaud's definition makes evident. That is, Marchaud derivative avoids applying the classical derivative after an integration in order to define the fractional operator. In a sense, this approach recalls the one that has characterized the Sobolev's approach (see, for instance [36], and, in in a sense, it could be considered as precursive of the notion of a weak solution to a PDE). In fact, roughly speaking, we recall that Sobolev's approach is based on the integration of both sides of an equation. In this way, we reduced looking for functions that satisfy the obtained integral equation.

In this order of ideas, in the Riemann–Liouville definition, the classical derivation still appears. On the contrary, in the Marchaud derivative, we simply recognize a singular integral where the reminiscence of the derivative is given by the kernel that multiplies the difference between the values of the function in two points. On the other hand, Marchaud's definition includes the Riemann–Liouville's one when the initial point is $-\infty$ and the functions are sufficiently smooth. We come back to this aspect later on in the section. From a philosophical point of view, the Marchaud derivative seems to make evident its non-local character. On the contrary, in the initial historical approach described by the Riemann–Liouville derivative, the classical derivative operator, which is a local object, still remains.

For instance, by considering a function defined in all of $\mathbb{R}$ and having a minimal smoothness, we, in principle, can modify its definition locally, for example simply changing its derivative in a *small* set of points. Nevertheless, the remaining part of the function is not affected by this modification. On the contrary, the Marchaud derivative, but also the Riemann–Liouville one, even if in a spurious way, determines a quantity that heavily depends on the modified function. This fact is evident thanks to the presence of the integral operator. Summarizing, by modifying the given function even only in a small set, the value of the fractional derivative will change, in general, in all the points where this fractional derivative will be evaluated.

Now, we comment separately on the Marchaud derivative and Weyl derivative.

6.1. Marchaud Derivative

The Marchaud derivative acts like an operator that associates to a function a new function that in general does not maintain local properties like the differential (of the function) do far away to the set where the function has been modified. Nevertheless, this operator, the fractional one, in a sense, still contains the classical derivative. Indeed, the classical derivative materializes as a particular (let say like an exception) case who realizes when the order of the fractional derivative goes to an integer. This focusing phenomenon is particularly interesting.

In order to clarify this remark, let us consider the Definition (1), in the case $p = 2$ and $\alpha = 1$. Then, we obtain:

$$\mathbf{D}^{\alpha} f(x) \int_0^{\infty} s^{-2}(1-e^{-s})^2 ds = \int_0^{\infty} s^{-\alpha-1} \left(f(x) - f(x-s) + f(x-2s)\right) ds, \tag{5}$$

and since

$$\int_0^{\infty} s^{-2}(1-e^{-s})^2 ds = 2\log 2,$$

we get:

$$\mathbf{D} f(x) = \frac{1}{2\log 2} \int_0^{\infty} \frac{f(x) - 2f(x-s) + f(x-2s)}{s^2} ds.$$

It is worth saying that here we have the value of the classical derivative in a point represented via an integral! Let us say: from the global to the local. How to explain this fact? We remark that, if f is a C^2 function with compact support or even $f \in \mathcal{S}(\mathbb{R})$, then

$$\begin{aligned} &\mathcal{F}\left(\int_0^{\infty} \frac{f(x) - 2f(x-s) + f(x-2s)}{s^2} ds\right) = \mathcal{F}f(\xi) \int_0^{\infty} \frac{(1-e^{-is\xi})^2}{s^2} ds \\ &= 2\log 2 (i\xi)\mathcal{F}f(\xi) = 2\log 2 \mathcal{F}(f'). \end{aligned}$$

This implies, recalling that the Fourier transform is invertible on Schwartz space $\mathcal{S}(\mathbb{R})$, that, for every $f \in \mathcal{S}(\mathbb{R})$, Formula (5) truly gives a representation of the derivative of a function in a point. We shall come back in Section 7 on this fact. On the other hand, the relationship (5) is correctly defined in a larger space of functions with respect to $\mathcal{S}(\mathbb{R})$.

In the case $p = 1$, $\alpha < 1$

$$\mathbf{D}^{\alpha} f(x) \int_0^{\infty} s^{-1-\alpha}(1-e^{-s}) ds = \int_0^{\infty} s^{-\alpha-1} \left(f(x) - f(x-s)\right) ds,$$

but

$$\int_0^{\infty} s^{-1-\alpha}(1-e^{-s}) ds = [-\alpha^{-1} s^{-\alpha}(1-e^{-s})]_{s=0}^{s=\infty} + \alpha^{-1} \int_0^{\infty} s^{-\alpha} e^{-s} ds = \alpha^{-1}\Gamma(1-\alpha).$$

As a consequence,

$$\mathbf{D}^{\alpha} f(x) = \frac{\alpha}{\Gamma(1-\alpha)} \int_0^{\infty} \frac{f(x) - f(x-s)}{s^{1+\alpha}} ds \tag{6}$$

and, even in this case, the easier case among Marchaud derivatives concerning the function f for $\alpha \in (0,1)$, we can read the non-locality of this definition and, in addition, for sufficiently smooth functions,

$$\lim_{\alpha \to 1^-} \mathbf{D}^{\alpha} f(x) = Df(x).$$

In this case, an important role is played by the normalizing constant $\frac{\alpha}{\Gamma(1-\alpha)}$ that multiplies the integral in the definition of Marchaud derivative.

The fact that, for sufficiently "good" functions, the fractional derivative $\mathbf{D}^\alpha f$ coincides with the Riemann–Liouville derivative

$$\mathcal{D}^\alpha f(x) = \frac{1}{\Gamma(1-\alpha)} \frac{d}{dx} \int_{-\infty}^{x} \frac{f(t)}{(x-t)^\alpha}$$

can be checked straightforwardly. Moreover, the definition given by Marchaud can be applied even for functions that may grow at infinity less than α. On the contrary, the definition of the Liouville derivative is less flexible since it does not admit (see p. XXXIII [3]) being applied to constant functions.

Let us check that the Marchaud derivative $\mathbf{D}_+^\alpha f$ coincides with the Riemann–Louville derivative from the right. In fact, since

$$\mathcal{D}_+^\alpha f(x) = \frac{1}{\Gamma(1-\alpha)} \frac{d}{dx} \int_0^{+\infty} \frac{f(x-t)}{t^\alpha} dt,$$

and supposing that $f \in C^1(\mathbb{R})$ and $f = o(|x|^{\alpha-1-\epsilon})$, $x \to +\infty$ for $\epsilon > 0$, then by Lebesgue dominated convergence theorem first and then integrating by parts, we get:

$$\begin{aligned}
\mathcal{D}_+^\alpha f(x) &= \frac{1}{\Gamma(1-\alpha)} \int_0^{+\infty} \frac{f'(x-t)}{t^\alpha} dt = \frac{\alpha}{\Gamma(1-\alpha)} \int_0^{+\infty} f'(x-t) \left(\int_t^\infty \tau^{-\alpha-1} d\tau \right) dt \\
&= \frac{\alpha}{\Gamma(1-\alpha)} \lim_{\epsilon \to 0^+} \left\{ [-f(x-t) \left(\int_t^{+\infty} \tau^{-\alpha-1} d\tau \right)]_{t=\epsilon}^{+\infty} - \int_\epsilon^{+\infty} \frac{f(x-t)}{t^{1+\alpha}} dt \right\} \\
&= \frac{\alpha}{\Gamma(1-\alpha)} \lim_{\epsilon \to 0^+} \left\{ f(x-\epsilon) \left(\int_\epsilon^{+\infty} \tau^{-\alpha-1} d\tau \right) - \int_\epsilon^{+\infty} \frac{f(x-t)}{t^{1+\alpha}} dt \right\} \qquad (7) \\
&= \frac{\alpha}{\Gamma(1-\alpha)} \lim_{\epsilon \to 0^+} \left\{ \int_\epsilon^{+\infty} \frac{f(x)-f(x-t)}{t^{1+\alpha}} dt + (f(x-\epsilon) - f(x)) \int_\epsilon^{+\infty} \tau^{-\alpha-1} d\tau \right\} = \\
&= \frac{\alpha}{\Gamma(1-\alpha)} \int_0^{+\infty} \frac{f(x)-f(x-t)}{t^{1+\alpha}} dt = \mathbf{D}_+^\alpha f(x)
\end{aligned}$$

because there exists $\eta \in]x-\epsilon, x[$ such that, as $\epsilon \to 0$:

$$|\alpha(f(x-\epsilon) - f(x)) \int_\epsilon^{+\infty} \tau^{-\alpha-1} d\tau| = |\alpha f'(x-\eta)| \leq \sup_{\tau \in [x-\epsilon, x]} |f'(\tau)| \epsilon^{1-\alpha} \to 0.$$

Thus, from this point of view, the Marchaud derivative is a sort of weaker version of the Riemann–Liouville derivative.

For example, constants satisfy $\mathbf{D}_+^\alpha f(x) = 0$ in the Marchaud sense, even if we can not consider, in all of $\mathbb{R}$, the Riemann–Liouville derivative of a constant. In fact, the parallel integral is divergent. This is, of course, absolutely unpleasant! Indeed, both Marchaud and Weyl were motivated also from this fact in order for looking for a different type of definition of fractional derivative.

We also think that the Marchaud derivative as some further properties that have to be better understood in its application. In order to focus on one of these aspects, we remark (see also [26]), that the sum of the two Marchaud derivatives ($\mathbf{D}_+^\alpha f$ and $\mathbf{D}_-^\alpha f$) gives, in a sense, the Riesz derivative in one dimension, namely the fractional Laplace operator in dimension 1. More precisely:

$$\mathbf{D}_+^\alpha f(x) + \mathbf{D}_-^\alpha f(x) = \frac{\alpha}{\Gamma(1-\alpha)} \int_0^{+\infty} \frac{2f(x) - f(x-t) - f(x+t)}{t^{1+\alpha}} dt$$

or

$$\begin{aligned}
\mathbf{D}_{+}^{\alpha} f(x)+\mathbf{D}_{-}^{\alpha} f(x) &= \frac{\alpha}{\Gamma(1-\alpha)}\left(\int_{0}^{+\infty} \frac{f(x)-f(x-t)}{t^{1+\alpha}} dt+\int_{0}^{+\infty} \frac{f(x)-f(x+t)}{t^{1+\alpha}} dt\right), \\
&= \frac{\alpha}{\Gamma(1-\alpha)}\left(\int_{0}^{+\infty} \frac{f(x)-f(x-t)}{t^{1+\alpha}} dt+\int_{-\infty}^{0} \frac{f(x)-f(x-t)}{|t|^{1+\alpha}} dt\right), \\
&= \frac{\alpha}{\Gamma(1-\alpha)} \int_{-\infty}^{+\infty} \frac{f(x)-f(x-t)}{|t|^{1+\alpha}} dt=\frac{\alpha}{\Gamma(1-\alpha)} \int_{-\infty}^{+\infty} \frac{f(x)-f(\xi)}{|x-\xi|^{1+\alpha}} d\xi, \\
&= \frac{\alpha}{c(1, \frac{\alpha}{2}) \Gamma(1-\alpha)}(-\frac{d^{2}}{dx^{2}})^{\frac{\alpha}{2}} f(x),
\end{aligned}$$

where $c(1,\frac{\alpha}{2})$ is the normalizing constant associated with the fractional Laplace operator $(-\frac{d^2}{dx^2})^{\frac{\alpha}{2}}$, and whose value we will recall later on in this paper. This fact was implicitly remarked in [3] and it seems that it can be connected with the different type of variable considered. In case of $\mathbf{D}_{+}^{\alpha} f$ and $\mathbf{D}_{-}^{\alpha} f$, the only one variable in $\mathbb{R}$ has a privileged direction in the two definitions of fractional derivatives. For instance, we can think of it as the time variable. On the contrary, considering the fractional Laplace operator

$$(-\frac{d^2}{dx^2})^{\frac{\alpha}{2}} f$$

in $\mathbb{R}$ (the same also in $\mathbb{R}^n$), there is not any privileged direction. Namely, the space (in this case $\mathbb{R}$) is homogeneous so that the previous connection is particularly interesting.

6.2. Weyl Derivative

As far as Weyl's approach is concerned, the relationship between spectral theory and fractional derivative is explicit. Indeed, supposing that working with a 2π-periodic function as having a zero average, it is well known that the associated Fourier series is

$$\sum_{k=-\infty}^{+\infty} c_k e^{ikx},$$

where, of course, $\{c_k\}_{k\in\mathbb{Z}}$ denotes the sequence of Fourier coefficients.

Then, by computing formally the derivative of this series, we obtain

$$\sum_{k=-\infty}^{+\infty} c_k (ik) e^{ikx}.$$

It is obvious that defining a new function for a fixed $\alpha < 1$, as

$$\sum_{k=-\infty}^{+\infty} \frac{c_k}{(ik)^\alpha} e^{ikx},$$

we formally obtain taking then a derivative we obtain

$$D\left(\sum_{k=-\infty}^{+\infty} \frac{c_k}{(ik)^\alpha} e^{ikx}\right) = \sum_{k=-\infty}^{+\infty} \frac{c_k}{(ik)^{\alpha-1}} e^{ikx}. \tag{8}$$

In this way, Weyl defines the parallel fractional integral so that it is natural to define the fractional derivative of f as

$$\sum_{k=-\infty,}^{+\infty} c_k (ik)^\alpha e^{ikx}.$$

On the other hand, we recall that, given two periodic functions f, g, the new function

$$\frac{1}{2\pi}\int_0^{2\pi} g(t)f(x-t)dt$$

is represented by the Fourier series

$$\sum_{k=-\infty}^{\infty} g_k c_k e^{ikx},$$

where $\{g_k\}_{k\in\mathbb{Z}}$ and $\{c_k\}_{k\in\mathbb{Z}}$ are the respective Fourier coefficients.

As a consequence, considering

$$\sum_{k=-\infty}^{+\infty} \frac{c_k}{(ik)^\alpha} e^{ikx},$$

as representing the Fourier series of an integral like the following one:

$$\frac{1}{2\pi}\int_0^{2\pi} g(t)f(x-t)dt,$$

we deduce that previous integral has to be written in the following form:

$$\frac{1}{2\pi}\int_0^{2\pi} f(x-t)\left(\sum_{k=-\infty,k\neq 0}^{+\infty} \frac{e^{ikt}}{(ik)^\alpha}\right) dt.$$

Since it can prove that (see [3]) that

$$\sum_{k=-\infty,k\neq 0}^{+\infty} \frac{e^{ikt}}{(ik)^\alpha} = 2\sum_{k=1}^{\infty} \frac{\cos(kt-\alpha\frac{\pi}{2})}{k^\alpha}.$$

Then, denoting the kernel

$$\psi_+^\alpha(t) := \sum_{k=-\infty,k\neq 0}^{+\infty} \frac{e^{ikt}}{(ik)^\alpha},$$

Weyl obtains the fractional integral

$$I_+^{(\alpha)} f(x) = \frac{1}{2\pi}\int_0^{2\pi} f(x-t)\psi_+^\alpha(t)dt.$$

At this point, by recalling (8), Weyl defines the fractional derivative as

$$\mathcal{D}_+^{(\alpha)}(x) = D\left(I_+^{(1-\alpha)} f\right)(x).$$

This definition corresponds to the Weyl-Riemann-Lioville version of this derivative, see [3] for the details. Then taking formally the derivative Weyl obtains the Weyl-Marchaud derivative, see also (4), discussed in Section 5:

$$\mathbf{D}_+^{(\alpha)} f(x) = \frac{1}{2\pi}\int_0^{2\pi} (f(x)-f(x-t))\frac{d}{dt}\psi_+^{1-\alpha}(t)dt.$$

Of course, the case concerning $\mathbf{D}_-^{(\alpha)} f$ is analogous to the one just described for $\mathbf{D}_+^{(\alpha)} f$.

7. General Setting of Marchaud Derivative and Some Further Remarks

The definition of Marchaud derivative, as it is known since [3], can be extended to all $\alpha > 0$ in the following way, see [3,37]. Let $l \in \mathbb{N}, l \geq 1$ and $\alpha < l$. We define for every $f \in \mathcal{S}(\mathbb{R})$

$$\mathbf{D}^{\alpha}_{\pm} f(x) = \frac{1}{\chi(\alpha, l)} \int_0^{+\infty} \frac{\Delta^l_{\pm\tau} f(x)}{\tau^{1+\alpha}} d\tau,$$

where

$$\chi(\alpha, l) = \Gamma(-\alpha) A_l(\alpha) = \int_0^{+\infty} \frac{(1-e^{-t})^l}{t^{1+\alpha}} dt,$$

$$A_l(\alpha) = \sum_{k=0}^{l} (-1)^k \binom{l}{k} k^{\alpha}, \tag{9}$$

and

$$\Delta^l_{\pm\tau} f(x) = \sum_{k=0}^{l} (-1)^k \binom{l}{k} f(x \mp k\tau).$$

Of course, this definition can be generalized to the case of functions with several variables having a nice behavior both at infinity and locally, simply considering for every $\xi \in \mathbb{R}^n$, $\xi \neq 0$ and for every $\alpha > 0$ and $l \in \mathbb{N}, l \geq 1$ such that $\alpha < l$ and defining:

$$\mathcal{D}^{\alpha)}_{\pm,\xi} f(x) = \frac{1}{\chi(\alpha, l)} \int_0^{+\infty} \frac{\Delta^l_{\pm\tau,\xi} f(x)}{\tau^{1+\alpha}} d\tau,$$

where

$$\Delta^l_{\pm\tau,\xi} f(x) = \sum_{k=0}^{l} (-1)^k \binom{l}{k} f(x \mp k\tau\xi).$$

It is worth saying that $\lim_{\alpha \to l^-} \mathcal{D}^{\alpha)}_{\pm,\xi} f(x) = \pm D^l_{\xi} f(x)$, in the local (classical sense) and $\lim_{\alpha \to (l-1)^+} \mathcal{D}^{\alpha)}_{\pm,\xi} f(x) = \pm D^{l-1}_{\xi} f(x)$, where $\mathcal{D}^{0)}_{\xi} = I$, whenever f is sufficiently smooth (for example in $\mathcal{S}(\mathbb{R}^n)$).

In this way, it is possible to consider interesting representation of local operators. For example, denoting by e_i the vector of the canonic base of $\mathbb{R}^n$, for every $i = 1, \ldots, n$, we get

$$\frac{\partial f(x)}{\partial x_i} = \frac{1}{\chi(1,2)} \int_0^{+\infty} \frac{\Delta^2_{\tau,e_i} f(x)}{\tau^2} d\tau \tag{10}$$

and

$$\frac{\partial^2 f(x)}{\partial x_i^2} = \frac{1}{\chi(2,3)} \int_0^{+\infty} \frac{\Delta^3_{\tau,e_i} f(x)}{\tau^3} d\tau. \tag{11}$$

As a consequence for every $f \in \mathcal{S}(\mathbb{R}^n)$:

$$\sum_{i=1}^{n} \frac{\partial}{\partial x_i} \frac{\partial f(x)}{\partial x_i} = \Delta f(x) = \frac{1}{\chi(1,2)} \int_0^{+\infty} \frac{\sum_{i=1}^n \left(\frac{\partial f(x)}{\partial x_i} + \sum_{k=1}^2 (-1)^{k+1} k \binom{2}{k} \tau \frac{\partial f(x-k\tau\xi)}{\partial x_i} \right)}{\tau^2} d\tau \tag{12}$$

and

$$\Delta f(x) = \sum_{i=1}^{n} \frac{\partial^2 f(x)}{\partial x_i^2} = \frac{1}{\chi(2,3)} \int_0^{+\infty} \frac{\sum_{i=1}^n \Delta^3_{\tau,e_i} f(x)}{\tau^3} d\tau, \tag{13}$$

that is

$$\Delta f(x) = \frac{1}{\chi(2,3)} \int_0^{+\infty} \frac{\sum_{k=0}^{3}(-1)^k \binom{3}{k} \sum_{i=1}^{n} f(x - k\tau e_i)}{\tau^3} d\tau. \tag{14}$$

From the Liouville Theorem, it is well known that there exists a unique function $f \in \mathcal{S}(\mathbb{R}^n)$ such that $\Delta f(x) = 0$ in $\mathbb{R}^n$ that is $f = 0$. Thus, the unique function $f \in \mathcal{S}(\mathbb{R}^n)$ that satisfies

$$\frac{1}{\chi(2,3)} \int_0^{+\infty} \frac{\sum_{i=1}^{n} \Delta^3_{\tau,e_i} f(x)}{\tau^3} d\tau = 0$$

has to be $f = 0$.

About the properties of the Marchaud derivative, we like to remind readers that, for every function $f \in \mathcal{S}(\mathbb{R})$,

$$\mathbf{D}f(x) = \frac{1}{2\log 2} \int_0^{\infty} \frac{f(x) - 2f(x-s) + f(x-2s)}{s^2} ds.$$

On the other hand, for every $f, g \in \mathcal{S}(\mathbb{R})$, $fg \in \mathcal{S}(\mathbb{R})$, so that

$$\mathbf{D}(fg)(x) = \frac{1}{2\log 2} \int_0^{\infty} \frac{f(x)g(x) - 2f(x-s)g(x-s) + f(x-2s)g(x-2s)}{s^2} ds.$$

On the other hand, we know that $\mathbf{D}(fg)(x) = \mathbf{D}(f)(x)g(x) + \mathbf{D}(g)(x)f(x)$. Then, as a by-product, we obtain the following formula for every $f, g \in \mathcal{S}(\mathbb{R})$

$$\begin{aligned} &\int_0^{\infty} \frac{f(x)g(x) - 2f(x-s)g(x-s) + f(x-2s)g(x-2s)}{s^2} ds \\ &= \int_0^{\infty} \frac{f(x) - 2f(x-s) + f(x-2s)}{s^2} ds g(x) + \int_0^{\infty} \frac{g(x) - 2g(x-s) + g(x-2s)}{s^2} ds f(x). \end{aligned}$$

Nevertheless, for instance, for every $\alpha \in (0,1)$ and for every $f, g \in \mathcal{S}(\mathbb{R})$, we get that $fg \in \mathcal{S}(\mathbb{R})$ and

$$\begin{aligned} \mathbf{D}^\alpha(fg)(x) &= \frac{\alpha}{\Gamma(1-\alpha)} \int_0^{\infty} \frac{f(x)g(x) - f(x-t)g(x-t)}{t^{1+\alpha}} dt \\ &= \frac{\alpha}{\Gamma(1-\alpha)} \int_0^{\infty} \frac{f(x) - f(x-t)}{t^{1+\alpha}} dt g(x) + \frac{\alpha}{\Gamma(1-\alpha)} \int_0^{\infty} \frac{f(x-t)(g(x) - g(x-t))}{t^{1+\alpha}} dt \\ &= \mathbf{D}^\alpha f(x)g(x) + \mathbf{D}^\alpha g(x)f(x) - \frac{\alpha}{\Gamma(1-\alpha)} \int_0^{\infty} \frac{(f(x) - f(x-t))(g(x) - g(x-t))}{t^{1+\alpha}} dt. \end{aligned}$$

This remark implies that the usual differential rule for the product of two functions does not hold. Nevertheless,

$$\frac{\alpha}{\Gamma(1-\alpha)} \int_0^{\infty} \frac{(f(x) - f(x-t))(g(x) - g(x-t))}{t^{1+\alpha}} dt \to 0$$

whenever $\alpha \to 1^-$. In fact,

$$\begin{aligned}
&\frac{\alpha}{\Gamma(1-\alpha)}\int_0^\infty \frac{(f(x)-f(x-t))(g(x)-g(x-t))}{t^{1+\alpha}}dt \\
&= \frac{\alpha}{\Gamma(1-\alpha)}\int_0^\eta \frac{(f(x)-f(x-t))(g(x)-g(x-t))}{t^{1+\alpha}}dt \\
&+ \frac{\alpha}{\Gamma(1-\alpha)}\int_\eta^\infty \frac{(f(x)-f(x-t))(g(x)-g(x-t))}{t^{1+\alpha}}dt \\
&= \frac{\alpha}{\Gamma(1-\alpha)}Df(x)Dg(x)\int_0^\eta t^{1-\alpha}dt + o(\eta^{2-\alpha}) \\
&+ \frac{\alpha}{\Gamma(1-\alpha)}\int_\eta^\infty \frac{(f(x)-f(x-t))(g(x)-g(x-t))}{t^{1+\alpha}}dt \to 0,
\end{aligned}$$

whenever $\alpha \to 1^-$, because $\frac{\alpha}{\Gamma(1-\alpha)} \to 0$, and there exists a positive constant such that

$$\left|\int_\eta^\infty \frac{(f(x)-f(x-t))(g(x)-g(x-t))}{t^{1+\alpha}}dt\right| \leq M$$

uniformly for every $\alpha \in (0,1)$ and for every fixed $\eta > 0$.

In this way, we obtain one more time the classical rule for the usual derivative of order one because $\mathbf{D}^\alpha(fg)(x) \to \mathbf{D}(fg)(x)$ and $\mathbf{D}^\alpha f(x)g(x) + \mathbf{D}^\alpha g(x)f(x) \to \mathbf{D}f(x)g(x) + \mathbf{D}g(x)f(x)$ if $\alpha \to 1^-$.

This behavior is heuristically clear thinking to the fractional operator as a nonlocal object. That is, the fractional derivative in a point *measures* something that depends on all the values of the function before that point. Thus, it is in a sense expected that, for this type of operator, a term depending on the interplay of the quantity associated with the fractional derivative of the functions acting, has to appear. In the special case of $\alpha \to 1^-$, this third term appears with value 0 thanks to the locality of the quantity expressed by the classical derivative of order one. The Marchaud derivative of order α rescales with the law λ^α. In fact, we have that, for every $f \in \mathcal{S}(\mathbb{R})$, the function $x \to f(\lambda x) = f_\lambda(x)$ has the following behavior with respect to the Marchaud fractional derivative:

$$\begin{aligned}
\mathbf{D}^\alpha f_\lambda(x) &= \frac{\alpha}{\Gamma(1-\alpha)}\int_0^\infty \frac{f(\lambda x)-f(\lambda(x-t))}{t^{1+\alpha}}dt \\
&= \frac{\alpha}{\Gamma(1-\alpha)}\int_0^\infty \frac{f(\lambda x)-f(\lambda x-\lambda t))}{t^{1+\alpha}}dt = \lambda^\alpha \frac{\alpha}{\Gamma(1-\alpha)}\int_0^\infty \frac{f(\lambda x)-f(\lambda x-\tau)}{\tau^{1+\alpha}}d\tau \\
&= \lambda^\alpha \mathbf{D}^\alpha f(\lambda x).
\end{aligned}$$

Remark also that, with respect to a different representation of the Marchaud fractional derivative, letting us see the case $\alpha < 2$, we get the same rescaling law:

$$\begin{aligned}
\mathbf{D}^\alpha f_\lambda(x) &= \frac{1}{\chi(\alpha,2)}\int_0^\infty \frac{f(\lambda x)-2f(\lambda(x-t))+f(\lambda(x-2t))}{t^{1+\alpha}}dt \\
&= \frac{1}{\chi(\alpha,2)}\int_0^\infty \frac{f(\lambda x)-2f(\lambda x-\lambda t))+f(\lambda x-2\lambda t))}{t^{1+\alpha}}dt \\
&= \lambda^\alpha \frac{1}{\chi(\alpha,2)}\int_0^\infty \frac{f(\lambda x)-2f(\lambda x-\tau)+f(\lambda x-2\tau)}{\tau^{1+\alpha}}d\tau \\
&= \lambda^\alpha \mathbf{D}^\alpha f(\lambda x).
\end{aligned}$$

The definition of the Marchaud derivative makes sense for a larger class of functions, with respect to the set $\mathcal{S}(\mathbb{R})$. For instance, all the constants have Marchaud derivative zero. The exponential function $x \to e^{\lambda x}$ does not belong to $\mathcal{S}(\mathbb{R})$. Nevertheless, for $\lambda \geq 0$ (here, we are using the Marchaud derivative

$\mathbf{D}^{\alpha}_{+}$, but, for the sake of simplicity, we omit writing the sign +), the function $e^{\lambda x}$ has Marchaud derivative and

$$\begin{aligned}\mathbf{D}^{\alpha} e^{\lambda x} &= \frac{\alpha}{\Gamma(1-\alpha)} \int_0^{\infty} \frac{e^{\lambda x} - e^{\lambda(x-t)}}{t^{1+\alpha}} dt = e^{\lambda x} \frac{\alpha}{\Gamma(1-\alpha)} \int_0^{\infty} \frac{1-e^{-\lambda t}}{t^{1+\alpha}} dt \\ &= \lambda^{\alpha} e^{\lambda x} \frac{\alpha}{\Gamma(1-\alpha)} \int_0^{\infty} \frac{1-e^{-\tau}}{\tau^{1+\alpha}} d\tau = \lambda^{\alpha} e^{\lambda x}\end{aligned}$$

because, integrating by parts, we obtain:

$$\int_0^{\infty} \frac{1-e^{-\tau}}{\tau^{1+\alpha}} d\tau = \frac{1}{\alpha} \int_0^{\infty} \frac{e^{-\tau}}{\tau^{\alpha}} d\tau = \frac{\Gamma(1-\alpha)}{\alpha}.$$

As a consequence, $e^{\lambda x}$ is solution of the fractional differential equation

$$\mathbf{D}^{\alpha} f(x) = \lambda^{\alpha} f(x).$$

8. Fractional Laplace Operator

The fractional Laplace operator can be represented in several ways. We should have to cite the contribution of many authors. We recall, for instance, [38–43]. Using the Fourier transform, for every $s \in (0,1)$ and for every $u \in \mathcal{S}(\mathbb{R}^n)$, the fractional Laplace operator is usually defined as

$$(-\Delta)^s u = \mathcal{F}^{-1}(||\xi||^{2s} \mathcal{F})u.$$

As a consequence, for every $u \in L^2(\mathbb{R}^n)$ if $||\xi||^{2s}\mathcal{F}u \in L^2(\mathbb{R}^n)$, then the fractional Laplace operator is defined by $\mathcal{F}^{-1}(||\xi||^{2s}\mathcal{F})u$.

On the other hand, for every $u \in \mathcal{S}(\mathbb{R}^n)$ and $s \in (0,1)$, we can define the operator

$$\mathcal{L}_s u(x) = c(\alpha, n) \int_{\mathbb{R}^n} \frac{f(x) - f(y)}{||x-y||^{n+2s}} dy := \lim_{\epsilon \to 0} c(s,n) \int_{\mathbb{R}^n \setminus B_\epsilon(x)} \frac{f(x)-f(y)}{||x-y||^{n+2s}} dy,$$

where $c(\alpha, n)$ is a normalizing constant, then $\mathcal{L}_s = (-\Delta)^s$ and

$$c(s,n) = \left(\int_{\mathbb{R}^n} \frac{1-\cos(\xi_1)}{||\xi||^{n+2s}} d\xi \right)^{-1}.$$

In addition, see [44], if $n > 1$, we get:

$$\lim_{s \to 1^-} \frac{\omega_{n-1} c(s,n)}{4ns(1-s)} = 1$$

and

$$\lim_{s \to 0^+} \frac{\omega_{n-1} c(s,n)}{2s(1-s)} = 1.$$

In addition in Lemma 5, [14], the previous constant has been surprisingly computed in a precise way so that it results as:

$$c_{s,n} = \frac{4^s \Gamma(\frac{n}{2}+s)}{\pi^{\frac{n}{2}} \Gamma(-\sigma)}.$$

We recall that in [3] a different expression of the fractional Laplace operator has been given, introducing a different constant of normalization and considering a more general situation. In fact, for every $f \in \mathcal{S}(\mathbb{R}^n)$ and $\alpha > 0, l \in \mathbb{N}, n \geq 1, \alpha < l$, we may define the following operator:

$$(-S)^{\frac{\alpha}{2}} f(x) = \frac{\sin(\alpha \frac{\pi}{2})}{\beta_n(\alpha) A_l(\alpha)} \int_{\mathbb{R}^n} \frac{\Delta_y^l f(x)}{||y||^{n+\alpha}} dy,$$

where $A_l(\alpha)$ is defined in (9),

$$\beta_n(\alpha) = \frac{\pi^{1+\frac{n}{2}}}{2^\alpha \Gamma(1+\frac{\alpha}{2})\Gamma(\frac{n+\alpha}{2})}$$

and

$$\Delta_y^l f(x) = \sum_{k=0}^{l} (-1)^k \binom{l}{k} f(x-ky)$$

denote the non-centered differences. Then, in [3], see Lemma 25.3, it is possible to find the proof that $(-S)^{\alpha/2} = (-\Delta)^{\alpha/2}$ in $\mathcal{S}(\mathbb{R}^n)$.

Another way of introducing the fractional Laplace operator can be done considering if $U : \mathbb{R}^n \times]0,+\infty[\to \mathbb{R}$ solution of the following nonlocal problem,

$$\begin{cases} \operatorname{div}_{(x,y)}(y^{1-2s}\nabla U(x,y)) = 0, & \text{in} \quad \mathbb{R}^n \times]0,+\infty[, \\ U(\cdot,0) = u, & x \in \mathbb{R}^n. \end{cases}$$

Then, defining

$$\mathcal{N}_s u := \lim_{y\to 0} y^{1-2s}\frac{\partial U(\cdot,y)}{\partial y},$$

it results, possibly up to a multiplicative factor depending only on s and n to $\mathcal{N}_s$, that $(-\Delta)^s = (-S)^s = \mathcal{L}_s = \mathcal{N}_s$ for every $u \in \mathcal{S}(\mathbb{R}^n)$. Among the application of this extension approach, we have the application to Carnot groups (see [14,15]).

In the next Section 9, we shall discuss the relationship of the Marchaud derivative with respect to the previous representation of the fractional Laplace operator. We recall, however, that, for the sake of completeness, the fractional Laplace operator may be represented also defining the operator

$$\mathcal{A}_s = -\frac{s}{\Gamma(1-s)}\int_0^{+\infty} (e^{t\Delta} - Id)\frac{dt}{t^{1+s}},$$

where $e^{t\Delta}$ denotes the heat semigroup generated by the Laplace operator Δ and it is also well known that defining the operator

$$\mathcal{B}_s = c(s,n)\int_0^{+\infty} \lambda^s dE(\lambda),$$

where $\{E(\lambda)\}_{\lambda\in[0,\infty[}$ is, as usual, the family of spectral projectors of the Laplace operator, we can conclude that, at least in $\mathcal{S}(\mathbb{R}^n)$, $(-\Delta)^s = (-S)^s = \mathcal{L}_s = \mathcal{N}_s = \mathcal{A}_s = \mathcal{B}_s$. We conclude this section recalling [24], where the semigroup method has been introduced and [45]. In [14], this approach has been developed and then generalized in [25] to a very large class of operators. The fractional Laplace operator in its representation via an extension has been applied in [46] for facing the regularity of the thin obstacle (see also [47]). In particular, we point out that this approach opened the way to a large number of papers in which this idea applied to many other problems. Other applications of the fractional calculus to the geometric measure theory can be found, for instance, in [48–50] and also coming to a very recently result [51,52], where the definition of nonlocal (fractional) perimeters is discussed. For further insights to the properties of the fractional Laplace operator, in addition to [3,38,44,45,53,54], we point out also the very recent preprint [55].

9. Extension Approach for Marchaud Derivative

We described here the simplest case given by $s = 1/2$ as follows. Let $\varphi : \mathbb{R} \to \mathbb{R}$ be a function in $\mathcal{S}(\mathbb{R})$ and U be a solution of the problem

$$\begin{cases} \dfrac{\partial U}{\partial t} = \dfrac{\partial^2 U}{\partial x^2}, & (x,t) \in (0,\infty)\times\mathbb{R}, \\ U(0,t) = \varphi(t), & t \in \mathbb{R}. \end{cases} \tag{15}$$

It is worth remarking that this is not the usual Cauchy problem for the heat operator. It is a heat conduction problem.

Without extra assumptions, we can not expect to have a unique solution of problem (15), see [56], Chapter 3.3. Anyhow, if we denote by $T_{1/2}$ the operator that associates with φ the partial derivative $\partial U/\partial x$, whenever U is sufficiently regular, we have that

$$T_{1/2}T_{1/2}\varphi = \frac{d\varphi}{dt}.$$

That is, $T_{1/2}$ acts like an half derivative, indeed

$$\frac{\partial}{\partial x}\frac{\partial U}{\partial x}(x,t) = \frac{\partial U}{\partial t}(x,t) \underset{x\to 0^+}{\longrightarrow} \frac{d\varphi(t)}{dt}.$$

The solution of problem (15) under the reasonable assumptions that φ is bounded and Hölder continuous, is explicitly known (check [56], Chapter 3.3) to be

$$\begin{aligned} U(x,t) &= cx\int_{-\infty}^{t} e^{-\frac{x^2}{4(t-\tau)}}(t-\tau)^{-\frac{3}{2}}\varphi(\tau)\,d\tau \\ &= cx\int_0^{\infty} e^{-\frac{x^2}{4\tau}}\tau^{-\frac{3}{2}}\varphi(t-\tau)\,d\tau, \end{aligned}$$

where the last line is obtained with a change of variables. Using $t = x^2/(4\tau)$ and the integral definition of the Gamma function we have that

$$\int_0^{\infty} xe^{-\frac{x^2}{4\tau}}\tau^{-\frac{3}{2}}\,d\tau = 2\int_0^{\infty} e^{-t}t^{-\frac{1}{2}}\,dt = 2\Gamma\left(\frac{1}{2}\right).$$

As a consequence,

$$\frac{U(x,t)-U(0,t)}{x} = c\int_0^{\infty} e^{-\frac{x^2}{4\tau}}\tau^{-\frac{3}{2}}\left(\varphi(t-\tau)-\varphi(t)\right)d\tau,$$

choosing c that takes into account the right normalization. In addition, by passing to the limit, we obtain

$$-\lim_{x\to 0^+}\frac{U(x,t)-U(0,t)}{x} = c\int_0^{\infty}\frac{\varphi(t)-\varphi(t-\tau)}{\tau^{\frac{3}{2}}}d\tau.$$

Hence, with the right choice of the constant, we get exactly $\mathbf{D}^{1/2}\varphi$ i.e., the Marchaud derivative of order $1/2$ of φ.

In [16], and independently also in [17], has been proved the following result.

Theorem 4. *Let $s \in (0,1)$ and $\bar{\gamma} \in (s,1]$ be fixed. Let $\varphi \in C^{\bar{\gamma}}(\mathbb{R})$ be a bounded function and let $U\colon [0,\infty)\times\mathbb{R}\to\mathbb{R}$ be a solution of the problem*

$$\begin{cases} \dfrac{\partial U}{\partial t}(x,t) = \dfrac{1-2s}{x}\dfrac{\partial U}{\partial x}(x,t) + \dfrac{\partial^2 U}{\partial x^2}(x,t), & (x,t)\in(0,\infty)\times\mathbb{R}, \\ U(0,t) = \varphi(t), & t\in\mathbb{R}, \\ \lim\limits_{x\to+\infty} U(x,t) = 0. \end{cases} \tag{16}$$

Then, U defines the extension operator for φ, such that

$$\mathbf{D}^s\varphi(t) = -\lim_{x\to 0^+} c_s x^{-2s}(U(x,t)-\varphi(t)), \quad \textit{where} \quad c_s = 4^s\Gamma(s).$$

An interesting application that follows from this extension procedure is a Harnack inequality for Marchaud stationary functions in an interval $J \subseteq \mathbb{R}$, namely for functions that satisfy $\mathbf{D}^s\varphi = 0$ in J. This fact is not obvious, indeed the set of functions determined by fractional-stationary functions (on an interval) is nontrivial, see e.g., [57].

Theorem 5. *Let $s \in (0,1)$. There exists a positive constant γ such that, if $\mathbf{D}^s\varphi = 0$ in an interval $J \subseteq \mathbb{R}$ and $\varphi \geq 0$ in $\mathbb{R}$, then*

$$\sup_{[t_0-\frac{3}{4}\delta, t_0-\frac{1}{4}\delta]} \varphi \leq \gamma \inf_{[t_0+\frac{3}{4}\delta, t_0+\delta]} \varphi$$

for every $t_0 \in \mathbb{R}$ and for every $\delta > 0$ such that $[t_0-\delta, t_0+\delta] \subset J$.

The previous result can be deduced from the Harnack inequality proved in [58] for some degenerate parabolic operators (see also [59] for the elliptic setting) that however are local operators. In particular, the constant γ used in Theorem 5 is the same that appears in the parabolic case in [58]. Concerning the Harnack inequality for the Riemann–Liouville fractional derivative, we also point out [60,61]. In concluding this section, we also remark that, as far as in the case of the fractional Laplace case, the result is true if $\varphi \geq 0$ in all of $\mathbb{R}$ (see [62] for a counterexample for the fractional Laplace case).

We end this section remarking that, concerning the numerical computation of the fractional operators, there exist many contributions. Among them, we point out [63] and the recent handbook [64].

10. Relationship between Marchaud Derivative and the Fractional Laplace Operator

In the end, we discuss here some relationships between Marchaud derivative and fractional Laplace operators. An application to this approach can be find in [26] in the first Heisenberg group case. By the way, we would like to point out that recently a major and renewed attention to fractional calculus and operators similar to Marchaud derivative has been testified by the application described in [20].

In order to explain how fractional Laplace and Marchaud derivative are linked, we fix our attention to the case $0 < \alpha < 1$ by considering

$$\mathbf{D}^{\alpha}_{\pm,\xi} f(x) = \frac{\alpha}{\Gamma(1-\alpha)} \int_0^{+\infty} \frac{f(x) - f(x \mp t\xi)}{t^{1+\alpha}} dt,$$

where $\xi \in \mathbb{S}^{n-1}$ and $f \in \mathcal{S}(\mathbb{R}^n)$. For clarity, we define a new operator as it follows: for every $f \in \mathcal{S}(\mathbb{R}^n)$,

$$\mathcal{M}_{\frac{\alpha}{2}} f(x) = \int_{\partial B_1(0)} \mathbf{D}^{\alpha}_{\xi} f(x) d\mathcal{H}^{n-1}(\xi).$$

Then, switching the order of integration, we get

$$\begin{aligned} \int_{\partial B_1(0)} \mathbf{D}^{\alpha}_{\xi} f(x) d\mathcal{H}^{n-1}(\xi) &= \frac{\alpha}{\Gamma(1-\alpha)} \int_0^{+\infty} \left(\int_{\partial B_1(0)} \frac{f(x) - f(x-t\xi)}{t^{1+\alpha}} d\mathcal{H}^{n-1}(\xi) \right) dt \\ &= \frac{\alpha}{\Gamma(1-\alpha)} \int_0^{+\infty} \left(\int_{\partial B_t(x)} \frac{f(x) - f(y)}{|x-y|^{1+\alpha}} d\mathcal{H}^{n-1}(y) \right) \frac{dt}{t^{n-1}} \\ &= \frac{\alpha}{\Gamma(1-\alpha)} \int_0^{+\infty} \left(\int_{\partial B_t(x)} \frac{f(x) - f(y)}{|x-y|^{n+\alpha}} d\mathcal{H}^{n-1}(y) \right) dt = \frac{\alpha}{\Gamma(1-\alpha)} \frac{\beta_n(\alpha)}{\sin(\alpha\frac{\pi}{2})} (-\Delta)^{\frac{\alpha}{2}} f(x). \end{aligned} \tag{17}$$

In general, as already remarked in Lemma 26.2, [3], and recalling previous Section 8 for the definition of the constants $\chi(\alpha, l)$, for every $\alpha > 0$, $l \in \mathbb{N}$, $l \geq 0$, $\alpha < l$, denoting

$$\mathcal{D}^{\alpha)}_{\xi} f(x) = \frac{1}{\chi(\alpha, l)} \int_0^{+\infty} \frac{\Delta^l_{\xi} f(x)}{\tau^{1+\alpha}} d\tau,$$

we obtain, for every $f \in \mathcal{S}(\mathbb{R}^n)$:

$$(-\Delta)^{\frac{\alpha}{2}} f(x) = -\frac{\Gamma(-\alpha)\sin(\alpha\frac{\pi}{2})}{\beta_n(\alpha)} \int_{\partial B_1(0)} \mathcal{D}_{\xi}^{\alpha)} f(x) d\mathcal{H}^{n-1}(\xi). \tag{18}$$

In particular, if $\alpha \in]0,1[$, and $l = 1$

$$\mathbf{D}^{\alpha,+} f(t) = \frac{1}{C_{\alpha,1}} \int_0^{+\infty} \frac{f(t) - f(t-s)}{s^{1+\alpha}} ds,$$

where

$$C_{\alpha,1} = \frac{\Gamma(1-\alpha)}{\alpha}$$

and

$$\mathbf{D}^{\alpha,-} f(t) = \frac{1}{C_{\alpha,1}} \int_0^{+\infty} \frac{f(t) - f(t+s)}{s^{1+\alpha}} ds,$$

where

$$C_{\alpha,1} = \frac{\Gamma(1-\alpha)}{\alpha}.$$

Thus,

$$(\mathbf{D}^{\alpha,+} + \mathbf{D}^{\alpha,-}) f(t) = \frac{1}{C_{\alpha,1}} \int_0^{+\infty} \frac{2f(t) - f(t+\tau) - f(t-\tau)}{\tau^{1+\alpha}} d\tau,$$

and for every $e \in \partial B_1(0)$ and for every $f \in \mathcal{S}(\mathbb{R}^n)$, we have:

$$(\mathbf{D}_e^{\alpha,+} + \mathbf{D}_e^{\alpha,-}) f(x) = \frac{1}{C_{\alpha,1}} \int_0^{+\infty} \frac{2f(x) - f(x+e\tau) - f(x-e\tau)}{\tau^{1+\alpha}} d\tau.$$

As a consequence, integrating on $\partial B_1(0)$, we obtain, as we already remarked in one variable only:

$$\begin{aligned}
&\int_{\partial B_1(0)} (\mathbf{D}_e^{\alpha,+} + \mathbf{D}_e^{\alpha,-}) f(x) d\mathcal{H}^{n-1}(e) \\
&= \frac{1}{C_{\alpha,1}} \int_{\partial B_1(0)} \left(\int_0^{+\infty} \frac{2f(x) - f(x+e\tau) - f(x-e\tau)}{\tau^{1+\alpha}} d\tau \right) d\mathcal{H}^{n-1}(e) \\
&= \frac{1}{C_{\alpha,1}} \int_0^{+\infty} \left(\int_{\partial B_1(0)} \frac{2f(x) - f(x+e\tau) - f(x-e\tau)}{\tau^{1+\alpha}} d\mathcal{H}^{n-1}(e) \right) d\tau \\
&= \frac{1}{C_{\alpha,1}} \int_0^{+\infty} \left(\int_{\partial B_\tau(0)} \frac{2f(x) - f(x+\xi) - f(x-\xi)}{\tau^{n+\alpha}} d\mathcal{H}^{n-1}(\xi) \right) d\tau \\
&= \frac{1}{C_{\alpha,1}} \int_0^{+\infty} \left(\int_{\partial B_\tau(0)} \frac{2f(x) - f(x+\xi) - f(x-\xi)}{|x-\xi|^{n+\alpha}} d\mathcal{H}^{n-1}(\xi) \right) d\tau \\
&= \frac{1}{C_{\alpha,1}} \int_{\mathbb{R}^n} \frac{2f(x) - f(x+\xi) - f(x-\xi)}{|x-\xi|^{n+\alpha}} d\xi = \frac{2}{C_{\alpha,1} c(\frac{\alpha}{2}, n)} (-\Delta)^{\frac{\alpha}{2}} f(x).
\end{aligned}$$

Acknowledgments: The author wishes to thank Salvatore Coen and Salomon Ofman for sharing with him their information about the life of French mathematicians between the two world wars of the 20th century and Bruno Franchi for telling him the role of some public officers in the French University system. Moreover, the author also thanks Catherine Goldstein for having told him her opinion about some questions concerning the life of A. P. Marchaud and for pointing out Leloup thesis, [23].

The author is supported by MURST (Ministero dell'Università e della Ricerca Scientifica e Tecnologica), Italy, University of Bologna and the INDAM-GNAMPA project 2017: *Regolarità delle soluzioni viscose per equazioni a derivate parziali non lineari degeneri.*

Conflicts of Interest: The author declares no conflict of interest.

References

1. Weyl, H. Bemerkungen zum Begriff des Differentialquotienten gebrochener Ordnung. *Zürich. Naturf. Ges.* **1917**, *62*, 296–302.
2. Marchaud, A. Sur Les Dérivées et Sur les Différences des Fonctions de Variables Réelles. *Numdam Thèses del L'entre-Deux-Guerres* **1927**, *78*, 98.
3. Samko, S.G.; Kilbas, A.A.; Marichev, O.I. Fractional integrals and derivatives. In *Theory and Applications;* Edited and with a foreword by Nikol'skiĭ, S.M. (1993); Translated from the 1987 Russian original, Revised by the authors; Elsevier Science B.V.: Amsterdam, The Netherlands, 2006.
4. Marchaud, A. Sur les dérivées et sur les différences des fonctions de variables réelles. *J. Math. Pures Appl.* **1927**, *9*, 337–425.
5. Baleanu, D.; Diethelm, K.; Scalas, E.; Trujillo, J. Fractional Calculus. In *Models and Numerical Methods*, 2nd ed.; Series on Complexity, Nonlinearity and Chaos, 5; World Scientific Publishing Co. Pte. Ltd.: Hackensack, NJ, USA, 2017.
6. Butzer, P.L.; Westphal, U. An introduction to fractional calculus. *Applications of Fractional Calculus in Physics;* World Science Publishing: River Edge, NJ, USA, 2000; pp. 1–85.
7. Kilbas, A.A.; Srivastava, H.M.; Trujillo, J.J. *Theory and Applications of Fractional Differential Equations, Volume 204 of North-Holland Mathematics Studies;* Elsevier Science Inc.: New York, NY, USA, 2006.
8. Ross, B. A brief history and exposition of the fundamental theory of fractional calculus. In Proceedings of the International Conference held at the University of New Haven, West Haven, CT, USA, 15–16 June 1974. Edited by Bertram Ross. Lecture Notes in Mathematics. Springer-Verlag: Berlin, Germany, 1975; Volume 457.
9. Abel, N.H. Solution de Quelques Problèmes à L'aide D'integrales Definies. In *Gesammelte Mathematische Werke;* Teubner: Leipzig, Germany, 1823; pp. 11–27.
10. Liouville, J. Memoire sur quelques questions de geometrie et de mecanique, et sur un noveau genre pour respondre ces questions. *J. École Polytech.* **1832**, *13*, 1–69.
11. Riemann, B. Versuch einer allgemeinen Auffassung der Integration und Differentiation. In *Gesammelte Mathematische Werke und Wissenschaftlicher Nachlass;* Teubner: Leipzig, Germany, 1876; Dover, NY, USA, 1953; pp. 331–344.
12. Caffarelli, L.; Silvestre, L. An extension problem related to the fractional Laplacian. *Commun. Part. Differ. Equ.* **2007**, *32*, 1245–1260.
13. Molčanov, S.A.; Ostrovskiĭ, E. Symmetric stable processes as traces of degenerate diffusion processes (Russian English summary). *Teor. Verojatnost. i Primenen.* **1969**, *14*, 127–130.
14. Stinga, P.R.; Torrea, J. Extension problem and Harnack's inequality for some fractional operators. *Commun. Part. Differ. Equ.* **2010**, *35*, 2092–2122.
15. Ferrari, F.; Franchi, B. Harnack inequality for fractional sub-Laplacians in Carnot groups. *Math. Z.* **2015**, *279*, 435–458.
16. Bernardis, A.; Martín-Reyes, F.J.; Stinga, P.R.; Torrea, J.L. Maximum principles, extension problem and inversion for nonlocal one-sided equations. *J. Differ. Equ.* **2016**, *260*, 6333–6362.
17. Bucur, C.; Ferrari, F. An extension problem for the fractional derivative defined by Marchaud. *Fract. Calc. Appl. Anal.* **2016**, *19*, 867–887.
18. Gorenflo, R.; Luchko, Y.; Yamamoto, M. Time-fractional diffusion equation in the fractional Sobolev spaces. *Fract. Calc. Appl. Anal.* **2015**, *18*, 799–820.
19. Luchko, Y.; Gorenflo, R. An operational method for solving fractional differential equations with the Caputo derivatives. *Acta Math. Vietnam.* **1999**, *24*, 207–233.
20. Allen, M.; Caffarelli, L.; Vasseur, A. Porous medium flow with both a fractional potential pressure and fractional time derivative. *Chin. Ann. Math. Ser. B* **2017**, *38*, 45–82.
21. Bell, J.L.; Korté, H. Hermann Weyl. The Stanford Encyclopedia of Philosophy, 2016th ed.; Zalta, E.N., Ed. Available online: https://plato.stanford.edu/archives/win2016/entries/weyl/ (accessed on 22 November 2017).
22. Condette, J.F. Marchaud André Paul [note bibliographique] Histoire biographique de l'enseignement Année 2006, Volume 12, Numéro 2, pp. 272–273. In *Les Recteurs D'académie en France de 1808 à 1940;* Tome II, Dictionnaire biographique; Histoire biographique de l'enseignement, 12; Institut National de Recherche Pédagogique: Paris, France, 2006; p. 418.

23. Leloup, J. L'entre-Deux-Guerres Mathématique à Travers Les Thèses Soutenues en France. Mathématiques Université Pierre et Marie Curie-Paris VI. 2009. Ph.D. Thesis, 2009. Français. tel-00426604. Available online: https://tel.archives-ouvertes.fr/tel-00426604/document (accessed on 22 November 2017).
24. Balakrishnan, A.V. Fractional powers of closed operators and the semigroups generated by them. *Pac. J. Math.* **1960**, *10*, 419–437.
25. Galé, J.; Miana, P.; Stinga, P.R. Extension problem and fractional operators: Semigroups and wave equations. *J. Evol. Equ.* **2013**, *13*, 343–368.
26. Ferrari, F. Some nonlocal operators in the first Heisenberg group. *Fractal Fract.* **2017**, *1*, 15, doi:10.3390/fractalfract1010015.
27. Grünwald, A.K. Uber "begrenzte" derivationen und deren anwen-dung. *Zeit. Fur Mathematik und Physik* **1867**, *12*, 441–480.
28. Letnikov, A.V. Theory of differentiation with an arbitrary index (Russian). *Mat. Sb.* **1868**, *3*, 1–66.
29. Rogosin, S.; Dubatovskaya, M. Letnikov vs. Marchaud: A Survey on Two Prominent Constructions of Fractional Derivatives *Mathematics* **2018**, *6*, 3, doi:10.3390/math6010003.
30. Chapman, S. On Non-Integral Orders of Summability of Series and Integrals. *Proc. Lond. Math. Soc.* **1911**, *9*, 369–409.
31. Abadias, L.; de León-Contreras, M.; Torrea, J. Non-local fractional derivatives. Discrete and continuous. *J. Math. Anal. Appl.* **2017**, *449*, 734–755.
32. Zygmund, A. Trigonometrical series. In *Warszava-Lwow, Monografje Matematyczne*; Springer: Berlin, Germany, 1935; Volume V.
33. Zygmund, A. *Trigonometric Series*; Cambridge University: Cambridge, UK, 2002; Volume I–II.
34. Roncal, L.; Stinga, P.R. Transference of fractional Laplacian regularity. Special Functions, Partial Differential Equations, and Harmonic Analysis, 203–212. In *Proceedings Mathematical Sciences*; Springer: Cham, Switzerland, 2014; Volume 108.
35. Montel, P. Sur les polynomes d'approximation. *S.M.F. Bull.* **1918**, *46*, 151–192.
36. Sobolev, S.L. On a theorem of functional analysis. *Math. Sb. N.S.* **1938**, *4*, 471–497 [Russian]; English Translation in *Am. Math. Soc. Transl.* **1963**, *34*, 39–68.
37. Samko, S.G. *Hypersingular Integrals and Their Applications, Volume 5 of Analytical Methods and Special Functions*; Taylor & Francis, Ltd.: London, UK, 2002.
38. Landkof, N.S. *Foundations of Modern Potential Theory, Die Grundlehren der Mathematischen Wissenschaften*; Springer: New York, NY, USA; Heidelberg, Germany, 1972; Volume 180.
39. Riesz, M. Potentiels de divers ordres et leurs fonctions de Green. *C. R. Congr. Internat. Math. Oslo* **1936**, *2*, 62–63.
40. Riesz, M. Intégrale de Riemann–Liouville et solution invariantive du probléme des ondes. *C. R. Congr. Internat. Math. Oslo* **1936**, *2*, 44–45.
41. Riesz, M. Intégrale de Riemann–Liouville et potentiels. *Acta Litt. Sci. Univ. Szeged. Sect. Sci. Math.* **1938**, *9*, 1–42.
42. Riesz, M. Intégrale de Riemann–Liouville et le problème de Cauchy. *Acta Math.* **1949**, *81*, 1–223.
43. Stein, E.M. Singular integrals and differentiability properties of functions. In *Princeton Mathematical Series*; Princeton University Press: Princeton, NJ, USA, 1970
44. Di Nezza, E.; Palatucci, G.; Valdinoci, E. Hitchhiker's Guide to the Fractional Sobolev Spaces. *Bull. Sci. Math.* **2012**, *136*, 521–573.
45. Yosida, K. *Functional Analysis*, 6th ed.; Springer: Berlin, Germany, 1980.
46. Caffarelli, L.; Salsa, S.; Silvestre, L. Regularity estimates for the solution and the free boundary of the obstacle problem for the fractional Laplacian. *Invent. Math.* **2008**, *171*, 425–461.
47. Salsa, S. The problems of the obstacle in lower dimension and for the fractional Laplacian; In Regularity estimates for nonlinear elliptic and parabolic problems. In *Lecture Notes in Mathematics*; 2045, Fond. CIME/CIME Found. Subser.; Springer: Heidelberg, Germany, 2012; pp. 153–244.
48. Ambrosio, L.; de Philippis, G.; Martinazzi, L. Gamma-convergence of nonlocal perimeter functionals. *Manuscr. Math.* **2011**, *134*, 377–403.
49. Caffarelli, L.; Savin, O.; Valdinoci, E. Minimization of a fractional perimeter-Dirichlet integral functional. *Annales de l'Institut Henri Poincaré C Analyse non Linéaire* **2015**, 32, 901–924.

50. Caffarelli, L.; Valdinoci, E. Uniform estimates and limiting arguments for nonlocal minimal surfaces. *Calc. Var. Part. Differ. Equ.* **2011**, *41*, 203–240.
51. Cinti, E.; Franchi, B.; Gonzàlez, M.d.M. Γ-convergence of variational functionals with boundary terms in Stein manifolds. *Calc. Var. Part. Differ. Equ.* **2017**, *56*, 52.
52. Ferrari, F.; Miranda, M., Jr.; Pallara, D.; Pinamonti, A.; Sire, Y. Fractional Laplacians, perimeters and heat semigroups in Carnot groups, accepted paper, to appear in Discrete Contin. *Dyn. Syst. Ser. S* **2018**, *11*, 477–491, doi:10.3934/dcdss.2018026.
53. Bucur, C.; Valdinoci, E. Nonlocal diffusion and applications. In *Lecture Notes of the Unione Matematica Italiana, 20*; Springer: Cham, Switzerland; Unione Matematica Italiana: Bologna, Italy, 2016.
54. Bisci, G.M.; Radulescu, V.; Servadei, R. Variational methods for nonlocal fractional problems. With a foreword by Jean Mawhin. In *Encyclopedia of Mathematics and its Applications, 162*; Cambridge University Press: Cambridge, UK, 2016.
55. Garofalo, N. Fractional thoughts. *arXiv* **2017**, arXiv:1712.03347.
56. Tichonov, A.N.; Samarskij, A.A.; Budak, B.M. In *Problemi Della Fisica Matematica*; Mir: Bergen, Norway, 1982.
57. Bucur, C. Local density of Caputo-stationary functions in the space of smooth functions. *ESAIM COCV* **2017**, *23*, 1361–1380.
58. Chiarenza, F.; Serapioni, R. A remark on a Harnack inequality for degenerate parabolic equations. *Rend. Sem. Mat. Univ. Padova* **1985**, *73*, 179–190.
59. Fabes, E.B.; Kenig, C.E.; Serapioni, R.P. The local regularity of solutions of degenerate elliptic equations. *Commun. Part. Differ. Equ.* **1982**, *7*, 77–116.
60. Zacher, R. A weak Harnack inequality for fractional differential equations. *J. Integral Equ. Appl.* **2007**, *19*, 209–232.
61. Zacher, R. The Harnack inequality for the Riemann–Liouville fractional derivation operator (English summary). *Math. Inequal. Appl.* **2011**, *14*, 35–43.
62. Kassmann, M. A new formulation of Harnack's inequality for nonlocal operators. *C. R. Math. Acad. Sci. Paris* **2011**, *349*, 637–640.
63. Diethelm, K.; Ford, N.J.; Freed, A.D.; Luchko, Y. Algorithms for the fractional calculus: A selection of numerical methods. *Comput. Methods Appl. Mech. Eng.* **2005**, *194*, 743–773.
64. Li, C.; Zeng, F. Numerical methods for fractional calculus. *Chapman & Hall/CRC Numerical Analysis and Scientific Computing*; CRC Press: Boca Raton, FL, USA, 2015.

mathematics

MDPI

Article

Letnikov vs. Marchaud: A Survey on Two Prominent Constructions of Fractional Derivatives

Sergei Rogosin * and Maryna Dubatovskaya

Department of Economics, Belarusian State University, 4, Nezavisimosti ave, 220030 Minsk, Belarus; dubatovska@bsu.by

* Correspondence: rogosinsv@gmail.com; Tel.: +375-17-220-22-84

Received: 22 November 2017; Accepted: 20 December 2017; Published: 25 December 2017

Abstract: In this survey paper, we analyze two constructions of fractional derivatives proposed by Aleksey Letnikov (1837–1888) and by André Marchaud (1887–1973), respectively. These derivatives play very important roles in Fractional Calculus and its applications.

Keywords: fractional integrals and derivatives; Grünwald–Letnikov approach; Marchaud approach; fractional differences; Hadamard finite part

MSC: primary 26A33; secondary 34A08; 34K37; 35R11; 39A70

1. Introduction

The aim of the paper is to present the essence of two important approaches in Fractional Calculus, namely, those developed by Letnikov (or by Grünwald–Letnikov) and by Marchaud. We collect here the most important results for the corresponding fractional derivatives, compare these constructions and highlight their role in Fractional Calculus and its applications.

In his master thesis (see [1], p. 37) Letnikov (When the thesiswas ready for defence, Letnikov discovered the paper by Grünwald [2] in which the same approach was realised.) defined the left-sided fractional derivative on the interval $[x_0, x]$ via the following limit:

$$y^{\alpha)}(x) := \lim_{h\to+0} \frac{\sum\limits_{k=0}^{n} (-1)^k \left(\begin{array}{c} \alpha \\ k \end{array} \right) y(x-kh)}{h^\alpha}, \tag{1}$$

where $nh = x - x_0$. Similarly, the right-sided fractional derivative is defined:

$$y^{\alpha)}(x) := \lim_{h\to+0} \frac{\sum\limits_{k=0}^{n} (-1)^k \left(\begin{array}{c} \alpha \\ k \end{array} \right) y(x+kh)}{h^\alpha}. \tag{2}$$

These definitions can be used at points on semi-axes $[x_0, +\infty)$ (or $(-\infty, x_0]$) whenever the function $y(x)$ is defined there. In this case, we formally have to replace the finite sum for an infinite one and remove any restriction on the increment h.

The aim of the work by Letnikov was to correct the definition of Liouville [3,4], who supposed that his construction of the fractional derivative is a general one. More exactly, Letnikov tried to overcome Liouville's assumption that the general definition of fractional derivative can be applied only to functions represented in the form convergent Dirichlet type series (see Representation (6)). Detailed description of the difficulties that bring such an assumption is given in [5] (see also Section 2.1 below). (In [6], Letnikov transformed his fractional derivative to the form which coincides with the Riemann's formula if one removes from the later so called "additional function". Note that at that time Letnikov did not know the work by Riemann since it was published only later in the first edition of Riemann's collected works in 1876 [7].)

The doctoral thesis by Marchaud was published in complete form in the Journal de Mathématiques Pures et Appliquées in 1927 [8] and reprinted in the series "Théses de L'Entre-Deux-Guerres" in 1965 [9]. His main idea was to generalize the Riemann–Liouville approach. Replacing positive parameter α in the Riemann–Liouville fractional integral:

$$I_a^\alpha = \frac{1}{\Gamma(\alpha)} \int_a^x \frac{f(\tau)d\tau}{(x-\tau)^{1-\alpha}} = \frac{1}{\Gamma(\alpha)} \int_0^{x-a} t^{\alpha-1} f(x-t)dt$$

by negative one, he considered the divergent integral:

$$\frac{1}{\Gamma(-\alpha)} \int_0^{x-a} t^{-\alpha-1} f(x-t)dt.$$

In order to regularize this definition, Marchaud made some transformations in the Riemann–Liouville fractional integral.

For arbitrary values of α, $\mathrm{Re}\,\alpha > 0$, the definition of the Marchaud derivative reads ([10], Section 5.6)

$$\mathbb{D}_\pm^\alpha f(x) = -\frac{1}{\Gamma(-\alpha)A_l(\alpha)} \int_0^\infty \frac{(\Delta_{\pm t}^l f)(x)}{t^{1+\alpha}} dt, \quad l > \mathrm{Re}\,\alpha > 0, \tag{3}$$

where

$$A_l(\alpha) = \sum_{k=0}^\infty (-1)^{k-1} \binom{l}{k} k^\alpha, \quad (\Delta_{\pm t}^l f)(x) = \sum_{k=0}^\infty (-1)^k \binom{l}{k} f(x \mp kt).$$

If $0 < \alpha < 1$, then the left- and right-sided are defined, respectively (see, e.g., [10], Section 5.4),

$$\mathbb{D}_+^\alpha f(x) = \frac{\alpha}{\Gamma(1-\alpha)} \int_0^\infty \frac{f(t) - f(x-t)}{t^{1+\alpha}} dt, \quad \mathbb{D}_-^\alpha f(x) = \frac{\alpha}{\Gamma(1-\alpha)} \int_0^\infty \frac{f(t) - f(x+t)}{t^{1+\alpha}} dt. \tag{4}$$

Since the integral in the right-side formula (3) is in general diverging, then the Marchaud derivative can be defined via the limit of the truncated derivative (if it exists):

$$\mathbb{D}_\pm^\alpha f(x) = \lim_{\varepsilon \to +0} \mathbb{D}_{\pm,\varepsilon}^\alpha f(x) = \lim_{\varepsilon \to +0} -\frac{1}{\Gamma(-\alpha)A_l(\alpha)} \int_\varepsilon^\infty \frac{(\Delta_{\pm t}^l f)(x)}{t^{1+\alpha}} dt. \tag{5}$$

It is seen from (1), (3) that there exists a formal relationship of the Grünwald–Letnikov and Marchaud derivatives. It will be discussed in more detail below. It should be noted that both constructions (by Letnikov and by Marchaud) are applicable to a much wider class of the functions than Liouville's construction (though all of them coincide on "good functions", see [10]).

Nowadays, both derivatives attract more attention among experts from different branches of Science. These derivatives are suitable for numerical analysis of the corresponding fractional models since they are defined based on discretization (see, e.g., [11,12]). Among applications of such derivatives, we have to mention those in different physical models (see, e.g., [13–21] and an extending survey in [22], Chapter 8), as well as in the study of certain problems of operator theory (see, e.g., [23–25]) and results describing interrelation between fractional calculus and fractal geometry (see, e.g., [26,27]). Since the Marchaud derivative is less known than Grünwald–Letnikov, Riemann–Liouville or Caputo derivatives, we have to mention some features of this construction that are important in application. First, Marchaud construction allows more freedom for the behavior at infinity (e.g., to the exogene variable X in the accelerator-multiplied model with memory as in [28]). Second, the Marchaud idea to regularize the Liouville fractional derivative by using finite differences met applications at the study of fractional differential equations as in [29]. Third, the Marchaud derivative possesses a simple and straightforward generalization to multi-dimensional case (see [10], Section 5.24), which is useful for models involving fractional powers of operators.

In this article, we analyze the construction and nature of the Grünwald–Letnikov and Marchaud derivatives starting from the original ideas of their creators. Section 2 is devoted to the Letnikov's contribution and Section 3 deals with Marchaud's approach. Some notes on common features and differences between these two constructions are presented in Section 4.

2. Letnikov Contribution to Fractional Calculus

Here, we describe the construction of so-called "general differentiation" (The words utilized by Letnikov for his derivative.) (or differentiation of arbitrary order) proposed by Alexej Vasil'evich Letnikov in the 1860s. It is based on certain results by Liouville that Letnikov considered as the most important results on fractional differentiation.

2.1. Preliminaries

Let us briefly describe some results of predecessors to the work by Letnikov. In spite of the fact that the idea of fractional derivative goes back to the end of the 17th century (see [30]), the real results in the area were made by Liouville [3,4].

Liouville applied his construction to the function representing in the form of the following (convergent!) series:

$$y(x) = \sum_{k=1}^{\infty} A_k e^{m_k x}. \tag{6}$$

For such functions, he used Leibnitz's idea of an arbitrary order differentiation of the exponential function:

$$\frac{d^p y}{dx^p} := \sum_{k=1}^{\infty} A_k m_k^p e^{m_k x}. \tag{7}$$

Liouville considered definition (7) as only possible, stressing (see [1], p. 14): "...it is impossible to get an exact and complete understanding of the nature of the arbitrary order derivative without taken certain series representation of the function".

Letnikov noted that Liouville's construction being deeply justified has an essential drawback. It follows from the definition (7) that it can be used only for functions whose derivatives (of all positive integer orders) are vanishing at infinity. Liouville himself met the first difficulty trying to apply his definition to power function with positive exponent $x^m, m > 0$, which does not satisfy the above condition. To overcome this difficulty he started with the function $y(x) = \frac{1}{x^m}, m > 0$. It follows from the definition of the Γ-function that the following relation holds:

$$\frac{1}{x^m} = \frac{1}{\Gamma(m)} \int_0^{\infty} e^{-zx} z^{m-1} dz.$$

The right-hand side of this formula can be considered as an expansion of the type (6), namely, $\sum A_n e^{-nx}$ with A_n being sufficiently small as $x \to \infty$. Thus, formula (7) applied in this case leads to the following result:

$$\frac{d^p \frac{1}{x^m}}{dx^p} = \frac{1}{\Gamma(m)} \int_0^{\infty} e^{-zx}(-z)^p z^{m-1} dz = \frac{(-1)^p}{\Gamma(m) x^{m+p}} \int_0^{\infty} e^{-t} t^{p+m-1} dz = \frac{(-1)^p \Gamma(m+p)}{\Gamma(m) x^{m+p}}, \tag{8}$$

which coincides with the celebrated Euler formula for the arbitrary order derivative of the power function.

Note that Liouville considered only the case when $m > 0$ and $m + p > 0$. (At this time, the Legendre-Gauss definition of Γ-function of the complex argument was not known.) For the remaining cases, he used the notion of so called additional functions in order to correct the above definition. These are the functions whose derivative of order $(-p)$ is equal to zero. Liouville gave a proof that additional functions should have the form:

$$A_0 + A_1 x + \ldots + A_n x^n,$$

with certain finite power n and arbitrary constant coefficients A_j. (This proof was not considered satisfactory by many mathematicians even in Liouville's time.)

During the next 30 years, several attempts to correct Liouville's construction were made (the most known results were due to Peacock, Kelland, Tardi, Roberts, see ([1], pp. 16–23), see also Chapter 1 by Butzer and Westphal in [18]). Anyway, only in 1867–1868, it was proposed by A.K. Grünwald [2] and by A.V. Letnikov [5] a really general definition of the fractional derivative. Both constructions (which differ only in few details) are based on the following formula for representation of the derivative of an arbitrary positive integer order via finite differences:

$$\frac{d^p f(x)}{dx^p} = \lim_{h \to +0} \frac{f(x) - \binom{p}{1} f(x-h) + \binom{p}{2} f(x-2h) - \ldots + (-1)^n \binom{p}{n} f(x-nh)}{h^p}, \tag{9}$$

where n is an arbitrary positive number, $n \geq p$.

2.2. Letnikov's Construction

In his construction [5] (see also [1] where the article [5] is reprinted practically without any changes.), Letnikov started with the now known formula (9) of the derivative of an integer order $p = m$. By assuming p being an arbitrary positive number and n arbitrary sufficiently large positive integer number, he supposed by definition:

$$f^{p)}(x) := \lim_{h \to 0} \frac{f(x) - \binom{p}{1} f(x-h) + \binom{p}{2} f(x-2h) - \ldots + (-1)^n \binom{p}{n} f(x-nh)}{h^p}, \tag{10}$$

where $\binom{k}{r} = \frac{k(k-1)\ldots(k-r+1)}{1 \cdot 2 \cdot \ldots \cdot r}$. It was noted that if p is a positive integer, $p \leq n$, then formula (10) leads to the derivative of p-th order.

Letnikov also considered an expression:

$$f^{-p)}(x) := \lim_{h \to 0} h^p \left[f(x) + \binom{p}{1} f(x-h) + \binom{p}{2} f(x-2h) + \ldots + \binom{p}{n} f(x-nh) \right].$$

If p is a positive integer in the last formula, then $f^{-p)}(x)$ is vanishing whenever n is finite. Hence, it is interesting to consider the case when n tends to infinity as $h \to 0$. Assuming $h = (x - x_0)/n$ Letnikov defined the following object (called the derivative of a negative order in finite limits by him)

$$\left[D^{-p} f(x) \right]_{x_0}^{x} := \lim_{h \to 0} f^{-p)}(x). \tag{11}$$

It was shown that, for p being a positive integer, the following relation holds whenever the integral on the right-hand side exists:

$$\left[D^{-p}f(x)\right]_{x_0}^{x} = \lim_{h\to 0}\sum_{r=0}^{n} h^p \left(\begin{array}{c} p \\ r \end{array}\right) f(x-rh) = \frac{1}{1\cdot 2\cdot \ldots \cdot (p-1)} \int_{x_0}^{x} (x-\tau)^{p-1} f(\tau)d\tau. \quad (12)$$

It follows from (12) that if this formula is valid for certain positive integer p, then it is valid for $p+1$ too. Moreover, the above introduced object $\left[D^{-p}f(x)\right]_{x_0}^{x}$ means the function whose p-th derivative coincides with $f(x)$:

$$\frac{d^p}{dx^p}\left[D^{-p}f(x)\right]_{x_0}^{x} = f(x),$$

and

$$\frac{d^j}{dx^j}\left[D^{-p}f(x)\right]_{x_0}^{x}\Big|_{x=x_0} = 0, \quad \forall j = 0,1,\ldots,p-1.$$

The above definition and its properties constitute the base for further generalizations. Thus, by using the properties of binomial coefficients, Letnikov proved that for any function f continuous on $[x_0, x]$ there exist the following limits:

$$\lim_{n\to\infty}\frac{\sum\limits_{r=0}^{n}(-1)^r \left(\begin{array}{c} p \\ r \end{array}\right) f(x-rh)}{h^p}, \quad \lim_{n\to\infty}\sum_{r=0}^{n} h^p \left(\begin{array}{c} p \\ r \end{array}\right) f(x-rh),$$

where $h = (x-x_0)/n$ and $p \in \mathbb{C}$, $\operatorname{Re} p > 0$. The values of these limits denoted by him $\left[D^{p}f(x)\right]_{x_0}^{x}$ and $\left[D^{-p}f(x)\right]_{x_0}^{x}$, respectively, are formally equal in this case to:

$$\left[D^{p}f(x)\right]_{x_0}^{x} = \frac{1}{\Gamma(-p)}\int_{x_0}^{x}\frac{f(\tau)d\tau}{(x-\tau)^{p+1}}, \quad (13)$$

$$\left[D^{-p}f(x)\right]_{x_0}^{x} = \frac{1}{\Gamma(p)}\int_{x_0}^{x}\frac{f(\tau)d\tau}{(x-\tau)^{1-p}}. \quad (14)$$

Formula (13) gives the derivative of arbitrary order $p \in \mathbb{C}$, $\operatorname{Re} p > 0$, and formula (14) gives the integral of arbitrary order $p \in \mathbb{C}$, $\operatorname{Re} p > 0$. As it was already mentioned, the integral in (14) exists whenever f is continuous, though it is not the case for the integral in (13). To overcome this difficulty, Letnikov transformed the right-hand side of (13) to the form:

$$\left[D^{p}f(x)\right]_{x_0}^{x} = \sum_{k=0}^{m}\frac{f^{(k)}(x_0)(x-x_0)^{-p+k}}{\Gamma(-p+k+1)} + \frac{1}{\Gamma(-p+m+1)}\int_{x_0}^{x}\frac{f^{(m+1)}(\tau)d\tau}{(x-\tau)^{p-m}}. \quad (15)$$

Surely, an assumption of existence and continuity of all derivatives up to the order $m+1$ is sufficient for this representation. It is suitable here to take $m = [\operatorname{Re} p]$.

Integration by parts in (14) leads to an analogous formula for $\left[D^{-p}f(x)\right]_{x_0}^{x}$, which is valid under the same conditions for any integer positive m:

$$\left[D^{-p}f(x)\right]_{x_0}^{x} = \sum_{k=0}^{m}\frac{f^{(k)}(x_0)(x-x_0)^{p+k}}{\Gamma(p+k+1)} + \frac{1}{\Gamma(p+m+1)}\int_{x_0}^{x}\frac{f^{(m+1)}(\tau)d\tau}{(x-\tau)^{p-1+m}}. \quad (16)$$

It can be considered also a special case when $x_0 \to +\infty$. It follows from (14):

$$\left[D^{-p}f(x)\right]_{+\infty}^{x} = \frac{(-1)^p}{\Gamma(p)}\int_{0}^{+\infty}\tau^{p-1}f(x+\tau)d\tau, \quad (17)$$

and from (15) we obtain:

$$[D^p f(x)]_{+\infty}^{x} = \frac{(-1)^{m+1-p}}{\Gamma(-p+m+1)} \int_{0}^{+\infty} \tau^{m-p} f^{(m+1)}(x+\tau)d\tau. \quad (18)$$

Letnikov noted that the integrals on the right-hand side of (18) and (17) converge in particular if $\lim_{x_0 \to +\infty} f^{(k)}(x_0) = 0$, $k = 0, \ldots, m$. This is exactly the class of functions considered by Liouville and formulas (17) and (18) that coincide with the corresponding formulas presented in [3,4].

2.3. Further Results on Grünwald–Letnikov Fractional Derivative/Integral

After the death of Letnikov, his theory did not get a serious development neither in Russia nor abroad. We have to mention the work by Nekrasov [31] who tried to overcome additional assumptions by Letnikov ensuring existence of the integrals appeared after passing to a limit. In his work, Nekrasov used the technique of Complex Analysis, in particular, the behaviour of complex valued function in a neighbourhood of branch-points.

Only much later, the Grünwald–Letnikov approach was retranslated into the modern language. The most essential improvement was presented in the article by Butzer and Westphal [32] and by Bugrov [33]. Historical remarks describing this development is presented in ([10], §23). Here we present the most essential results on the Grünwald–Letnikov fractional derivative.

In [10], the Grünwald–Letnikov fractional derivative is defined via differences of arbitrary order for functions from the space $X = X(\mathbb{R}^1)$ (it is either $L_p(\mathbb{R}^1)$ or $C(\mathbb{R}^1)$) or for 2π-periodic functions from the same type of spaces denoting $X_{2\pi} = X(0, 2\pi)$. Such differences of arbitrary order are defined as follows:

$$(\Delta_{\pm t}^{\alpha} f)(x) = \sum_{k=0}^{\infty} (-1)^k \left(\begin{array}{c} \alpha \\ k \end{array} \right) f(x \mp kt). \quad (19)$$

Left-sided/right-sided differences correspond to upper/lower signs. Letnikov proved convergence of differences (19) for any $\alpha > 0$ and each bounded continuous function. In fact, absolute and uniform convergence follows already for any bounded function. If the function f belongs to one of the above spaces, then the convergence takes place in the norm of the corresponding space. Classical definition of the *Grüwald–Letnikov fractional derivative* remains the same as in original papers [2,5]:

$$f_{\pm}^{\alpha)}(x) = \lim_{h \to +0} \frac{(\Delta_{\pm t}^{\alpha} f)(x)}{h^{\alpha}}, \quad \alpha > 0. \quad (20)$$

If the convergence in (20) is understood in the norm of the corresponding space, then $f_{\pm}^{\alpha)}(x)$ is called the strong Grüwald–Letnikov fractional derivative.

Existence of the strong Grüwald–Letnikov fractional derivative in the periodic case gives the following:

Theorem 1. *(Theorem 20.1 [10]) Let $f \in X_{2\pi}$. The strong Grüwald–Letnikov fractional derivative exists in the corresponding space iff there exists a function $\varphi_{\pm} \in X_{2\pi}$ such that the following representation holds*

$$f(x) = I_{\pm}^{(\alpha)} \varphi_{\pm} + f_0, \quad f_0 = \frac{1}{2\pi} \int_{0}^{2\pi} f(x)dx, \quad (21)$$

where $I_{\pm}^{(\alpha)}$ are the Liouville fractional integrals:

$$I_{+}^{(\alpha)} \phi(x) = \frac{1}{\Gamma(\alpha)} \int_{-\infty}^{x} \frac{\phi(\tau)d\tau}{(x-\tau)^{1-\alpha}}, \quad I_{-}^{(\alpha)} \phi(x) = \frac{1}{\Gamma(\alpha)} \int_{x}^{+\infty} \frac{\phi(\tau)d\tau}{(\tau-x)^{1-\alpha}}.$$

If $f_{\pm}^{(\alpha)}(x)$ *exists, then* $\varphi_{\pm}(x) = f_{\pm}^{(\alpha)}(x)$.

In a nonperiodic case, an existence (in sense of L_p-norm, $p > 1$) of the strong Grünwald–Letnikov fractional derivative of the function $f \in L_r(\mathbb{R}^1)$ implies that f can be represented by the (Liouville) fractional integral of this fractional derivative.

2.4. Short Biography of A.V. Letnikov

Alexey Vasil'evich Letnikov was born in Moscow on 1 January 1837 in the family of a rather poor nobleman. His father died when Alexey was 8 years old. His mother tried to give an education to Alexey and his sister. His mother sent Alexey to grammar school in 1847. In spite of his evident abilities, he was not too successful in education. Therefore, he was moved to Konstantin's land surveyors institute (full-time provisional military type institute). This was a second rank educational establishment. Anyway, the director of the Institute tried to do the best for good students. He discovered Alexey's big interest in mathematics and supported his growth in the subject. The director decided to prepare him for a career of a teacher in mathematics in Konstantin's land surveyors institute. To get the corresponding position, Letnikov was sent to Moscow University and studied mathematics there for two years (1856–1858) as an extern student. After graduation, he was sent to Paris in order to extend his knowledge at the most well-known mathematical center for two years and to study the structure and the content of the technical education in France.

In Paris, Letnikov attended the lectures of many well-known mathematicians (Liouville among them) in the Ecole Polytechnique, College de France and Sorbonne. Returning from Paris in December 1860, he was appointed as a teacher in the engineering class of Konstantin's land surveyors institute and started to teach Probability Theory. Letnikov actively participated in mathematical life in Moscow. In particular, he was among the founders of Moscow Mathematical Society in 1864.

In 1863, a new Statute of Higher Education was approved. Among other regulations, it was supposed to enlarge a number of chairs at universities and to recruit new university teachers. To get a position in university, one had either to pass the graduation gymnasium's exams or to receive the degree at a foreign university. Letnikov decided to use the second option. In 1867, he defended a Ph.D. "Über die Bedingungen der Integrabilität einiger Differential-Gleichungen" at Leipzig university.

In 1868, Konstantin's land surveyors institute moved from the military ministry to a civilian one and its teachers had to leave the military service. Therefore, Letnikov searched for another position as a teacher of mathematics. At that time, the Imperial Technical College (now Bauman's Technical University) was reopened after being transformed from the Moscow Industrial School and Letnikov got a position there. He was working at this College up to 1883 when he moved to Alexandrov's Commercial College sharing this job with a part-time teacher at Konstantin's land surveyors institute and at the Imperial Technical College. It was an active time for him, and he was awarded the degree of a state councillor, received the order of Saint Stanislav and was appointed in 1884 as a corresponding member of St.-Petersburg Academy of Sciences (by recommendation of Imshenetsky, Bunyakovsky and Backlund).

Letnikov's main scientific results were obtained during the 20 years starting from the middle of the 1860s. In 1866, he began to work on the theme "differentiation of arbitrary order". When the work was almost ready (and he supposed to defend it as a Master's thesis), he found the article by a professor from Prague university, A.K. Grünwald [2], who exploited the same idea. Only due to a professor Davidov from Moscow University (who knew the research of Letnikov and Grünwald and recommended to Letnikov to complete the presentation) was the Master's thesis defended and published at Mathematical Sbornik in 1868 [5]. (This work as well as the work [34] was reprinted in [1].) Letnikov's theory was not completely recognized at that time. Thus, in 1972, Sonin (future academician) published a paper [35] in which he criticized some statements of Letnikov's theory. Letnikov answered [34] in the same year with an explanation of the basic results of his theory recovering mistakes made by Sonin. In 1874, Letnikov published a paper [6] in which he corrected the models taken by Liouville in his works [3] and showed how one can use the new theory of fractional differentiation to attack the problems arising. It should be noted that Letnikov published all his results in Russian at Mathematical Sbornik. This is why these works are not too familiar for mathematicians outside Russia.

The personal life of A.V. Letnikov was quite heavy. He married in 1863 and had a large family. Because of that, he had to work hard teaching at several institutions. At the end of the 1880s, Letnikov should have received a state pension and was supposed to leave teaching and concentrate on research. He was dreaming of getting a position at Moscow University, but it did not happen, since, at the opening ceremony of a new building of Alexandrov's Commercial College, he caught a cold. He had no serious illness before and continued to deliver lectures. However, this time his disease was serious. Moreover, it was complicated by typhoid fever and he died on 27 February 1988.

3. Marchaud Contribution to Fractional Calculus

3.1. Main Constructions

In his thesis (published in 1927 [8] and reprinted in [9]), André Paul Marchaud (1887–1973) started from the Riemann–Liouville construction and tried to generalize it by using new knowledge on the properties of the functions of real variables. By examination of the main properties of the Riemann-Lioville integral $\left(I_{a+}^{(\alpha)}f\right)(x) = \frac{1}{\alpha}\int\limits_{a}^{x-a} \tau^{\alpha-1}f(x-\tau)d\tau$ and derivative $\left(D_{a+}^{(\alpha)}f\right)(x) = \frac{d^n}{dx^n}I_{a+}^{n-\alpha}f(x), n > \alpha$, he noted that $\left(I_{a+}^{(\alpha)}f\right)$ is well-defined for any bounded function (respectively, bounded and integrable in the case of Liouville fractional integral with integration along semi-axes). As for the Riemann–Liouville fractional derivative, it can be unbounded at the lower end of integration as it follows, e.g., from the simple example:

$$\left(D_{a+}^{(\alpha)}(1)\right)(x) = \frac{(x-a)^{-\alpha}}{\Gamma(1-\alpha)}.$$

Marchaud studied such situation, taking into account the following consideration. If the function f admits a bounded derivative of an integer order r on the interval (a, a_1), then there exists the derivative $\left(D_{a+}^{(\alpha')}(f)\right)(x)$ of any order $\alpha' < r$. Such derivatives can be defined via formula:

$$\left(D_{a+}^{(\alpha')}(f)\right)(x) = \left(I_{a+}^{(r-\alpha')}f^{(r)}\right)(x) + \sum_{j=0}^{r-1}\frac{(x-a)^{j-\alpha'}}{\Gamma(j+1-\alpha')}f^{(j)}(x). \tag{22}$$

Any derivative $D_{a+}^{(\alpha')}(f)$ is unbounded at $x = a$, but it is bounded on each subinterval $(a', a_1) \subset (a, a_1)$.

Marchaud proved the following relation (valid in the above conditions with noninteger α and $r = [\alpha]$)

$$f(x) = \left(I_{a'+}^{(\alpha)}D_{a+}^{(\alpha)}f\right)(x) + \sum_{j=1}^{r}\frac{(x-a')^{j+\alpha-r-1}}{\Gamma(j+\alpha-r)}F^{(j)}(a') + \frac{(x-a')^{\alpha-r}}{\Gamma(\alpha-r)\Gamma(r+1-\alpha)}\int\limits_{a}^{a'}\frac{(a'-t)^{r-\alpha}f(t)dt}{(x-t)}, \tag{23}$$

where

$$F(x) := \left(I_{a+}^{(r+1-\alpha)}f\right)(x).$$

It gives, in particular, that if all derivatives $\left(D_{a'+}^{(\alpha')}f\right)(x)$ $(\alpha' < \alpha)$ exist and are continuous on the interval (a', a_1), then there exists the derivative $\left(D_{a+}^{(\alpha)}f\right)(x)$, which is bounded on (a', a_1). Such result was obtained by Montel (see [36]) under additional conditions that $\left(D_{a+}^{(\alpha)}f\right)(x)$ is bounded on the interval (a, a_1). Montel proved:

$$\left(D_{a+}^{(\alpha')}f\right)(x) = \left(I_{a+}^{(\alpha-\alpha')}D_{a+}^{(\alpha)}f\right)(x) + \sum_{j=1}^{r}\frac{(x-a)^{j+\alpha-\alpha'-r-1}}{\Gamma(j+\alpha-\alpha'-r)}F^{(j)}(a), \ \ \alpha' < \alpha.$$

Thus, sufficient conditions for continuity of these derivatives at $x = a$ is vanishing of all derivatives $F^{(j)}(a)$ and $\lim_{x\to a}(x-a)^{-\alpha+1}f(x) = 0$.

Marchaud noted that relation (23) can not be taken for a definition of the derivative of a noninteger order since the function f remains in the last term on the right-hand side. Therefore, he tried to use another approach to the definition taking into account the above consideration on the behaviour of the function at the initial point $x = a$. The starting point for this definition was a formula obtained by the formal replacement of positive α in the definition of fractional integral for a negative parameter:

$$\left(D_{a+}^{(\alpha)}f\right)(x) = \frac{1}{\Gamma(-\alpha)}\int_0^{x-a}\frac{f(x-t)dt}{t^{\alpha+1}}. \tag{24}$$

Since the integral on the right-hand side is divergent, this definition needs some transformation. For the function defined on the interval (a, a_1), it was proposed to extend it by zero on $(-\infty, a]$ and to denote the corresponding fractional integral by $f_{(\alpha)}$ and fractional derivative by $f^{(\alpha)}$.

In order to find a proper transformation, Marchaud started to regularize fractional integrals. In the following formula (valid, e.g., for exponential function):

$$f_{(\alpha)}(x)\int_0^\infty t^{\alpha-1}e^{-t}dt = \int_0^\infty t^{\alpha-1}f(x-t)dt,$$

he replaced t for $k_j t$ ($j = 0, 1, \ldots, p$) and made a linear combination of the obtained equalities with unknown coefficients $C_j, j = 0, 1, \ldots, p$. It leads to the following formula:

$$f_{(\alpha)}(x)\int_0^\infty t^{\alpha-1}\psi(t)dt = \int_0^\infty t^{\alpha-1}\varphi(x,t)dt, \tag{25}$$

where

$$\psi(t) = \sum_{j=0}^{p}C_j e^{-k_j t}, \quad \varphi(x,t) = \sum_{j=0}^{p}C_j f(x-k_j t). \tag{26}$$

Then, replacing α for $-\alpha$, we arrive at the following formal relation, which needs some extra conditions for its validity:

$$f^{(\alpha)}(x)\int_0^\infty t^{-\alpha-1}\psi(t)dt = \int_0^\infty t^{-\alpha-1}\varphi(x,t)dt. \tag{27}$$

If the integral:

$$\gamma(\alpha) = \int_0^\infty t^{-\alpha-1}\psi(t)dt$$

has a sense for certain $\alpha = \alpha_0$, then γ_α is defined and continuous for all $\alpha \le \alpha_0$.

Denoting $r = [\alpha]$, one can see that $\gamma(\alpha)$ has a sense whenever $\psi(t)$ has the order of infinitesimality equal to $r+1$. It is required:

$$\sum_{j=0}^{p}k_j^s C_j = 0, \quad s = 0, 1, \ldots, r.$$

r-times integration by parts gives:

$$\gamma(\alpha) = \Gamma(-\alpha) \sum_{j=0}^{p} k_j^{\alpha} C_j \quad (\alpha \neq r); \quad \gamma(r) = \lim_{\alpha \to r} \gamma(\alpha) = \frac{(-1)^{r+1}}{r!} \sum_{j=0}^{p} k_j^r C_j \log k_j. \tag{28}$$

The most simple function $\psi(t)$ that has an order of infinitesimality equal to p is the following one:

$$\psi(t) = e^{-t}\left(1 - e^{-t}\right)^p = e^{-t} - \begin{pmatrix} p \\ 1 \end{pmatrix} e^{-2t} + \begin{pmatrix} p \\ 2 \end{pmatrix} e^{-3t} + \dots. \tag{29}$$

In this case:

$$\varphi(x,t) = f(x-t) - \begin{pmatrix} p \\ 1 \end{pmatrix} f(x-2t) + \begin{pmatrix} p \\ 2 \end{pmatrix} f(x-3t) + \dots. \tag{30}$$

The simple properties of the Marchaud derivatives are similar to that of the Liouville derivative, e.g., permutability with the operators of reflection, translation and scaling (see [10], Section 2.5), composition formulas with the singular integral operator S and relation between $\mathbb{D}_-^{\alpha}$ and $\mathbb{D}_+^{\alpha}$:

$$\left(\mathbb{D}_-^{\alpha} f\right) = \cos\alpha\pi \left(\mathbb{D}_+^{\alpha} f\right) - \sin\alpha\pi \left(S\mathbb{D}_+^{\alpha} f\right).$$

Among characteristic properties, we have to point out the vanishing of the Marchaud derivative on the constant function:

$$\mathbb{D}_{\pm}^{\alpha} const \equiv 0.$$

For all $\alpha > 0$, the Marchaud derivative $\mathbb{D}_{\pm}^{\alpha}$ is defined on all bounded functions $f \in C^{[\alpha]}(\mathbb{R}^1)$ satisfying the following condition:

$$|f^{([\alpha])}(x+h) - f^{([\alpha])}(x)| \leq A(x)|h|^{\lambda}.$$

3.2. Condition on Convergence

In the case of "small" parameter α (see (4)), the Marchaud derivative is defined, for instance, in the class of the locally Hölder continuous bounded functions $H^{\lambda}_{loc}(\mathbb{R})$ with Hölder exponent $\lambda > \alpha$. Further analysis of convergence was done by Marchaud himself (see [8], Chapter 2). He proposed to introduce the fractional derivative $f^{(\alpha)}$ by formula (27) (with a proper choice of the constant $\gamma(\alpha)$). He gave the following condition for the existence of such derivative: *the necessary and sufficient condition for a function f continuous on (a, a_1) to admit the fractional derivative $D^{\alpha'}_{a+} f$ of any order $\alpha' \leq \alpha$ is the uniform convergence of the integral $\int\limits_{\varepsilon}^{\infty} t^{-\alpha-1}\varphi(x,t)dt$ on any subinterval $(a', a_1) \subset (a, a_1)$.*

It was noted that the order of infinitesimality of:

$$\left(\Delta_h^n f\right)(x) := f(x+nh) - \begin{pmatrix} n \\ 1 \end{pmatrix} f(x+(n-1)h) + \begin{pmatrix} n \\ 2 \end{pmatrix} f(x+(n-2)h) + \dots$$

depends on the behaviour of the function f at a point x. For the function f defined and bounded on $(0,1)$, Marchaud (following [37]) considered the properties of the generalized modulus of continuity defined as:

$$\omega_n(\delta) = \omega_n(\delta; f(x)) := \sup_{|h| \leq \delta \leq \frac{1}{n}} \left(\Delta_h^n f\right)(x) \quad (0 \leq x \leq 1, 0 \leq x+nh \leq 1). \tag{31}$$

Then, the following statement is valid: *letting the integral* $\int_0 t^{-\alpha-1}\omega_n(t)dt$ *be convergent for certain* α *and* n $(\alpha < n)$*, then the function* f *admits the derivative of any order* $\leq \alpha$ *given by formula* (27).

For instance,

$$f'(x) = \frac{1}{2\log 2}\int_0^\infty t^{-2}\left[f(x) - 2f(x-t) + f(x+2t)\right] dt,$$

and, for $n = 1, 0 < \alpha < 1$,

$$f^{(\alpha)} = \frac{\alpha}{\Gamma(1-\alpha)}\int_0^\infty t^{-\alpha-1}\left[f(x) - f(x-t)\right] dt.$$

Finally, Marchaud made the following remark ([8], p. 396) concerning the differences of noninteger order (citing the paper by Liouville):

$$\left(\Delta_h^\alpha f\right)(x) := f(x+\alpha h) - \begin{pmatrix} \alpha \\ 1 \end{pmatrix} f(x + (\alpha-1)h) + \begin{pmatrix} \alpha \\ 2 \end{pmatrix} f(x + (\alpha-2)h) + \ldots$$

He proved that if the following limit exists:

$$\lim_{\delta \to +0} \delta^{-\alpha}\Delta_\delta^\alpha f(x) = g(x), \tag{32}$$

for certain $\alpha > 0$, then the following relation holds (Thus, $g(x)$ can be considered as a fractional derivative of the function $f(x)$ of order α):

$$f(x) = I^{(\alpha)}g(x).$$

3.3. Further Results on the Marchaud Derivative

As it was already mentioned, the Marchaud derivative appeared due to a formal replacement of the positive parameter α for negative one $-\alpha$ in the definition of the Liouville fractional integral. The appeared object is not well-defined. In order to give a sense to the integral in (3), one can use Hadamard's finite part (p.f.) (see [10], Lemma 5.2):

$$\left(\mathbb{D}_\pm^\alpha f\right)(x) = p.f.\left(I_\pm^{(-\alpha)} f\right)(x).$$

An application of this approach gives, in particular, the following result (see [10], Lemma 5.3): let $f \in C_{loc}^{m,\lambda}(\mathbb{R}), 0 \leq \lambda < 1, m = [\alpha], \alpha > 0, \alpha \neq 1, 2, \ldots$. Then, the following representation holds:

$$p.f.\int_0^\infty \frac{f(x-t)dt}{t^{1+\alpha}} = \int_0^\infty \frac{f(x-t) - \sum\limits_{k=0}^{m} \frac{(-1)^k f^{(k)}(x)}{k!}}{t^{1+\alpha}} dt.$$

The Marchaud derivative is a suitable object for representation of the fractional powers of the operators. Thus, for instance,

$$\left(\frac{d}{dx}\right)^\alpha = \frac{1}{\Gamma(-\alpha)}\int_0^\infty \frac{\phi(x-t) - \phi(x)}{t^{1+\alpha}} dt = \left(\mathbb{D}_+^{(\alpha)}\phi\right)(x),$$

where $\phi \in \left\{\phi(t) : \phi'(t) \in L_{p,\omega}\right\}, \omega > 0, 0 < \alpha < 1$.

Marchaud's ideas is applied in the study of another constructions, such as the Hadamard type derivative in [29]:

$$\left(\mathcal{D}_{0+,\mu}^{(\alpha)} f\right) = x^{1-\mu} \frac{d}{dx} \frac{1}{\Gamma(1-\alpha)} \int_0^x \left(\log \frac{x}{t}\right)^{-\alpha} \frac{f(t)dt}{t^{1-\mu}} \quad (0 < \alpha < 1).$$

It was shown that, for any function $f \in X_c^p(\mathbb{R}_+) = \left\{ g : \left(\int_0^\infty |t^c g(t)|^p \frac{dt}{t} \right)^{1/p} < +\infty \right\}$, the following relation holds: $\left(\mathbf{D}_{0+,\mu}^{\alpha} f\right)(x) = \left(\mathcal{D}_{0+,\mu}^{(\alpha)} f\right)$, $0 < \alpha < 1$, where

$$\left(\mathbf{D}_{0+,\mu}^{\alpha} f\right)(x) = \frac{\alpha}{\Gamma(1-\alpha)} \int_0^\infty e^{-\mu t} \frac{f(x) - f(xe^{-t})}{t^{1+\alpha}} dt + \mu^\alpha f(x) =$$

$$= \frac{\alpha}{\Gamma(1-\alpha)} \int_0^x \left(\frac{t}{x}\right)^\mu \left(\log \frac{x}{t}\right)^{-\alpha-1} \frac{f(x) - f(t)}{t} dt + \mu^\alpha f(x).$$

The difference between the Marchaud derivatives on semi-axes and the Liouville derivatives (e.g., for small values of parameter α, $0 < \alpha < 1$)

$$(\mathcal{D}_+^\alpha f)(x) = \frac{1}{\Gamma(1-\alpha)} \frac{d}{dx} \int_{-\infty}^x \frac{f(t)dt}{(x-t)^\alpha}, \quad (\mathcal{D}_-^\alpha f)(x) = -\frac{1}{\Gamma(1-\alpha)} \frac{d}{dx} \int_{-\infty}^x \frac{f(t)dt}{(x-t)^\alpha}$$

is related to invertibility of fractional integral. Thus, the identity:

$$\mathcal{D}_\pm^\alpha I_\pm^\alpha f \equiv f$$

is valid only for $f \in L_1(\mathbb{R}^1)$, but the identity:

$$\mathbb{D}_\pm^\alpha I_\pm^\alpha f \equiv f$$

is valid for all $f \in L_p(\mathbb{R}^1), 1 \le p < 1/\alpha$. Analogous results are valid for the Marchaud derivatives on a finite interval and the Riemann–Liouville fractional derivative: if $f = I_{a+}^\alpha \varphi$, $\varphi \in L_1(a,b)$, then the Marchaud derivatives on the interval (a,b) and the Riemann–Liouville derivatives coincide almost everywhere (see [10], Theorem 13.1). In particular, these derivatives coincide (a.e.) for all functions f absolutely continuous on (a,b).

Since the Marchaud fractional derivative is defined via finite differences, then it follows that sufficient conditions on a function f to have the Marchaud fractional derivative of any order $\le \alpha$ is the convergence of the integral (see [8]):

$$\int_0 \frac{\omega_n(t)dt}{t^{\alpha+1}},$$

where ω_n is the finite difference of f of order $n = [\alpha + 1]$. Vice versa, if such integral diverges for certain $\alpha > 0$, then the function f does not have a bounded fractional Marchaud derivative of any order greater than α. One of the possible ways of using the Marchaud derivative in a discrete case is discussed in [28].

Marchaud fractional derivative has a form similar to Riesz fractional derivative (or hypersingular integral) and thus possesses multidimensional generalization (see [10], Section 26). One of the most prominent constructions for certain domains $\Omega \subseteq \mathbb{R}^n$ is given in [38]:

$$\left(\mathbb{D}^{\alpha}_{\Omega} f\right)(\mathbf{x}) = c(\alpha)\left[a_{\Omega}(\mathbf{x})f(\mathbf{x}) + \int_{\Omega} \frac{f(\mathbf{x}) - f(\mathbf{y})}{|\mathbf{x}-\mathbf{y}|^{n+\alpha}} d\mathbf{y}\right], \quad \mathbf{x} \in \Omega,$$

where

$$a_{\Omega}(\mathbf{x}) = \int_{\mathbb{R}^n \setminus \Omega} \frac{d\mathbf{y}}{|\mathbf{x}-\mathbf{y}|^{n+\alpha}}, \quad c(\alpha) = \frac{2^{\alpha}\Gamma(1+\alpha/2)\Gamma((n+\alpha)/2)\sin(\pi\alpha)/2}{\pi^{1+n/2}}.$$

3.4. Short Biography of A.P. Marchaud

André Paul Marchaud was born (Here, we follow the biographical paper [39].) in Saintes (Charente-Maritime) on 27 April 1887 in a religious family (Protestants). He received his primary education at Saintes and Paris, and then returned to secondary school in Saintes. At the age 17, he got ill with pleurisy. He obtained his Bachelor's degree in 1906, and then studied in College Chaptal preparing for entering École normal supérieure or École polytechnique. Marchaud studied at the École normal supérieure from 1909–1912, got his Master's degree in 1911. After graduating from École normal supérieure, he obtained an aggregation in mathematics in 1912. He was a teacher in mathematics at the high school in Montauban (from 1 October 1913 until 2 October 1919).

In 1914, Marchaud was mobilized as a second lieutenant of the 344 infantry regiment of Bordeaux. He was captured on 20 August 1914, and transferred to Germany. He was a prisoner for more than two years. In 1917, Marchaud became ill and was moved to Switzerland by the Red Cross. He returned to France on 6 July 1918.

In France, Marchaud continued his work as a teacher in mathematics at high schools (in Montauban and in Monpellier). In March, 1927, he defended a Doctor of Science thesis "On derivatives and differences of functions of real variables" (chairman of the committee Montel, examiners Denjoy and Chazy). In the period from October 1927 until October 1938, he occupied different positions at the mathematical faculty in Marseille. A.P. Marchaud was a rector of the Academy Clermont (1938–1944), Academy of Bordeaux (1944–1950) and a member of rectorate of the city University, Paris (1950–1957). He retired in August 1957 and passed away on 15 October 1973.

4. Conclusions

In this paper, we describe two approaches to the definition of fractional derivatives. The aim of both constructions is to define a fractional derivative that can be applied to a wide class of functions.

Letnikov's (1) and Marchaud's (3) derivatives are based on the use of differences. Thus, Letnikov studied the properties of differences of an arbitrary order and applied them to define his derivative. Marchaud proposed the regularization of the divergent integral by using finite differences of an integer order. The functions to which these constructions can be applied are not differentiable in general. One more common feature of these fractional derivatives is that both of them are started from the Riemann–Liouville fractional derivative, which needs to be simplified in order to be applied to a rather wide class of functions.

We also have to mention the following result on coincidence of Letnikov's and Marchaud's derivatives.

Theorem 2. *(Theorem 20.4 [10]) Let $f \in L_r(\mathbb{R})$ for certain $1 \le r < \infty$. Then, the following limits $f^{\alpha)}_{\pm} = \lim_{h\to+0,(L_p(\mathbb{R}))} \frac{(\Delta^{\alpha}_{\pm h} f)(x)}{h^{\alpha}}$ and $\left(\mathbb{D}^{\alpha}_{\pm} f\right)(x) = \lim_{\varepsilon\to+0,(L_p(\mathbb{R}))} \left(\mathbb{D}^{\alpha}_{\pm,\varepsilon} f\right)(x)$ exist simultaneously and coincide (for the same choice of signs and $1 < p < 1/\alpha$).*

Letnikov's idea is briefly mentioned in the main paper by Marchaud in spite of the fact that Marchaud never read the paper by Letnikov or Grünwald. Since 50 years separate Letnikov's and Marchaud's proposals, we can see that Marchaud's paper already explored the results from the theory functions, the theory of sets and the theory of integration developed at the beginning of the 20th century by French and Russian schools.

In conclusion, we can say that both constructions are prominent and we suppose that many mathematicians would apply them in their work.

Acknowledgments: The work has received funding from the People Programme (Marie Curie Actions) of the European Union's Seventh Framework Programme FP7/2007–2013/ under the REAGrant Agreement PIRSES-GA-2013-610547—TAMER and supported by the Belarusian Fund for Fundamental Scientific Research through the grant F17MS-002.

Author Contributions: The authors made equal contributions in the article.

Conflicts of Interest: The authors declare no conflict of interest.

References

1. Letnikov, A.V.; Chernykh, V.A. *The Foundation of Fractional Calculus (with Applications to the Theory of Oil and Gas Production, Underground Hydrodynamics and Dynamics of Biological Systems)*; Neftegaz: Moscow, Russia, 2011. (In Russian)
2. Grünwald, A.K. Über "begrentze" Derivationen und deren Anwendung. *Z. Angew. Math. Phys.* **1867**, *12*, 441–480.
3. Liouville, J. Mémoire sur quelques questiones de géométrie et de mécanique, et nouveau gendre de calcul pour résoudre ces questiones. *J. Ecole Roy. Polytéchn.* **1832**, *13*, 1–69.
4. Liouville, J. Mémoire sur le calcul des différentielles a indices quelconques. *J. Ecole Roy. Polytéchn.* **1832**, *13*, 71–162.
5. Letnikov, A.V. Theory of differentiation of an arbitrary order. *Mat. Sb.* **1868**, *3*, 1–68. (In Russian)
6. Letnikov, A.V. An investigation related to the theory of integrals of the form $\int\limits_a^x (x-u)^{p-1} f(u) du$. *Mat. Sb.* **1874**, *7*, 5–205. (In Russian)
7. Riemann, B.; Weber, H . *Gesammelte Mathematische Werke und Wissenschaflicher Nachlass*; Teubner: Leipzig, Germany, 1876.
8. Marchaud, A.P. Sur les dérivées et sur les différences des fonctions de variablr réelles. *J. Math. Pure Appl.* **1927**, *6*, 337–426. (In French)
9. Marchaud, A.P. *Sur les Dérivées et sur les Différences des Fonctions de Variablr Réelles*; Thésis de Entre-Deux-Guerres; Numdam, The French Digital Mathematics Library: Paris, France, 1965. (In French)
10. Samko, S.G.; Kilbas, A.A.; Marichev, O.I. *Fractional Integrals and Derivatives: Theory and Applications*; Breach Science Publishers: London, UK, 1993.
11. Baleanu, D.; Diethelm, K.; Scalas, E.; Trujillo, J.J. *Fractional Calculus: Models and Numerical Methods*, 2nd ed.; Series on Complexity, Nonlinearity and Chaos; World Scientific: Singapore, 2016.
12. Almeida, R.; Pooseh, S.; Torres D. *Computational Methods in the Fractional Calculus of Variations*; Imperial College Press and World Scientific: Singapore, 2015.
13. Uchaikin, V. *Fractional Derivatives for Physicists and Engineers*; Springer: Berlin, Germany; Higher Education Press: Beijing, China, 2013.
14. Atanackovic, T.; Philipović, S., Stanković, B., Zorica, D. *Fractional Calculus with Applications in Mechanics: Vibrations and Diffusion Processes*; Wiley: London, UK, 2014.
15. Gorenflo, R.; Kilbas, A.; Mainardi, F.; Rogosin, S. *Mittag-Leffler Functions, Related Topics and Applications*; Springer: Berlin/Heidelberg, Germany, 2014.
16. Podlubny, I. *Fractional Differential Equations*; Academic Press: New York, NY, USA, 1999.
17. Caponetto, R.; Dongola, G.; Fortuna, L.; Petráš, I. *Fractional Order Systems: Modeling and Control Applications*; World Scientific: Singapore, 2010.
18. Hilfer, R. (Ed.) *Applications of Fractional Calculus in Physics*; World Scientific: Singapore, 2000.
19. Mainardi, F. *Fractional Calculus and Waves in Linear Viscoelasticity*; Imperial College Press and World Scientific: Singapore, 2010.
20. Hille, E.; Tamarkin, J.D. On the theory of linear integral equations. *Ann. Math.* **1930**, *31*, 479–528.
21. Miller, K.S.; Ross, B. *An Introduction to the Fractional Calculus and Fractional Differential Equations*; John Wiley and Sons: New York, NY, USA, 1993.
22. Kilbas, A.A.; Srivastava, H.M.; Trujillo, J.J. *Theory and Applications of Fractional Differential Equations*; Elsevier: Amsterdam, The Netherlands, 2006.
23. Allen, M.; Caffarelli, L.; Vasseur, A. A parabolic problem with a fractional time derivative, *Arch. Rat. Mech. Anal.* **2016**, *221*, 603–630.

24. Bergounioux, M.; Leaci, A.; Nardi, G.; Tomarelli, F. Fractional Sobolev spaces and functions of bounded variation of one variable. *Frac. Calc. Appl. Anal.* **2017**, *20*, 936–962.
25. Bucur, C.; Ferrari, F. An extesion problem for the fractional derivative defined by Marchaud. *Frac. Calc. Appl. Anal.* **2016**, *19*, 867–887.
26. West, B. J.; Bologna, M.; Grigolini, P. *Physics of Fractal Operators*; Springer: New York, NY, USA, 2003.
27. Yao, K.; Su, W.-Y.; Zhou, S.P. The fractional derivative of a fractal functions. *Acta Math. Sinica. Engl. Ser.* **2006**, *22*, 719–722.
28. Tarasova, V.V.; Tarasov, V.E. Exact discretization of economic accelerator and multiplier with memory. *Fractal Fract.* **2017**, *1*, 6.
29. Kilbas, A.A.; Titioura, A.A. Nonlinear differential equations with Marchaud-Hadamard-type fractional derivative in the weighted spaces of summable functions. *Math. Model. Anal.* **2007**, *12*, 343–356.
30. Tenreiro Machado, J.; Kiryakova, V.; Mainardi, F. A poster about the old history of fractional calculus. *Frac. Calc. Appl. Anal.* **2010**, *13*, 447–454.
31. Nekrasov, P.A. General differentiation. *Mat. Sb.* **1888**, *14*, 45–168. (In Russian)
32. Butzer, P.L.; Westphal, U. An access to fractional differentiation via fractional difference quotients. *Lect. Notes Math.* **1975**, *457*, 116–145.
33. Bugrov, J.S. Fractional difference operators and classes of functions. *Trudy Matematicheskogo Instituta imeni VA Steklova* **1985**, *172*, 60–70. (In Russian)
34. Letnikov, A.V. On explanation of the main provisions of the theory of differentiation of arbitrary order. *Mat. Sb.* **1872**, *6*, 413–445. (In Russian)
35. Sonin, N.Y. On differentiation of arbitrary order. *Mat. Sb.* **1872**, *6*, 1–38. (In Russian)
36. Montel, P. Sure les polynomes d'approximation. *Bull. Soc. Math. Fr.* **1918**, *46*, 151–192.
37. De la Vallee Poussin, C.-J. Over differentie quotienten en differential quotienten. *K. Ak. Wetensch. Amsterdam* **1908**, *17*, 38–45.
38. Rafeiro, H.; Samko, S. On multidimensional analogue of Marchaud formula for fractional Riesz-type derivatives in domains in $\mathbb{R}^n$. *Frac. Calc. Appl. Anal.* **2005**, *8*, 393–401.
39. Condette, J.-F. Marchaud André Paul. In *Les Recteurs d'académie en France de 1808 à 1940*, Tom II; Dictionnaire biographique; Institut National de Recherche Pédagogique: Paris, France, 2006; pp. 272–273.

mathematics

MDPI

Article

Generalized Langevin Equation and the Prabhakar Derivative

Trifce Sandev [1,2,3]

[1] Radiation Safety Directorate, Partizanski Odredi 143, P.O. Box 22, 1020 Skopje, Macedonia; trifce.sandev@drs.gov.mk; Tel.: +389-230-990-30

[2] Institute of Physics, Faculty of Natural Sciences and Mathematics, Ss. Cyril and Methodius University in Skopje, P.O. Box 162, 1001 Skopje, Macedonia

[3] Research Center for Computer Science and Information Technologies, Macedonian Academy of Sciences and Arts, Bul. Krste Misirkov 2, 1000 Skopje, Macedonia

Received: 15 October 2017; Accepted: 15 November 2017; Published: 20 November 2017

Abstract: We consider a generalized Langevin equation with regularized Prabhakar derivative operator. We analyze the mean square displacement, time-dependent diffusion coefficient and velocity autocorrelation function. We further introduce the so-called *tempered* regularized Prabhakar derivative and analyze the corresponding generalized Langevin equation with friction term represented through the tempered derivative. Various diffusive behaviors are observed. We show the importance of the three parameter Mittag-Leffler function in the description of anomalous diffusion in complex media. We also give analytical results related to the generalized Langevin equation for a harmonic oscillator with generalized friction. The normalized displacement correlation function shows different behaviors, such as monotonic and non-monotonic decay without zero-crossings, oscillation-like behavior without zero-crossings, critical behavior, and oscillation-like behavior with zero-crossings. These various behaviors appear due to the friction of the complex environment represented by the Mittag-Leffler and tempered Mittag-Leffler memory kernels. Depending on the values of the friction parameters in the system, either diffusion or oscillations dominate.

Keywords: generalized Langevin equation; regularized Prabhakar derivative; tempered memory kernel; Mittag-Leffler functions

---

1. Introduction

The Langevin equation for a Brownian particle with mass $m = 1$ is represented by the following equaion [1–3]

$$\ddot{x}(t) + \gamma\dot{x}(t) = \xi(t), \quad \dot{x}(t) = v(t), \tag{1}$$

where $x(t)$ is the particle displacement, $v(t)$ is the particle velocity, γ is the friction coefficient, and $\xi(t)$ is a stationary random force with zero mean and correlation $\langle\xi(t)\xi(t')\rangle = 2\gamma k_B T\delta(t'-t)$, where $2\gamma k_B T$ is the so-called spectral density. By calculation of the mean square displacement (MSD) it has been shown that such process shows normal diffusive behavior, i.e., linear dependence of the MSD on time, $\langle x^2(t)\rangle = \frac{2k_B T}{\gamma}t$. The corresponding diffusion coefficient $D = \lim_{t\to\infty}\frac{\langle x^2(t)\rangle}{2t}$ for the Brownian motion is given by $D = \dfrac{k_B T}{\gamma}$, and the normalized velocity autocorrelation function (VACF) by $\dfrac{\langle v(t)v(0)\rangle}{\langle v^2(0)\rangle} = e^{-\gamma t}$ [1,3].

In the work by Lutz [4], a fractional Langevin equation describing non-Markovian stochastic process is introduced,

$$\ddot{x}(t) + \gamma_\mu \, {}_C D_t^\mu x(t) = \xi(t), \quad \dot{x}(t) = v(t), \tag{2}$$

where

$${}_C D_t^\mu f(t) = \frac{1}{\Gamma(1-\mu)} \int_0^t (t-t')^{-\mu} \frac{d}{dt'} f(t') \, dt' \tag{3}$$

is the Caputo fractional derivative of order $0 < \mu < 1$ [5], and γ_μ is the generalized friction coefficient. This equation is a special case of the generalized Langevin equation (GLE) with the power-law memory kernel $\gamma(t) = \gamma_\mu t^{-\mu}/\Gamma(1-\mu)$ (see Section 3). Additionally, if the noise is internal, the second fluctuation-dissipation relation

$$\langle \xi(t)\xi(t') \rangle = k_b T \bar{\gamma}_\mu \, |t-t'|^{-\mu} \tag{4}$$

holds true, where $\bar{\gamma}_\mu = \gamma_\mu/\Gamma(1-\mu)$. It has been shown that the MSD for the fractional Langevin Equation (2) is represented in terms of the two parameter Mittag-Leffler (M-L) function [4] (see Section 2 for details),

$$\left\langle x^2(t) \right\rangle = 2k_B T \, t^2 E_{2-\mu,3}\left(-\gamma_\mu \, t^{2-\mu}\right) \simeq_{t\to\infty} \frac{2k_B T}{\gamma_\mu} \frac{t^\mu}{\Gamma(1+\mu)}, \tag{5}$$

which is a proof of existence of anomalous diffusion in the system. Since $0 < \mu < 1$, it is a subdiffusive process [6].

In this paper we introduce the GLE with a friction term represented through the regularized Prabhakar derivative (see Section 2 for details), i.e.,

$$\ddot{x}(t) + \gamma_{\mu,\rho,\delta} \, {}_C\mathcal{D}_{\rho,-\nu,t}^{\delta,\mu} x(t) = \xi(t), \quad \dot{x}(t) = v(t), \tag{6}$$

where $0 < \mu, \delta < 1$, $0 < \mu/\delta < 1$, $0 < \mu/\delta - \rho < 1$, $\nu = \tau^{-\mu}$, τ is a time parameter, and $\gamma_{\mu,\rho,\delta}$ is the generalized friction coefficient. This equation is a generalization of the fractional Langevin Equation (2) which is recovered by setting $\delta = 0$. We further introduce a GLE with a tempered regularized Prabhakar friction term and analyze the normalized displacement correlation function in case of harmonic potential. Tempered fractional equations nowadays attract more and more attention due to their application in different systems [7–10].

This paper is organized as follows. In Section 2 we give definitions for the Prabhakar derivatives and integral, and related three parameter Mittag-Leffler function. We also introduce a so-called tempered regularized Prabhakar derivative and derive its Laplace transform. GLE for a free particle is considered in Section 3. The MSD, time-dependent diffusion coefficient and VACF are obtained explicitly. In Section 4 we introduce a GLE with friction term represented via tempered regularized Prabhakar derivative. Normal diffusive behavior in the long time limit is obtained due to the exponential truncation in the memory kernel. The case of harmonic oscillator is considered in Section 5, and the normalized displacement correlation function, which is experimentally measured quantity, is obtained. Different diffusion regimes have been observed, therefore the considered equations are of importance for description of anomalous diffusion in complex media. A summary is provided in Section 6.

2. Prabhakar Derivatives

The Prabhakar integral is defined by [11]

$$\left(\mathcal{E}^{\delta}_{\rho,\mu,-\nu,t} f\right)(t) = \int_0^t (t-t')^{\mu-1} E^{\delta}_{\rho,\mu}\left(-\nu(t-t')^{\rho}\right) f(t')\,dt', \tag{7}$$

where

$$E^{\delta}_{\alpha,\beta}(z) = \sum_{k=0}^{\infty} \frac{(\delta)_k}{\Gamma(\alpha k+\beta)} \frac{z^k}{k!}, \tag{8}$$

with $(\delta)_k = \Gamma(\gamma+k)/\Gamma(\gamma)$—the Pochhammer symbol, is the three parameter M-L function [11]. The Laplace transform, $\mathcal{L}[f(t)](s) = \int_0^{\infty} f(t)e^{-st}\,dt$, of the three parameter M-L function is given by

$$\mathcal{L}\left[t^{\beta-1}E^{\delta}_{\alpha,\beta}(-\nu t^{\alpha})\right](s) = \frac{s^{\alpha\delta-\beta}}{(s^{\alpha}+\nu)^{\delta}}, \quad \Re(s) > |\nu|^{1/\alpha}. \tag{9}$$

The functions $E_{\alpha,\beta}(z) = E^{1}_{\alpha,\beta}(z)$ and $E_{\alpha}(z) = E^{1}_{\alpha,1}(z)$ are the two parameter and one parameter M-L function, respectively. For $\delta = 0$ the Prabhakar integral becomes the Riemann-Liouville (R-L) fractional integral, defined by ${}_{\mathrm{RL}}I^{\mu}_t f(t) = \frac{1}{\Gamma(\mu)} \int_0^t (t-t')^{\mu-1} f(t')\,dt'$. Its Laplace transform is given by $\mathcal{L}\left[{}_{\mathrm{RL}}I^{\mu}_t f(t)\right](s) = s^{-\mu}\mathcal{L}\left[f(t)\right](s)$ [5].

The regularized Prabhakar derivative ${}_{\mathrm{C}}\mathcal{D}^{\gamma,\mu}_{\rho,-\nu,t}$ is defined as follows [12]

$${}_{\mathrm{C}}\mathcal{D}^{\delta,\mu}_{\rho,-\nu,t} f(t) = \left(\mathcal{E}^{-\delta}_{\rho,m-\mu,-\nu,t} \frac{d^m}{dt^m} f\right)(t), \tag{10}$$

where $\mu, \nu, \gamma, \rho \in C$, $\Re(\mu) > 0$, $\Re(\rho) > 0$, $m = [\mu] + 1$. For $\delta = 0$ one obtains the Caputo fractional derivative ${}_{\mathrm{C}}\mathcal{D}^{\mu}_t f(t) = {}_{\mathrm{RL}}I^{m-\mu}_t \frac{d^m}{dt^m} f(t)$ [5]. Here we note that the Prabhakar derivative in a form of R-L is given by ${}_{\mathrm{RL}}\mathcal{D}^{\delta,\mu}_{\rho,-\nu,t} f(t) = \frac{d^m}{dt^m}\left(\mathcal{E}^{-\delta}_{\rho,m-\mu,-\nu,t} f\right)(t)$ [12,13] (see also [14]). For $\delta = 0$ it becomes the R-L fractional derivative ${}_{\mathrm{RL}}\mathcal{D}^{\mu}_t f(t) = \frac{d^m}{dt^m} {}_{\mathrm{RL}}I^{m-\mu}_t f(t)$ [5]. These derivatives are applicable in the fractional Poisson process [12], for description of dielectric relaxation phenomena [15,16], in the fractional Maxwell model in the linear viscoelasticity [17], in mathematical modeling of fractional differential filtration dynamics [18], in fractional dynamical systems [19], generalized reaction-diffusion equations [20], in generalized model of particle deposition in porous media [21], etc.

In our paper we are particularly interested in the case with $0 < \mu < 1$, so one has

$${}_{\mathrm{C}}\mathcal{D}^{\delta,\mu}_{\rho,-\nu,t} f(t) = \left(\mathcal{E}^{-\delta}_{\rho,1-\mu,-\nu,t} \frac{d}{dt} f\right)(t) = \int_0^t (t-t')^{-\mu} E^{-\delta}_{\rho,1-\mu}\left(-\nu(t-t')^{\rho}\right) \frac{d}{dt'} f(t')\,dt'. \tag{11}$$

From here we conclude that the regularized Prabhakar derivative is a special case of the generalized derivative

$$\left({}_{\mathrm{C}}\mathbf{G}_{\gamma,t} f\right)(t) = \int_0^t \gamma(t-t') \frac{d}{dt'} f(t')\,dt', \tag{12}$$

which has been investigated in different contexts in [10,22,23], where

$$\gamma(t) = t^{-\mu} E^{-\delta}_{\rho,1-\mu}\left(-\nu t^{\rho}\right). \tag{13}$$

We note that the Prabhakar derivative in the R-L form for $0 < \mu < 1$ is a special case of the generalized derivative $\left({}_{\mathrm{RL}}\mathbf{G}_{\eta,t} f\right)(t) = \frac{d}{dt}\int_0^t \eta(t-t') f(t')\,dt'$, which has been investigated in [10,22,23], where the memory kernel is given by $\eta(t) = t^{-\mu} E^{-\delta}_{\rho,1-\mu}\left(-\nu t^{\rho}\right)$.

The Laplace transform of regularized Prabhakar derivative, Equation (11), is given by [12]

$$\begin{aligned}\mathcal{L}\left[{}_{\mathrm{C}}\mathcal{D}^{\delta,\mu}_{\rho,-\nu,t}f(t)\right](s) &= s^{\mu}\left(1+\nu s^{-\rho}\right)^{\delta}\mathcal{L}\left[f(t)\right](s) - s^{\mu-1}\left(1+\nu s^{-\rho}\right)^{\delta}f(0+) \\ &= s^{\mu-1}\left(1+\nu s^{-\rho}\right)^{\delta}\left[s\mathcal{L}\left[f(t)\right](s) - f(0+)\right],\end{aligned} \tag{14}$$

where $\Re(s) > |\nu|^{1/\rho}$ [12]. For $\delta = 0$, Prabhakar derivative corresponds to the Caputo fractional derivative of order $0 < \mu < 1$ with Laplace transform [5]

$$\mathcal{L}\left[{}_{\mathrm{C}}\mathcal{D}^{\mu}_{t}f(t)\right](s) = s^{\mu}\mathcal{L}\left[f(t)\right](s) - s^{\mu-1}f(0+) = s^{\mu-1}\left[s\mathcal{L}\left[f(t)\right](s) - f(0+)\right].$$

Here we note that the regularized Prabhakar derivative, Equation (11), is a special case of the Hilfer-Prabhakar derivative defined by [12]

$${}_{\mathrm{HP}}\mathcal{D}^{\delta,\mu,\bar{\nu}}_{\rho,\nu,t}f(t) = \left(\mathcal{E}^{-\delta\bar{\nu}}_{\rho,\bar{\nu}(1-\mu),-\nu,t}\frac{d}{dt}\left(\mathcal{E}^{-\delta(1-\bar{\nu})}_{\rho,(1-\bar{\nu})(1-\mu),-\nu,t}\right)f\right)(t), \quad 0<\mu<1, \quad 0\le\bar{\nu}\le 1, \tag{15}$$

for the case with $\bar{\nu} = 1$. Its Laplace transform reads [12]

$$\mathcal{L}\left[{}_{\mathrm{HP}}\mathcal{D}^{\delta,\mu,\bar{\nu}}_{\rho,-\nu,t}f(t)\right](s) = s^{\mu}\left(1+\nu s^{-\rho}\right)^{\delta}\mathcal{L}\left[f(t)\right](s) - s^{\bar{\nu}(\mu-1)}\left(1+\nu s^{-\rho}\right)^{\delta\bar{\nu}}\left[\mathcal{E}^{-\delta(1-\bar{\nu})}_{\rho,(1-\bar{\nu})(1-\mu),-\nu,t}f\right]\Big|_{t=0+}.$$

From this formula we see that the initial value term is given in an integral form. Only the case with $\bar{\nu} = 1$ yields the initial value in the natural form $f(0+)$. Therefore, the regularized Prabhakar derivative (11) is suitable for application in the GLE model. The case with $\delta = 0$ corresponds to the so-called Hilfer composite fractional derivative of order $0 < \mu < 1$ and type $0 \le \bar{\nu} \le 1$, which is given by $\mathcal{D}^{\mu,\bar{\nu}}_{t}f(t) = \left({}_{\mathrm{RL}}I^{\bar{\nu}(1-\mu)}_{t}\frac{d}{dt}({}_{\mathrm{RL}}I^{(1-\bar{\nu})(1-\mu)}_{t}f)\right)(t)$ [24]. This composite fractional derivative has been successfully applied in description of dielectric and viscoelastic phenomena [25,26].

Furthermore, in this paper we introduce the *tempered* regularized Prabhakar derivative in the following way

$${}_{\mathrm{TC}}\mathcal{D}^{\delta,\mu}_{\rho,-\nu,t}f(t) = \left({}_{\mathrm{T}}\mathcal{E}^{-\delta}_{\rho,1-\mu,-\nu,t}\frac{d}{dt}f\right)(t), \tag{16}$$

where

$$\left({}_{\mathrm{T}}\mathcal{E}^{\delta}_{\rho,\mu,-\nu,t}f\right)(t) = \int_0^t e^{-b(t-t')}(t-t')^{\mu-1}E^{\delta}_{\rho,\mu}\left(-\nu(t-t')^{\rho}\right)f(t')\,dt', \tag{17}$$

and $b > 0$. Other parameters are the same as in the regularized Prabhakar derivative (11). From the definition we see that this derivative is a special case of the generalized derivative (12) for

$$\gamma(t) = e^{-bt}t^{-\mu}E^{-\delta}_{\rho,1-\mu}\left(-\nu t^{\rho}\right). \tag{18}$$

Therefore, for the Laplace transform of the tempered regularized Prabhakar derivative we find

$$\mathcal{L}\left[{}_{\mathrm{TC}}\mathcal{D}^{\delta,\mu}_{\rho,-\nu,t}f(t)\right](s) = (s+b)^{\mu-1}\left(1+\nu(s+b)^{-\rho}\right)^{\delta}\left[s\mathcal{L}\left[f(t)\right](s) - f(0+)\right]. \tag{19}$$

We will use this derivative in the GLE and analyze the influence of the exponential truncation on the particle behavior.

On the other hand, one can introduce the so-called *tempered* Prabhakar derivative in the R-L form in a similar way, by introducing exponential truncation in the memory kernel, i.e., ${}_{\mathrm{TRL}}\mathcal{D}^{\delta,\mu}_{\rho,-\nu,t}f(t) = \frac{d}{dt}\int_0^t e^{-b(t-t')}(t-t')^{-\mu}E^{-\delta}_{\rho,1-\mu}\left(-(t-t')^{\rho}\right)f(t')\,dt'$.

Here we note that different fractional equations have been used for modeling anomalous diffusion in various systems, including fractional reaction-diffusion equations [27,28] and their application [29], fractional relaxation and diffusion equations [5,6,9,10,24–26], fractional cable equation [30], etc.

3. Free Particle

We showed in Section 2 that the regularized Prabhakar derivative (11) is a special case of the generalized derivative (12), therefore we conclude that the Langevin Equation (6) can be written in a form of the generalized Langevin equation (see [31])

$$\ddot{x}(t) + \int_0^t \gamma(t-t')\frac{d}{dt'}x(t')dt' = \xi(t), \quad \dot{x}(t) = v(t), \tag{20}$$

where

$$\gamma(t) = \gamma_{\mu,\rho,\delta}\, t^{-\mu} E_{\rho,1-\mu}^{-\delta}\left(-\left[\frac{t}{\tau}\right]^{\rho}\right). \tag{21}$$

By using the asymptotic expansion of the three parameter M-L function $E_{\alpha,\beta}^{\delta}(-z) \simeq \frac{z^{-\delta}}{\Gamma(\beta-\alpha\delta)}$, which follows from the formula [9] (for details of the three parameter M-L function see [32])

$$E_{\alpha,\beta}^{\delta}(-z) = \frac{z^{-\delta}}{\Gamma(\delta)} \sum_{k=0}^{\infty} \frac{\Gamma(\delta+k)}{\Gamma(\beta-\alpha(\delta+n))} \frac{(-z)^{-n}}{n!}, \tag{22}$$

for $0 < \alpha < 2$ and $z \to \infty$, one can show that the assumption $\lim_{t\to\infty} \gamma(t) = 0$ [33] is satisfied since $\mu > \rho\delta$. Additionally to this equation, we assume that the second fluctuation-dissipation relation

$$\langle \xi(t)\xi(t') \rangle = k_b T \gamma_{\mu,\rho,\delta}\, |t-t'|^{-\mu} E_{\rho,1-\mu}^{-\delta}\left(-\left[\frac{|t-t'|}{\tau}\right]^{\rho}\right), \tag{23}$$

holds true.

From the Laplace transform method one finds that

$$x(t) = \langle x(t) \rangle + \int_0^t G(t-t')\xi(t')\,dt', \quad v(t) = \langle v(t) \rangle + \int_0^t g(t-t')\xi(t')\,dt', \tag{24}$$

where $x_0 = x(0)$ and $v_0 = v(0)$ are the initial particle displacement and initial particle velocity, respectively,

$$\langle x(t) \rangle = x_0 + v_0 G(t), \quad \langle v(t) \rangle = v_0 g(t), \tag{25}$$

are the average particle displacement and the average particle velocity, respectively, and the Laplace pairs of the relaxation functions $g(t)$, $G(t)$ and $I(t)$ are given through $\mathcal{L}[\gamma(t)](s) = \hat{\gamma}(s)$ by

$$\hat{g}(s) = \frac{1}{s+\hat{\gamma}(s)}, \quad \hat{G}(s) = \frac{s^{-1}}{s+\hat{\gamma}(s)}, \quad \hat{I}(s) = \frac{s^{-2}}{s+\hat{\gamma}(s)}, \tag{26}$$

respectively. From the relaxation functions one can derive the MSD, time-dependent diffusion coefficient and VACF, for thermal initial conditions $x_0 = 0$ and $v_0 = k_B T$, and under the assumption $\lim_{t\to\infty} \gamma(t) = \lim_{s\to 0} s\hat{\gamma}(s) = 0$, as follows [33–35]

$$\left\langle x^2(t) \right\rangle = 2k_B T I(t), \tag{27}$$

$$D(t) = \frac{1}{2}\frac{d}{dt}\left\langle x^2(t) \right\rangle = k_B T G(t), \tag{28}$$

$$C_V(t) = \frac{\langle v(t)v(0)\rangle}{\langle v^2(0)\rangle} = g(t). \tag{29}$$

Therefore, for the MSD, $D(t)$ and VACF we find

$$\left\langle x^2(t)\right\rangle = 2k_BT\sum_{n=0}^{\infty}(-\gamma_{\mu,\rho,\delta})^n t^{(2-\mu)n+2}E_{\rho,(2-\mu)n+3}^{-\delta n}\left(-\left[\frac{t}{\tau}\right]^{\rho}\right), \tag{30}$$

$$D(t) = k_BT\sum_{n=0}^{\infty}(-\gamma_{\mu,\rho,\delta})^n t^{(2-\mu)n+1}E_{\rho,(2-\mu)n+2}^{-\delta n}\left(-\left[\frac{t}{\tau}\right]^{\rho}\right), \tag{31}$$

$$C_V(t) = \sum_{n=0}^{\infty}(-\gamma_{\mu,\rho,\delta})^n t^{(2-\mu)n}E_{\rho,(2-\mu)n+1}^{-\delta n}\left(-\left[\frac{t}{\tau}\right]^{\rho}\right), \tag{32}$$

respectively. Such series in three parameter M-L functions are convergent [36–38].

From Equation (22), for the long time limit we find

$$\left\langle x^2(t)\right\rangle \simeq 2k_BT\,t^2E_{2-\mu+\rho\delta,3}\left(-\bar{\gamma}t^{2-\mu+\rho\delta}\right) \simeq \frac{2k_BT}{\bar{\gamma}}\frac{t^{\mu-\rho\delta}}{\Gamma(1+\mu-\rho\delta)}, \tag{33}$$

$$D(t) \simeq k_BT\,tE_{2-\mu+\rho\delta,2}\left(-\bar{\gamma}t^{2-\mu+\rho\delta}\right) \tag{34}$$

$$C_V(t) \simeq E_{2-\mu+\rho\delta}\left(-\bar{\gamma}t^{2-\mu+\rho\delta}\right), \tag{35}$$

where $\bar{\gamma} = \gamma_{\mu,\rho,\delta}\tau^{-\rho\delta}$. Therefore, from the MSD we conclude that there exists subdiffusion $\left\langle x^2(t)\right\rangle \simeq t^{\alpha}$ with anomalous diffusion exponent $\alpha = \mu - \rho\delta$, where $0 < \alpha < \delta < 1$.

Graphical representation of the MSD (30) is given in Figure 1. From the figure we see that the MSD shows oscillation-like behavior for intermediate times which can be explained as a result of the cage effect of the environment represented by the M-L memory kernel (see also Section 5).

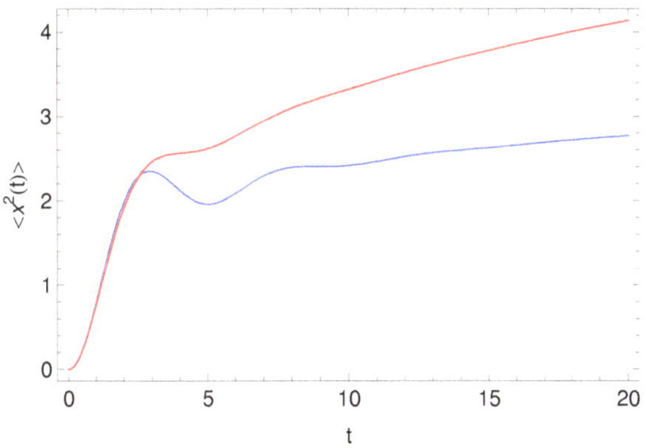

Figure 1. Graphical representation of the MSD (30) for $k_BT = 1$, $\gamma_{\mu,\rho,\delta} = 1$, $\tau = 1$, $\rho = 1/2$, $\delta = 3/4$, and $\mu = 1/2$ (blue line), $\mu = 5/8$ (red line).

Remark 1. *The case with high viscous damping, i.e., Equation (20) with vanishing second derivative term $\ddot{x}(t) = 0$, yields the following result*

$$\left\langle x^2(t)\right\rangle = \frac{2k_BT}{\gamma_{\mu,\rho,\delta}}\,t^{\mu}E^{\delta}_{\rho,\mu+1}\left(-\left[\frac{t}{\tau}\right]^{\rho}\right) \simeq \frac{2k_BT}{\gamma_{\mu,\rho,\delta}}\begin{cases}\frac{t^{\mu}}{\Gamma(\mu+1)}, & t\to 0\\ \frac{t^{\mu-\rho\delta}}{\tau^{-\rho\delta}\Gamma(1+\mu-\rho\delta)}, & t\to\infty.\end{cases} \tag{36}$$

Since the anomalous diffusion exponent from μ for the short time limit turns to $\mu-\rho\delta$ in the long time limit, we conclude that decelerating subdiffusion exists in the system.

4. Tempered Friction

We further consider the GLE with a friction term represented through the tempered regularized Prabhakar derivative. Therefore, we consider

$$\ddot{x}(t) + \gamma_{\mu,\rho,\delta}\;{}_{\mathrm{TC}}\mathcal{D}^{\delta,\mu}_{\rho,-\nu,t}x(t) = \xi(t), \quad \dot{x}(t) = v(t), \tag{37}$$

where $b>0$, and all the parameters are the same as in Equation (6). From definition (16) one may conclude that the memory kernel in GLE is given by

$$\gamma(t) = \gamma_{\mu,\rho,\delta}\,e^{-bt}t^{-\mu}E^{-\delta}_{\rho,1-\mu}\left(-\left[\frac{t}{\tau}\right]^{\rho}\right). \tag{38}$$

The second fluctuation-dissipation relation then reads

$$\langle\xi(t)\xi(t')\rangle = k_bT\gamma_{\mu,\rho,\delta}\,e^{-b|t-t'|}|t-t'|^{-\mu}E^{-\delta}_{\rho,1-\mu}\left(-\left[\frac{|t-t'|}{\tau}\right]^{\rho}\right). \tag{39}$$

Following the same procedure as previous, for the MSD we find

$$\left\langle x^2(t)\right\rangle = \sum_{n=0}^{\infty}\left(-\gamma_{\mu,\rho,\delta}\right)^n\;{}_{\mathrm{RL}}I_t^{n+3}\left(e^{-bt}t^{(1-\mu)n-1}E^{-\delta n}_{\rho,(1-\mu)n}\left(-\left[\frac{t}{\tau}\right]^{\rho}\right)\right), \tag{40}$$

where ${}_{\mathrm{RL}}I_t^{\alpha}$ is the R-L fractional integral. In the case of no truncation ($b=0$), from Equation (40), by using the formula ${}_{\mathrm{RL}}I_t^{\zeta}\left(t^{\beta-1}E^{\delta}_{\alpha,\beta}\left(-\nu t^{\alpha}\right)\right) = t^{\zeta+\beta-1}E^{\delta}_{\alpha,\zeta+\beta}\left(-\nu t^{\alpha}\right)$ [14], we recover the result (30) for the MSD.

Remark 2. *In the absence of the inertial term, $\ddot{x}(t)=0$ in Equation (37), we find the following result for the MSD*

$$\left\langle x^2(t)\right\rangle = \frac{2k_BT}{\gamma_{\mu,\rho,\delta}}\;{}_{\mathrm{RL}}I_t^{2}\left(e^{-bt}t^{\mu-2}E^{\delta}_{\rho,\mu-1}\left(-\left[\frac{t}{\tau}\right]^{\rho}\right)\right). \tag{41}$$

The short time limit yields subdiffusion $\left\langle x^2(t)\right\rangle \simeq \frac{2k_BT}{\gamma_{\mu,\rho,\delta}}t^{\mu}/\Gamma(1+\mu)$, and the long time limit normal diffusion $\left\langle x^2(t)\right\rangle \simeq t$. Therefore, there exists accelerating diffusion—from subdiffusion to normal diffusion. Such crossover from subdiffusion to normal diffusion has been observed, for example, in complex viscoelastic systems [39].

Graphical representation of the MSD (40) is given in Figure 2. From the figure one observes the influence of the truncation parameter b on the behavior of the MSD. The case with no truncation shows subdiffusive behavior (blue line), and the case with truncation (red and green lines) normal diffusion in the long time limit.

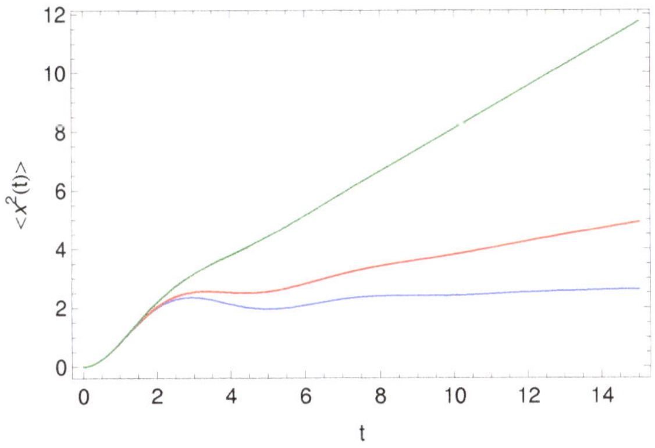

Figure 2. Graphical representation of the MSD (40), for $k_BT = 1$, $\gamma_{\mu,\rho,\delta} = 1$, $\tau = 1$, $\rho = 1/2$, $\mu = 1/2$, $\delta = 3/4$, and $b = 0$ (blue line), $b = 0.1$ (red line) and $b = 0.5$ (green line).

5. Harmonically Bounded Particle in Presence of Prabhakar Friction Term

At the end we consider GLE for a harmonic oscillator with tempered regularized Prabhakar friction, i.e.,

$$\ddot{x}(t) + \gamma_{\mu,\rho,\delta}\, {}_{\mathrm{TC}}\mathcal{D}^{\delta,\mu}_{\rho,-\nu,t}\, x(t) + \omega^2 x(t) = \xi(t), \quad \dot{x}(t) = v(t), \tag{42}$$

where ω is the frequency of the oscillator. By Laplace transform method we find exact result for the MSD. It is given by

$$\begin{aligned} \frac{\langle x^2(t)\rangle}{2k_BT} &= \sum_{n=0}^{\infty}\left(-\gamma_{\mu,\rho,\delta}\right)^n \int_0^t (t-t')^{n+2} E^{n+1}_{2,n+3}\left(-\omega^2(t-t')^2\right) e^{-bt'} t'^{(1-\mu)n-1} E^{-\delta n}_{\rho,(1-\mu)n}\left(-\left[\tfrac{t'}{\tau}\right]^{\rho}\right) dt' \\ &= \sum_{n=0}^{\infty}\left(-\gamma_{\mu,\rho,\delta}\right)^n \mathcal{E}^{n+1}_{2,n+3,-\omega^2,t}\left(e^{-bt} t^{(1-\mu)n-1} E^{-\delta n}_{\rho,(1-\mu)n}\left(-\left[\tfrac{t}{\tau}\right]^{\rho}\right)\right), \end{aligned} \tag{43}$$

where $\left(\mathcal{E}^{\delta}_{\alpha,\beta,-\omega^2,t} f\right)(t)$ is the Prabhakar integral (7). For the free particle case $\omega = 0$, the Prabhakar integral corresponds to the R-L fractional integral, therefore, from Equation (43) one finds the previously obtained result for free particle, Equation (30).

Here we are particularly interested in the normalized displacement correlation function which is experimental measured quantity related to the GLE [40],

$$C_X(t) = \frac{\langle x(t)x_0\rangle}{\langle x_0^2\rangle} = \frac{s + \hat{\gamma}(s)}{s^2 + s\hat{\gamma}(s) + \omega^2}, \tag{44}$$

under the conditions $x_0^2 = \frac{k_BT}{\omega^2}$, $\langle x_0 v_0\rangle = 0$, and $\langle \xi(t)x_0\rangle = 0$ [41]. From here one concludes that $C_X(t) = 1 - \omega^2 I(t)$, where $I(t) = \frac{\langle x^2(t)\rangle}{2k_BT}$. Therefore, it is given by

$$C_X(t) = 1 - \omega^2 \sum_{n=0}^{\infty}\left(-\gamma_{\mu,\rho,\delta}\right)^n \mathcal{E}^{n+1}_{2,n+3,-\omega^2,t}\left(e^{-bt} t^{(1-\mu)n-1} E^{-\delta n}_{\rho,(1-\mu)n}\left(-\left[\frac{t}{\tau}\right]^{\rho}\right)\right). \tag{45}$$

The case with no truncation yields

$$C_X(t) = 1 - \omega^2 \sum_{n=0}^{\infty}\left(-\gamma_{\mu,\rho,\delta}\right)^n \mathcal{E}^{n+1}_{2,n+3,-\omega^2,t}\left(t^{(1-\mu)n-1} E^{-\delta n}_{\rho,(1-\mu)n}\left(-\left[\frac{t}{\tau}\right]^{\rho}\right)\right). \tag{46}$$

Graphical representation of the $C_X(t)$, Equations (46) and (45), is given in Figures 3 and 4, respectively. From Figure 3 we see that different behaviors of $C_X(t)$ appear, monotonic or non-monotonic decay without zero crossings (for $\omega < 1.44$), critical behavior between the situations with and without zero crossings (at critical frequency $\omega \approx 1.44$), and oscillation-like behavior with zero crossings (for $\omega > 1.44$), which appear due to the cage effect of the environment [8,41]. The friction, depending on the memory kernel parameters, forces either diffusion or oscillations. In Figure 4 we see that with increasing of tempering, oscillation behavior with zero crossings appears. Therefore, in comparison to the standard harmonic oscillator (standard Langevin equation in presence of harmonic potential) where two types of motion are observed for the normalized displacement correlation function, either monotonic decay without zero crossings or oscillation-like behavior with zero crossings (that are separated at the critical frequency $\omega_c = \gamma/2$), in case of the GLE with Prabhakar memory kernel more different complex behaviors of the $C_X(t)$ are observed. Thus, the friction parameters contained in the tempered Prabhakar derivative, by their tuning, increase the versatility to fit complex experimental data.

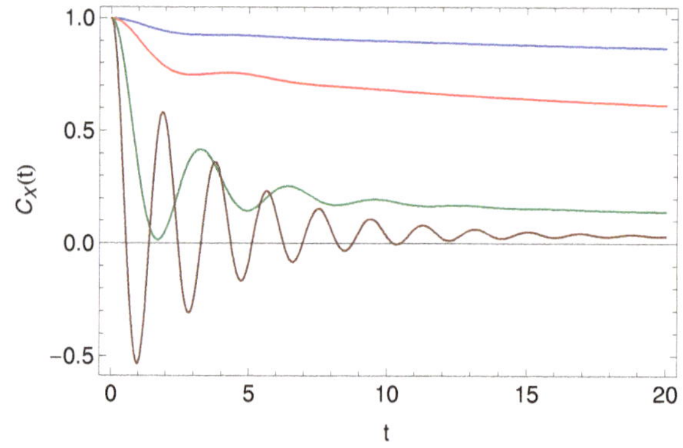

Figure 3. Graphical representation of the normalized displacement correlation function, Equation (46), for $\gamma_{\mu,\rho,\delta} = 1$, $\tau = 1$, $\rho = 1/5$ $\mu = 1/2$, $\delta = 3/4$, and $\omega = 0.25$ (blue line), $\omega = 0.5$ (red line), $\omega = 1.44$ (green line), $\omega = 3$ (brown line).

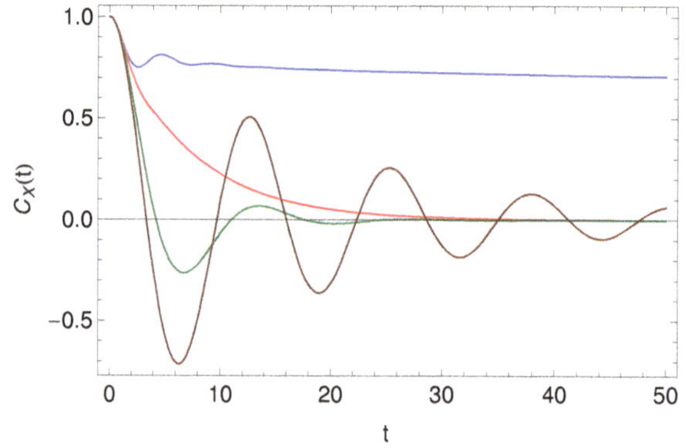

Figure 4. Graphical representation of the normalized displacement correlation function, Equation (45), for $\gamma_{\mu,\rho,\delta} = 1$, $\tau = 1$, $\rho = 1/2$ $\mu = 1/2$, $\delta = 3/4$, $\omega = 0.5$ and $b = 0$ (blue line), $b = 1$ (red line), $b = 10$ (green line), $b = 100$ (brown line).

6. Summary

We show that the generalized Langevin equation with a friction represented in terms of the regularized Prabhakar derivative generates decelerating subdiffusion. We also introduce a tempered regularized Prabhakar derivative in the friction term and we show that the system from subdiffusion switches to normal diffusion due to the exponential truncation in the memory kernel. Such model could be used in the description of diffusive processes in viscoelastic systems, where subdiffusion turns to normal diffusive behavior. We demonstrate the role of the three parameter Mittag-Leffler function and the Prabhakar integral in the anomalous dynamics theory. The obtained results are generalizations of those for the fractional Langevin equation. We also observe various behaviors of the normalized displacement correlation function in case of harmonic oscillator. The proposed models in this paper generate various anomalous diffusive realizations; therefore, by adjusting the memory kernel parameters one could better fit complex experimental data.

Acknowledgments: The author acknowledges funding from the Deutsche Forschungsgemeinschaft (DFG), project ME 1535/6-1 "Random search processes, Lévy flights, and random walks on complex networks".

Conflicts of Interest: The author declares no conflict of interest.

References

1. Coffey, W.T.; Kalmykov, Yu.P.; Waldron, J.T. *The Langevin Equation: With Applications to Stochastic Problems in Physics, Chemistry and Electrical Engineering*; World Scientific: Singapore, 2012.
2. Langevin, P. On the theory of Brownian motion. *Comptes Rendus* **1908**, *146*, 530–533.
3. Zwanzig, R. *Nonequilibrium Statistical Mechanics*; Oxford University Press: New York, NY, USA, 2001.
4. Lutz, E. Fractional Langevin equation. *Phys. Rev. E* **2001**, *64*, 051106.
5. Mainardi, F. *Fractional Calculus and Waves in Linear Viscoelesticity: An Introduction to Mathematical Models*; Imperial College Press: London, UK, 2010.
6. Metzler, R.; Jeon, J.-H.; Cherstvy, A.G.; Barkai, E. Anomalous diffusion models and their properties: Non-stationarity, non-ergodicity, and ageing at the centenary of single particle tracking. *Phys. Chem. Chem. Phys.* **2014**, *16*, 24128–24164.
7. Chen, Y.; Wang, X.; Deng, W. Localization and ballistic diffusion for the tempered fractional Brownian-Langevin motion. *J. Stat. Phys.* **2017**, *169*, 18–37.
8. Liemert, A.; Sandev, T.; Kantz, H. Generalized Langevin equation with tempered memory kernel. *Physica A* **2017**, *466*, 356–369.
9. Sandev, T.; Chechkin, A.; Kantz, H.; Metzler, R. Diffusion and Fokker-Planck-Smoluchowski equations with generalized memory kernel. *Fract. Calc. Appl. Anal.* **2015**, *18*, 1006–1038.
10. Sandev, T.; Sokolov, I.M.; Metzler, R.; Chechkin, A. Beyond monofractional kinetics. *Chaos Solitons Fractals* **2017**, *102*, 210–217.
11. Prabhakar, T.R. A singular integral equation with a generalized Mittag-Leffler function in the kernel. *Yokohama Math. J.* **1971**, *19*, 7–15.
12. Garra, R.; Gorenflo, R.; Polito, F.; Tomovski, A. Hilfer-Prabhakar derivatives and some applications. *Appl. Math. Comput.* **2014**, *242*, 576–589.
13. Polito, F.; Tomovski, Z. Some properties of Prabhakar-type fractional calculus operators. *Fract. Differ. Calc.* **2016**, *6*, 73–94.
14. Kilbas, A.A.; Saigo, M.; Saxena, R.K. Generalized Mittag-Leffler function and generalized fractional calculus operators. *Integral Transform. Spec. Funct.* **2004**, *15*, 31–49.
15. Garrappa, R. Grünwald-Letnikov operators for fractional relaxation in Havriliak-Negami models. *Commun. Nonlinear Sci. Numer. Simul.* **2016**, *38*, 178–191.
16. Garrappa, R.; Mainardi, F.; Maione, G. Models of dielectric relaxation based on completely monotone functions. *Fract. Calc. Appl. Anal.* **2016**, *19*, 1105–1160.
17. Giusti, A.; Colombaro, I. Prabhakar-like fractional viscoelasticity. *Commun. Nonlinear Sci. Numer. Simul.* **2018**, *56*, 138–143.

18. Bulavatsky, V.M. Mathematical modeling of fractional differential filtration dynamics based on models with Hilfer–Prabhakar derivative. *Cybern. Syst. Anal.* **2017**, *53*, 204–216.
19. Darani, M.A.; Derakhshan, M.H.; Ansari, A.; Khoshsiar, R. On asymptotic stability of Prabhakar fractional differential systems. *Comput. Methods Differ. Equ.* **2016**, *4*, 276–284.
20. Agarwal, R.; Jain, S.; Agarwal, R.P. Analytic solution of generalized space time fractional reaction diffusion equation. *Fract. Differ. Calc.* **2017**, *7*, 169–184.
21. Xu, J. Time-fractional particle deposition in porous media. *J. Phys. A Math. Theor.* **2017**, *50*, 195002.
22. Kochubei, A. General fractional calculus, evolution equations, and renewal processes. *Integral Equ. Oper. Theory* **2011**, *71*, 583–600.
23. Luchko, Y.; Yamamoto, M. General time fractional diffusion equation: Some uniqueness and existence results for the initial-boundary-value problems. *Fract. Calc. Appl. Anal.* **2016**, *19*, 676–695.
24. Hilfer, R. *Application of Fractional Calculus in Physics*; World Scientific: Singapore, 2000.
25. Hilfer, R. Experimental evidence for fractional time evolution in glass forming materials. *Chem. Phys.* **2002**, *284*, 399–408.
26. Sandev, T.; Metzler, R; Tomovski, Z. Fractional diffusion equation with a generalized Riemann-Liouville time fractional derivative. *J. Phys. A Math. Theor.* **2011**, *44*, 255203.
27. Saxena, R.K.; Mathai, A.M.; Haubold, H.J. Space-time fractional reaction-diffusion equations associated with a generalized Riemann-Liouville fractional derivative. *Axioms* **2014**, *3*, 320–334.
28. Saxena, R.K.; Mathai, A.M.; Haubold, H.J. Computational solutions of distributed order reaction-diffusion systems associated with Riemann-Liouville derivatives. *Axioms* **2015**, *4*, 120–133.
29. Haubold, H.J.; Mathai, A.M.; Saxena, R.K. Analysis of solar neutrino data from Super-Kamiokande I and II. *Entropy* **2014**, *16*, 1414–1425.
30. Saxena, R.K.; Tomovski, Z.; Sandev, T. Analytical solution of generalized space-time fractional cable equation. *Mathematics* **2015**, *3*, 153–170.
31. Kubo, R. The Fluctuation-Dissipation Theorem. *Rep. Prog. Phys.* **1966**, *29*, 255–284.
32. Garra, R.; Garrappa, R. The Prabhakar or three parameter Mittag–Leffler function: Theory and application. *Commun. Nonlinear Sci. Numer. Simul.* **2018**, doi:10.1016/j.cnsns.2017.08.018.
33. Despósito, M.A.; Viñales, A.D. Subdiffusive behavior in a trapping potential: Mean square displacement and velocity autocorrelation function. *Phys. Rev. E* **2009**, *80*, 021111.
34. Mainardi, F; Pironi, P. The fractional Langevin equation: Brownian motion revisited. *Extr. Math.* **1996**, *10*, 140–154.
35. Pottier, N. Aging properties of an anomalously diffusing particle. *Physica A* **2003**, *317*, 371–382.
36. Paneva-Konovska, J. Convergence of series in three parametric Mittag-Leffler functions. *Math. Slovaca* **2014**, *64*, 73–84.
37. Paneva-Konovska, J. *From Bessel to Multi-Index Mittag-Leffler Functions: Enumerable Families, Series in Them and Convergence*; World Scientific: Hackensack, NJ, USA, 2016.
38. Sandev, T.; Tomovski, Z.; Dubbeldam, J.L.A. Generalized Langevin equation with a three parameter Mittag-Leffler noise. *Physica A* **2011**, *390*, 3627–3636.
39. Jeon, J.-H.; Monne, H.M.-S.; Javanainen, M.; Metzler, R. Anomalous diffusion of phospholipids and cholesterols in a lipid bilayer and its origins. *Phys. Rev. Lett.* **2012**, *109*, 188103.
40. Min, W.; Luo, G.; Cherayil, B.J.; Kou, S.C.; Xie, X.S. Observation of a power-law memory kernel for fluctuations within a single protein molecule. *Phys. Rev. Lett.* **2005**, *94*, 198302.
41. Burov, S.; Barkai, E. Fractional Langevin equation: Overdamped, underdamped, and critical behaviors. *Phys. Rev. E* **2008**, *78*, 031112.

mathematics

MDPI

Article

A Note on Hadamard Fractional Differential Equations with Varying Coefficients and Their Applications in Probability

Roberto Garra [1,*], Enzo Orsingher [1] and Federico Polito [2]

1 Dipartimento di Scienze Statistiche, "Sapienza" Università di Roma, P. le A. Moro 5, 00185 Roma, Italy; enzo.orsingher@uniroma1.it

2 Dipartimento di Matematica "G. Peano", Università degli Studi di Torino, Via Carlo Alberto 10, 10123 Torino, Italy; f.polito@gmail.com

* Correspondence: roberto.garra@sbai.uniroma1.it

Received: 18 November 2017; Accepted: 26 December 2017; Published: 1 January 2018

Abstract: In this paper, we show several connections between special functions arising from generalized Conway-Maxwell-Poisson (COM-Poisson) type statistical distributions and integro-differential equations with varying coefficients involving Hadamard-type operators. New analytical results are obtained, showing the particular role of Hadamard-type derivatives in connection with a recently introduced generalization of the Le Roy function. We are also able to prove a general connection between fractional hyper-Bessel-type equations involving Hadamard operators and Le Roy functions.

Keywords: Hadamard fractional derivatives; COM-Poisson distributions; Modified Mittag–Leffler functions

MSC: 33E12; 34A08; 60G55

1. Introduction

The analysis of fractional differential equations involving Hadamard fractional derivatives has increased interest in mathematical analysis, as proved, for example, by the publication of the recent monograph [1]. On the other hand, few results regarding the applications of Hadamard fractional differential equations in mathematical physics (see, for example, [2] and probability (see [3]) exist. In [4], some analytical results regarding Hadamard fractional equations with time-varying coefficients have been pointed out. In particular, the following α-Mittag–Leffler function was introduced:

$$E_{\alpha;\nu,\gamma}(z) = \sum_{k=0}^{\infty} \frac{z^k}{[\Gamma(\nu k+\gamma)]^{\alpha}}, \quad z \in \mathbb{C}, \alpha, \nu, \gamma \in \mathbb{C}. \tag{1}$$

The α-Mittag–Leffler function is an entire function of the complex variable z if the parameters are such that $\Re(\nu) > 0$, $\gamma \in \mathbb{R}$ and $\alpha \in \mathbb{R}^+$ (see [5]). This function was independently introduced and studied by Gerhold in [6]. In [4], we called this special function α-Mittag–Leffler function, since it includes for $\alpha = 1$ the well-known two-parameters Mittag–Leffler function

$$E_{\nu,\gamma}(z) = \sum_{k=0}^{\infty} \frac{z^k}{\Gamma(\nu k+\gamma)}, \quad z \in \mathbb{C},\ \nu \in \mathbb{C},\ \Re(\nu) > 0,\ \gamma \in \mathbb{R}, \tag{2}$$

widely used in the theory of fractional differential equations (see the recent monograph [7] and the references therein). Moreover, the α-Mittag–Leffler function is a generalization of the so-called Le Roy function [8]

$$R_\rho(z) = \sum_{k=0}^{\infty} \frac{z^k}{[(k+1)!]^\alpha}, \quad z \in \mathbb{C},\ \alpha > 0. \tag{3}$$

In the more recent paper [5], the authors studied the asymptotic behavior and numerical simulation of this new class of special functions.

We observe that the Le Roy functions are used in probability in the context of the studies of COM-Poisson distributions [9], which are special classes of weighted Poisson distributions (see for example [10]). We show in Section 3, as a first probabilistic application, that the α-Mittag-Leffler functions can also be used in the construction of a new generalization of the COM-Poisson distribution that can be interesting for statistical applications and in physics in the context of generalized coherent states [11].

The aim of this paper is to study some particular classes of Hadamard fractional integral equations and differential equations whose solutions can be written in terms of the α-Mittag–Leffler function (1) and are somehow related to a fractional-type generalization of the COM-Poisson distribution.

A second application in probability is contained in Section 4. We observe that Imoto [12] has recently introduced the following generalization of the COM-Poisson distribution,

$$P\{N(t) = k\} = \frac{(\Gamma(\nu+k))^r\, t^k}{k!C(r,\nu,t)}, \quad k \in \mathbb{N}, \tag{4}$$

involving the normalizing function

$$C(r,\nu,t) = \sum_{k=0}^{\infty} \frac{(\Gamma(\nu+k))^r\, t^k}{k!}, \tag{5}$$

with $r < 1/2$, $t > 0$ and $1 > \nu > 0$ or $r = 1$, $\nu = 1$ and $|t| < 1$. The special function (5) is somehow related to the generalization of the Le Roy function, while the distribution (4) includes the COM-Poisson for $\nu = 1$ and $r = 1 - n$, $n \in \mathbb{R}^+$. In Section 4, we show that this function is related to integral equations with a time-varying coefficient involving Hadamard integrals.

In Section 5, we present other results concerning the relation between Le Roy-type functions and Hadamard fractional differential equations. We introduce a wide class of integro-differential equations extending the hyper-Bessel equations. This is a new interesting approach in the context of the mathematical studies of fractional Bessel equations (we refer for example to the recent paper [13] and the references therein).

In conclusion, the main aim of this paper is to establish a connection between some generalizations of the COM-Poisson distributions and integro-differential equations with time-varying coefficients involving Hadamard integrals or derivatives. As a by-product we suggest a possible application of the α-Mittag–Leffler function to build a generalized COM-Poisson distribution that in future should be investigated in more detail.

2. Preliminaries About Fractional Hadamard Derivatives and Integrals

Starting from the seminal paper by Hadamard [14], many papers have been devoted to the analysis of fractional operators with logarithmic kernels (we refer in particular to [15]). In this section, we briefly recall the definitions and main properties of Hadamard fractional integrals and derivatives and their Caputo-like regularizations recently introduced in the literature.

Definition 1. *Let $t \in \mathbb{R}^+$ and $\Re(\alpha) > 0$. The Hadamard fractional integral of order α, applied to the function $f \in L^p[a,b]$, $1 \le p < +\infty$, $0 < a < b < \infty$, for $t \in [a,b]$, is defined as*

$$\mathcal{J}^\alpha f(t) = \frac{1}{\Gamma(\alpha)} \int_a^t \left(\ln \frac{t}{\tau} \right)^{\alpha-1} f(\tau) \frac{\mathrm{d}\tau}{\tau}. \tag{6}$$

Before constructing the corresponding differential operator we must define the following space of functions.

Definition 2. *Let $[a,b]$ be a finite interval such that $-\infty < a < b < \infty$ and let $AC[a,b]$ be the space of absolutely continuous functions on $[a,b]$. Let us denote $\delta = t\frac{\mathrm{d}}{\mathrm{d}t}$ and define the space*

$$AC^n_\delta[a,b] = \left\{ f : t \in [a,b] \to \mathbb{R} \text{ such that } \left(\delta^{n-1} f \right) \in AC[a,b] \right\}. \tag{7}$$

Clearly $AC^1_\delta[a,b] \equiv AC[a,b]$ for $n = 1$.

Definition 3. *Let $\delta = t\frac{\mathrm{d}}{\mathrm{d}t}$, $\Re(\alpha) > 0$ and $n = [\alpha] + 1$, where $[\alpha]$ is the integer part of α. The Hadamard fractional derivative of order α applied to the function $f \in AC^n_\delta[a,b]$, $0 \le a < b < \infty$, is defined as*

$$\mathcal{D}^\alpha f(t) = \frac{1}{\Gamma(n-\alpha)} \left(t\frac{\mathrm{d}}{\mathrm{d}t} \right)^n \int_a^t \left(\ln \frac{t}{\tau} \right)^{n-\alpha-1} f(\tau) \frac{d\tau}{\tau} = \delta^n \left(\mathcal{J}^{n-\alpha} f \right)(t). \tag{8}$$

It has been proved (see e.g., Theorem 4.8 in [16]) that in the $L^p[a,b]$ space, $p \in [1,\infty)$, $0 \le a < b < \infty$, the Hadamard fractional derivative is the left-inverse operator to the Hadamard fractional integral, i.e.,

$$\mathcal{D}^\alpha \mathcal{J}^\alpha f(t) = f(t), \quad \forall t \in [a,b]. \tag{9}$$

Analogously to the Caputo fractional calculus, the regularized Caputo-type Hadamard fractional derivative is defined in terms of the Hadamard fractional integral in the following way (see, for example, [17])

$$\left(t\frac{\mathrm{d}}{\mathrm{d}t} \right)^\alpha f(t) = \frac{1}{\Gamma(n-\alpha)} \int_a^t \left(\ln \frac{t}{\tau} \right)^{n-\alpha-1} \left(\tau \frac{\mathrm{d}}{\mathrm{d}\tau} \right)^n f(\tau) \frac{d\tau}{\tau} = \mathcal{J}^{n-\alpha} \left(\delta^n f \right)(t), \tag{10}$$

where $t \in [a,b]$, $0 \le a < b < \infty$ and $n - 1 < \alpha \le n$, with $n \in \mathbb{N}$. In this paper, we will use the symbol $\left(t\frac{\mathrm{d}}{\mathrm{d}t} \right)^\alpha$ for the Caputo-type derivative in order to distinguish it from the Riemann-Liouville type definition (8) and also to underline the fact that essentially it coincides with the fractional power of the operator $\delta = t\frac{\mathrm{d}}{\mathrm{d}t}$. Moreover, by definition, when $\alpha = n$, $\left(t\frac{\mathrm{d}}{\mathrm{d}t} \right)^\alpha \equiv \delta^n$. The relationship between the Hadamard derivative (8) and the regularized Caputo-type derivative is given by ([17], Equation (12))

$$\left(t\frac{\mathrm{d}}{\mathrm{d}t} \right)^\alpha f(t) = \mathcal{D}^\alpha \left[f(t) - \sum_{k=0}^{n-1} \frac{\delta^k f(t_0)}{k!} \ln^k \left(\frac{t}{t_0} \right) \right], \quad \alpha \in (n-1, n], \; n = [\alpha] + 1. \tag{11}$$

In the sequel we will use the following useful equalities (that can be checked by simple calculations),

$$\left(t\frac{\mathrm{d}}{\mathrm{d}t} \right)^\alpha t^\beta = \beta^\alpha t^\beta, \tag{12}$$

$$\mathcal{J}^\alpha t^\beta = \beta^{-\alpha} t^\beta, \tag{13}$$

for $\beta \in (-1, \infty) \setminus \{0\}$ and $\alpha > 0$. It is immediate to see that

$$\left(t\frac{\mathrm{d}}{\mathrm{d}t}\right)^{\alpha} const = 0. \tag{14}$$

Finally, we observe that formally the relationship between Hadamard-type derivatives and Riemann-Liouville derivatives is given by the change of variable $t \to \ln(t)$, leading to the logarithmic kernel. According to this change of variable, the Hadamard derivative represents the counterpart of the fractional power of the operator $\frac{\mathrm{d}}{\mathrm{d}t}$, i.e., the fractional power of the operator $\delta = t\frac{\mathrm{d}}{\mathrm{d}t}$.

3. Generalized COM-Poisson Processes Involving the α-Mittag–Leffler Function and the Related Hadamard Equations

Here we show a possible application of the α-Mittag–Leffler function (1) in the context of fractional-type Poisson statistics and we discuss the relation with Hadamard equations.

We first recall that weighted Poisson distributions have been widely studied in statistics in order to consider over or under-dispersion with respect to the homogeneous Poisson process. The probability mass function of a weighted Poisson process $N^w(t)$, $t > 0$, is given by (see [10])

$$P\{N^w(t) = n\} = \frac{w(n)p(n,t)}{\mathbb{E}[w(N)]}, \quad n \geq 0, \tag{15}$$

where N is a Poisson distributed random variable

$$p(n,t) = \frac{t^n}{n!}e^{-t}, \quad n \geq 0,\ t > 0,$$

$w(\cdot)$ is a non-negative weight function with non-zero, finite expectation, i.e.,

$$0 < \mathbb{E}[w(N)] = \sum_{k=0}^{\infty} w(k)p(k,t) < +\infty. \tag{16}$$

The so-called COM-Poisson distribution introduced by Conway and Maxwell [9] in a queueing model belongs to this wide class of distributions. A random variable $N_\nu(t)$ is said to have a COM-Poisson distribution if

$$P\{N_\nu(t) = n\} = \frac{1}{f(t)}\frac{t^n}{n!^\nu}, \quad n \geq 0,\ t > 0, \tag{17}$$

where

$$f(t) = \sum_{k=0}^{\infty} \frac{t^k}{k!^\nu}, \qquad \nu > 0, \tag{18}$$

is the Le-Roy function.

This distribution can be viewed as a particular weighted Poisson distribution with $w(n) = 1/(n!)^{\nu-1}$. The Conway-Maxwell-Poisson distribution is widely used in statistics in order to take into account over-dispersion (for $0 < \nu < 1$) or under-dispersion (for $\nu > 1$) in data analysis. Moreover it represents a useful generalization of the Poisson distribution (that is obtained for $\nu = 1$) for many applications (see for example [18] and the references therein). Observe that when $t \in (0,1)$ and $\nu \to 0$ it reduces to the geometric distribution, while for $\nu \to +\infty$ it converges to the Bernoulli distribution.

Another class of interesting weighted Poisson distributions are the fractional Poisson-type distributions

$$P\{N_\alpha(t) = k\} = \frac{1}{E_{\alpha,1}(t)}\frac{t^k}{\Gamma(\alpha k + 1)}, \quad \alpha \in (0,1), \tag{19}$$

where

$$E_{\alpha,1}(t) = \sum_{k=0}^{\infty} \frac{t^k}{\Gamma(\alpha k + 1)}, \tag{20}$$

is the Mittag–Leffler function (see the recent monograph [7] for more details). In this case, the weight function is given by $w(k) = k!/\Gamma(\alpha k + 1)$. There is a recent wide literature about fractional generalizations of Poisson processes (see e.g., [19–27]) and we should explain in which sense Equation (19) can be regarded as a fractional-type Poisson distribution. The relationship of this kind of distributions with fractional calculus, first considered by Beghin and Orsingher in [28], is given by the fact that the probability generating function

$$G(u,t) = \sum_{k=0}^{\infty} u^k P\{N_\alpha(t) = k\} = \frac{E_{\alpha,1}(ut)}{E_{\alpha,1}(t)}, \quad \alpha \in (0,1], \tag{21}$$

satisfies the fractional differential equation

$$\frac{d^\alpha G(u^\alpha, t)}{du^\alpha} = tG(u^\alpha, t), \quad \alpha \in (0,1], \ |u| \leq 1, \tag{22}$$

where d^α/du^α denotes the Caputo fractional derivative w.r.t. the variable u (see [16] for the definition). Essentially, the connection between the distribution (19) and fractional calculus is therefore given by the normalizing function, since the Mittag–Leffler plays the role of the exponential function in the theory of fractional differential equations. In this context, we discuss a new generalization of the COM-Poisson distribution by using the generalized α-Mittag–Leffler function Equation (1), independently introduced by Gerhold in [6] and by Garra and Polito in [4]. This new fractional distribution includes a wide class of special distributions already studied in the recent literature: the fractional Poisson distribution Equation (19) considered by Beghin and Orsingher [28], the classical COM-Poisson distribution and so on.

In this paper, we introduce the following generalization, namely the fractional COM-Poisson distribution

$$P\{N_{\nu;\alpha,\gamma}(t) = n\} = \frac{(\lambda t)^n}{(\Gamma(\alpha n + \gamma))^\nu} \frac{1}{E_{\nu;\alpha,\gamma}(\lambda t)}, \tag{23}$$

where

$$E_{\nu;\alpha,\gamma}(t) = \sum_{k=0}^{\infty} \frac{t^k}{(\Gamma(\alpha k + \gamma))^\nu}, \quad \alpha > 0, \ \nu > 0, \ \gamma \in \mathbb{R} \tag{24}$$

is the α-Mittag–Leffler function, considered by Gerhold [6], Garra and Polito [4]. The distribution Equation (23) is a weighted Poisson distribution with weights $w(k) = k!/\left(\Gamma(\alpha k + \gamma)\right)^\nu$. Obviously the probability generating function is here given by a ratio of α-Mittag–Leffler functions, i.e.,

$$G(u,t) = \sum_{n=0}^{\infty} u^n P\{N_{\nu;\alpha,\gamma}(t) = n\} = \frac{E_{\nu;\alpha,\gamma}(\lambda u t)}{E_{\nu;\alpha,\gamma}(\lambda t)}, \quad |u| \leq 1. \tag{25}$$

With the next theorem we explain the reason why we consider the distribution Equation (23), a kind of fractional generalization of the COM-Poisson classical distribution related to Hadamard fractional equations with varying coefficients.

Theorem 1. *For $\nu \in (0,1)$, the function*

$$g(u,t) = u^{\nu-1} \frac{E_{\nu;1,\nu}(\lambda u t)}{E_{\nu;1,\nu}(\lambda t)} = u^{\nu-1} G(u,t),$$

satisfies the equation

$$\left(u\frac{\mathrm{d}}{\mathrm{d}u}\right)^{\nu} g(u,t) = \lambda t u g(u,t) + \frac{u^{\nu-1}}{\Gamma^{\nu}(\nu-1)}, \tag{26}$$

where $\left(u\frac{\mathrm{d}}{\mathrm{d}u}\right)^{\nu}$ *is the Caputo-Hadamard fractional derivative of order* ν.

Proof. By direct calculations we obtain that

$$\begin{aligned}\left(u\frac{\mathrm{d}}{\mathrm{d}u}\right)^{\nu} u^{\nu-1}\frac{E_{\nu;1,\nu}(\lambda u t)}{E_{\nu;1,\nu}(\lambda t)} &= \frac{1}{E_{\nu;1,\nu}(\lambda t)}\sum_{k=0}^{\infty}\frac{\lambda^k u^{k+\nu-1} t^k}{(\Gamma(k+\nu-1))^{\nu}} \\ &= \frac{1}{E_{\nu;1,\nu}(\lambda t)}\sum_{k=-1}^{\infty}\frac{\lambda^{k+1} u^{k+\nu} t^{k+1}}{(\Gamma(k+\nu))^{\nu}} = \lambda t u g(u,t) + \frac{u^{\nu-1}}{(\Gamma(\nu-1))^{\nu}},\end{aligned} \tag{27}$$

where we changed the order of summation with fractional differentiation since the ν-Mittag–Leffler function is, for $\nu \in (0,1)$, an entire function. □

A more detailed analysis about the statistical properties and possible applications of this new class of distributions is not the object of this paper and will be considered in a future work.

We finally observe that the fractional COM-Poisson distribution (23) includes as a special case, for $\nu = 1$ the fractional Poisson distribution (19) previously introduced by Beghin and Orsingher in [28] and recently treated by Chakraborty and Ong [29], Herrmann [21] and Porwall and Dixit [30]. Asymptotic results for the multivariate version of this distribution have been recently analyzed by Beghin and Macci in [19].

4. Hadamard Fractional Equations Related to the GCOM-Poisson Distribution

In a recent paper, Imoto [12] studied a different generalization of the Conway-Maxwell-Poisson distribution (in the following GCOM Poisson distribution), depending on two real parameters r and ν. According to [12], a random variable $X(t)$ has a GCOM Poisson distribution if

$$P\{X(t) = n\} = \frac{(\Gamma(\nu+n))^r\, t^n}{n! C(r,\nu,t)}, \quad t \geq 0,\ n = 0,1,\ldots, \tag{28}$$

where

$$C(r,\nu,t) = \sum_{k=0}^{\infty}\frac{(\Gamma(\nu+k))^r\, t^k}{k!}, \tag{29}$$

is the normalizing function which converges for $\nu > 0$, $t > 0$ and $r < 1/2$ (by applying the ratio criterion) or $r = 1$, $\nu > 0$ and $|t| < 1$. In the context of weighted Poisson distributions, it corresponds to the choice $w(k) = \Gamma^r(\nu + k)$.

This distribution includes as particular cases the COM-Poisson distribution for $\nu = 1$ and $r = 1 - n$, the geometric for $r = \nu = 1$, $|t| < 1$ and the homogeneous Poisson distribution for $r = 0$. The new parameter introduced in the distribution plays the role of controlling the length of tails within the framework of queueing processes. Indeed it was proved that, it displays over-dispersion for $r \in (0,1)$ and under-dispersion for $r < 0$. Moreover, from the ratios of successive probabilities, when $\nu > 1$ and $r \in (0,1)$ it emerges to be heavy tailed, while for $r < 0$ light-tailed.

We also underline that for $r = 1 - n$, with $n > 0$, the function (29) becomes a sort of generalization of the Wright function which can be written as

$$C(1-n,\nu,t) = \sum_{k=0}^{\infty}\frac{t^k}{k!\Gamma^n(\nu+k)}. \tag{30}$$

Recall that, in the general form, the Wright function is defined as follows (we refer to [31] for a good survey):

$$W_{\alpha,\beta}(t) = \sum_{k=0}^{\infty} \frac{t^k}{k!\Gamma(\alpha k + \beta)}, \quad \alpha > -1, \beta \in \mathbb{C}. \tag{31}$$

We show now the relation between the normalizing function Equation (29) and an Hadamard fractional differential equation with time-varying coefficients. Therefore, we can consider in some sense also the GCOM Poisson distribution as a fractional-type modification of the homogeneous distribution, in the sense that its probability generating function satisfies a fractional differential equation.

Proposition 1. *The function*

$$C(r,\nu,t) = \sum_{k=0}^{\infty} \frac{\Gamma^r(\nu+k)(\lambda t)^k}{k!}, \tag{32}$$

satisfies the integro-differential equation

$$\mathcal{J}^r \left(t^\nu \frac{\mathrm{d}f}{\mathrm{d}t} \right) = \lambda t^\nu f, \quad t > 0, \; \nu \in (0,1], \; r \in (0,1/2) \tag{33}$$

involving an Hadamard fractional integral $\mathcal{J}^r$ *of order* r.

Proof.

$$\begin{aligned}
\mathcal{J}^r \left(t^\nu \frac{\mathrm{d}C}{\mathrm{d}t} \right) &= \mathcal{J}^r \sum_{k=1}^{\infty} \frac{\Gamma^r(\nu+k)\lambda^k t^{k+\nu-1}}{(k-1)!} = \sum_{k=1}^{\infty} \frac{\Gamma^r(\nu+k)(k+\nu-1)^{-r}\lambda^k t^{k+\nu-1}}{(k-1)!} \\
&= \sum_{k=1}^{\infty} \frac{\Gamma^r(\nu+k-1)\lambda^k t^{k+\nu-1}}{(k-1)!} = \lambda t^\nu f,
\end{aligned} \tag{34}$$

where we used Equation (13). □

Corollary 1. *The probability generating function* $\mathcal{G}(u,t)$ *of the GCOM Poisson distribution Equation* (28)

$$\mathcal{G}(u,t) = \sum_{k=0}^{\infty} u^k P\{X(t) = k\} = \frac{C(r,\nu,ut)}{C(r,\nu,t)}, \quad |u| \le 1, \tag{35}$$

satisfies the integro-differential equation

$$\mathcal{J}^r \left(u^\nu \frac{df}{du} \right) = u^\nu f, \quad |u| \le 1. \tag{36}$$

5. Further Connections between Modified Mittag–Leffler Functions and Hadamard Fractional Equations

The analysis of fractional differential equations with non-constant coefficients is an interesting and non-trivial topic. In particular, the analysis of equations involving fractional-type Bessel operators (i.e., the fractional counterpart of singular linear differential operators of arbitrary order) has attracted the interest of many researchers (see e.g., [13,32] and references therein). In [4], some results about the connection between Le-Roy functions and equations with space-varying coefficients involving Hadamard derivatives and Laguerre derivatives have been obtained. Here we go further in the direction started in [4], showing other interesting applications of Hadamard fractional equations in the theory of hyper-Bessel functions. With the next theorem, we find the equation interpolating classical Bessel equations of arbitrary order possessing exact solution in terms of Le Roy functions. We remark, indeed, that Le Roy functions include as special cases hyper-Bessel functions.

Theorem 2. *The Le Roy function* $E_{\alpha;1,1}\left(t^{\alpha}/\alpha\right)$ *satisfies the equation*

$$\frac{1}{t^{\alpha}}\left(t\frac{\mathrm{d}}{\mathrm{d}t}\right)^{\alpha} f(x) = \alpha^{\alpha-1} f(t) \tag{37}$$

Proof. Recall that

$$\left(t\frac{\mathrm{d}}{\mathrm{d}t}\right)^{\alpha} t^{\beta} = \beta^{\alpha} t^{\beta}. \tag{38}$$

Therefore

$$\left(t\frac{\mathrm{d}}{\mathrm{d}t}\right)^{\alpha} E_{\alpha;1,1}\left(\frac{t^{\alpha}}{\alpha}\right) = \sum_{k=0}^{\infty} \frac{(\alpha k)^{\alpha} t^{\alpha k}}{\alpha^{k} k!^{\alpha}} = \alpha^{\alpha-1} t^{\alpha} E_{\alpha;1,1}\left(\frac{t^{\alpha}}{\alpha}\right). \tag{39}$$

□

Remark 1. *For* $\alpha = 2$*, we have that the function* $E_{2;1,1}(t^2/2)$ *is a solution of*

$$\left(t\frac{\mathrm{d}}{\mathrm{d}t}\right)^{2} f(t) = 2t^2 f(t). \tag{40}$$

For $\alpha = 3$*, the function* $E_{3;1,1}(t^3/3)$ *satisfies*

$$\left(t\frac{\mathrm{d}}{\mathrm{d}t}\right)^{3} f(t) = 9t^3 f(t), \tag{41}$$

and so forth for any integer value of α*. From this point of view, the solution of Hadamard Equation (37), for non integer values of* α*, leads to an interpolation between successive hyper-Bessel functions.*

Let us consider the operator

$$L_H = \frac{1}{t^{\alpha n}} \underbrace{\left(t\frac{\mathrm{d}}{\mathrm{d}t}\right)^{\alpha}\left(t\frac{\mathrm{d}}{\mathrm{d}t}\right)^{\alpha}\cdots\left(t\frac{\mathrm{d}}{\mathrm{d}t}\right)^{\alpha}}_{n \text{ times}}, \quad \alpha > 0, \tag{42}$$

where H stands for an Hyper-Bessel type operator involving Caputo-type Hadamard derivatives. We have the following general Theorem

Theorem 3. *A solution of the equation*

$$L_H f(t) = \alpha^{n\alpha - n} n^{n\alpha} f(t), \tag{43}$$

is given by

$$f(t) = E_{n\alpha;1,1}\left(t^{\alpha n}/\alpha^{n}\right)$$

To conclude this section, we restate Theorem 3.3 of [4], in view of the comments of Turmetov [33] in the case in which Caputo-type Hadamard derivatives appear in the governing equations.

Theorem 4. *The function* $E_{\beta;1,1}(t)$*, with* $\beta \in [1, \infty)$*,* $t \geq 0$*,* $\lambda \in \mathbb{R}$ *is an eigenfunction of the operator*

$$\underbrace{\frac{\mathrm{d}}{\mathrm{d}t}t\frac{\mathrm{d}}{\mathrm{d}t}\cdots\frac{\mathrm{d}}{\mathrm{d}t}t\frac{\mathrm{d}}{\mathrm{d}t}}_{r \text{ derivatives}}\left(t\frac{\mathrm{d}}{\mathrm{d}t}\right)^{\beta-r}, \quad r = 1, \ldots, n-1, \tag{44}$$

where $n = [\beta]$ *is the integer part of* β *and* $\left(t\frac{\mathrm{d}}{\mathrm{d}t}\right)^{\beta-r}$ *denotes the Caputo-type regularized Hadamard derivative of order* $\beta - r$.

The difference w.r.t. the previous version is simply given by the fact that, by using the regularized Caputo-type Hadamard derivative $\left(t\frac{\mathrm{d}}{\mathrm{d}t}\right)^{\alpha} t^0 = 0$, that is necessary for the correctness of the result.

Acknowledgments: The work of R.G. has been carried out in the framework of the activities of GNFM. F.P. has been supported by the projects *Memory in Evolving Graphs* (Compagnia di San Paolo/Università di Torino), *Sviluppo e analisi di processi Markoviani e non Markoviani con applicazioni* (Università di Torino), and by INDAM–GNAMPA.

Author Contributions: The authors contributed equally to this work.

Conflicts of Interest: The authors declare no conflict of interest.

References

1. Ahmad, B.; Alsaedi, A.; Ntouyas, S.K.; Tariboon, J. *Hadamard-Type Fractional Differential Equations, Inclusions and Inequalities*; Springer: Basel, Switzerland, 2017.
2. Garra, R.; Giusti, A.; Mainardi, F.; Pagnini, G. Fractional relaxation with time-varying coefficient. *Fract. Calc. Appl. Anal.* **2014**, *17*, 424–439.
3. Saxena, R.K.; Garra, R.; Orsingher, E. Analytical solution of space-time fractional telegraph-type equations involving Hilfer and Hadamard derivatives. *Integral Transform. Spec. Funct.* **2016**, *27*, 30–42.
4. Garra, R.; Polito, F. On Some Operators Involving Hadamard Derivatives. *Integral Transform Spec. Funct.* **2013**, *24*, 773–782.
5. Garrappa, R.; Mainardi, F.; Rogosin, S.V. On a generalized three-parameter Wright function of Le Roy-type. *Fract. Calc. Appl. Anal.* **2017**, *20*, 1196–1215.
6. Gerhold, S. Asymptotics for a variant of the Mittag–Leffler function. *Integral Transform Spec. Funct.* **2012**, *23*, 397–403.
7. Gorenflo, R.; Kilbas, A.A.; Mainardi, F.; Rogosin, S.V. *Mittag–Leffler Functions. Related Topics and Applications*; Springer Monographs in Mathematics; Springer: Berlin, Germany, 2014.
8. Le Roy, E. Valeurs asymptotiques de certaines séries procédant suivant les puissances entéres et positives d'une variable réelle. *Darboux Bull.* **1899**, *24*, 245–268. (In French).
9. Conway, R.W.; Maxwell, W.I. A queueing model with state dependent service rate. *J. Ind. Eng.* **1961**, *12*, 132–136.
10. Balakrishnan, N.; Kozubowski, T.J. A class of weighted Poisson processes. *Stat. Probab. Lett.* **2008**, *78*, 2346–2352.
11. Sixdeniers, J.M.; Penson, K.A.; Solomon, A.I. Mittag–Leffler coherent states. *J. Phys. A* **1999**, *32*, 7543–7563.
12. Imoto, T. A generalized Conway–Maxwell–Poisson distribution which includes the negative binomial distribution. *Appl. Math. Comput.* **2014**, *247*, 824–834.
13. Shishkina, E.L.; Sitnik, S.M. On fractional powers of Bessel operators. *J. Inequal. Spec. Funct.* **2017**, *8*, 49–67.
14. Hadamard, J. Essai sur l'etude des fonctions donnees par leur developpment de Taylor. *J. Math. Pures Appl.* **1892**, *8*, 101–186.
15. Kilbas, A.A. Hadamard-type fractional calculus. *J. Korean Math. Soc.* **2001**, *38*, 1191–1204.
16. Kilbas, A.A.; Srivastava, H.M.; Trujillo, J.J. *Theory and Applications of Fractional Differential Equations*; Elsevier: Amsterdam, The Netherlands, 2006; Volume 204.
17. Jarad, F.; Abdeljawad, T.; Baleanu, D. Caputo-type modification of the Hadamard fractional derivative. *Adv. Differ. Equ.* **2012**, 1–8, doi:10.1186/1687-1847-2012-142.
18. Rodado, A.; Bebbington, M.; Noble, A.; Cronin, S.; Jolly, G. On selection of analog volcanoes. *Math. Geosci.* **2011**, *43*, 505–519.
19. Beghin, L.; Macci, C. Asymptotic results for a multivariate version of the alternative fractional Poisson process. *Stat. Probab. Lett.* **2017**, *129*, 260–268.
20. Beghin, L.; Ricciuti, C. Time-inhomogeneous fractional Poisson processes defined by the multistable subordinator. *arXiv* **2016**, arXiv:1608.02224.

21. Herrmann, R. Generalization of the fractional Poisson distribution. *Fract. Calc. Appl. Anal.* **2006**, *19*, 832–842.
22. Laskin, N. Fractional Poisson process. *Commun. Nonlinear Sci. Numer. Simul.* **2003**, *8*, 201–213.
23. Mainardi, F.; Gorenflo, R.; Scalas, E. A fractional generalization of the Poisson processes. *Vietnam J. Math.* **2013**, *32*, 53–64.
24. Meerschaert, M.; Nane, E.; Vellaisamy, P. The fractional Poisson process and the inverse stable subordinator. *Electron. J. Probab.* **2011**, *16*, 1600–1620.
25. Orsingher, E.; Toaldo, B. Counting processes with Bernstein intertimes and random jumps. *J. Appl. Probab.* **2015**, *52*, 1028–1044.
26. Orsingher, E.; Polito, F. The space-fractional Poisson process. *Stat. Probab. Lett.* **2012**, *82*, 852–858.
27. Polito, F.; Scalas, E. A Generalization of the Space-Fractional Poisson Process and its Connection to some Lévy Processes. *Electron. Commun. Probab.* **2016**, *21*, 1–14.
28. Beghin, L.; Orsingher, E. Fractional Poisson processes and related planar random motions. *Electron. J. Probab.* **2009**, *14*, 1790–1826.
29. Chakraborty, S.; Ong, S. H. Mittag-Leffler function distribution—A new generalization of hyper-Poisson distribution. *J. Stat. Distrib. Appl.* **2017**, *4*, 8, doi:10.1186/s40488-017-0060-9.
30. Porwal, S.; Dixit, K.K. On Mittag–Leffler type Poisson distribution. *Afr. Mat.* **2017**, *28*, 29–34.
31. Gorenflo, R.; Luchko, Y.; Mainardi, F. Analytical properties and applications of the Wright function. *Fract. Calc. Appl. Anal.* **1999**, *2*, 383–414.
32. Kiryakova, V.S. *Generalized Fractional Calculus and Applications*; CRC Press: Boca Raton, FL, USA, 1993.
33. Turmetov, B.K. On certain operator method for solving differential equations. *AIP Conf. Proc.* **2015**, *1676*, doi:10.1063/1.4930520.

mathematics

MDPI

Article

On Some New Properties of the Fundamental Solution to the Multi-Dimensional Space- and Time-Fractional Diffusion-Wave Equation

Yuri Luchko †

Beuth University of Applied Sciences Berlin, 13353 Berlin, Germany; luchko@beuth-hochschule.de; Tel.: +49-30-4504-5295

† Current address: Beuth Hochschule für Technik Berlin, Fachbereich II Mathematik-Physik-Chemie, Luxemburger Str. 10, 13353 Berlin, Germany.

Received: 11 November 2017; Accepted: 4 December 2017; Published: 8 December 2017

Abstract: In this paper, some new properties of the fundamental solution to the multi-dimensional space- and time-fractional diffusion-wave equation are deduced. We start with the Mellin-Barnes representation of the fundamental solution that was derived in the previous publications of the author. The Mellin-Barnes integral is used to obtain two new representations of the fundamental solution in the form of the Mellin convolution of the special functions of the Wright type. Moreover, some new closed-form formulas for particular cases of the fundamental solution are derived. In particular, we solve the open problem of the representation of the fundamental solution to the two-dimensional neutral-fractional diffusion-wave equation in terms of the known special functions.

Keywords: multi-dimensional diffusion-wave equation; neutral-fractional diffusion-wave equation; fundamental solution; Mellin-Barnes integral; integral representation; Wright function; generalized Wright function

MSC: 26A33; 35C05; 35E05; 35L05; 45K05; 60E99

1. Introduction

Partial fractional differential equations are nowadays both an important research subject and a popular modeling approach. Despite of importance of mathematical models in two- and three-dimensions for applications, most of the recent publications devoted to fractional diffusion-wave equations have dealt with the one-dimensional case. The literature dealing with multi-dimensional partial fractional differential equations is still not numerous and can be divided into several groups, those devoted to Cauchy problems on the whole space, boundary-value problems on bounded domains, and of course to different types of equations, including single- and multi-term equations as well as equations of distributed order. Because the focus of this paper is on the Cauchy problem for a model linear time- and space-fractional diffusion-wave equation, we mention here only some important relevant publications.

The fundamental solution to the multi-dimensional time-fractional diffusion-wave equation with the Laplace operator was derived for the first time by Kochubei in [1] and Schneider and Wyss in [2] independently from each other in terms of the Fox H-function. We note that in [1], the Cauchy problem for the general fractional diffusion equation with the regularized fractional derivative (the Caputo fractional derivative in modern terminology) and the general second-order spatial differential operator was also investigated. In the series of publications [3–5], Hanyga considered mathematical, physical, and probabilistic properties of the fundamental solutions to multi-dimensional time-, space- and space-time-fractional diffusion-wave equations, respectively. Recently, Luchko and his co-authors started to employ the method of the Mellin-Barnes integral representation to derive further properties of the multi-dimensional space-time-fractional diffusion-wave equation (see, e.g., [6–10]). Still, the list of the properties, particular cases, integral and series representations, asymptotic formulas, and so forth known for the fundamental solution to the one-dimensional diffusion-wave equation (see, e.g., [11]) is essentially more expanded compared to the multi-dimensional case, and thus further investigations of the multi-dimensional case are required.

In this paper, some new properties and particular cases of the fundamental solution to the multi-dimensional space- and time-fractional diffusion-wave equation are deduced. In the second section, we recall the Mellin-Barnes representations of the fundamental solution that were derived in the previous publications of the author and his co-authors. In the third section, the Mellin-Barnes integral is used to obtain two new representations of the fundamental solution in the form of the Mellin convolution of the special functions of the Wright type. The fourth section is devoted to the derivation of some new closed-form formulas for the fundamental solution. In particular, the open problem of the representation of the fundamental solution to the two-dimensional neutral-fractional diffusion-wave equation in terms of the known elementary or special functions is solved.

2. Problem Formulation and Auxiliary Results

In this section, we present a problem formulation and some auxiliary results that are used in the rest of the paper.

2.1. Problem Formulation

In this paper, we deal with the multi-dimensional space- and time-fractional diffusion-wave equation in the following form:

$$D_t^{\beta} u(\mathrm{x},t) = -(-\Delta)^{\frac{\alpha}{2}} u(\mathrm{x},t), \quad \mathrm{x} \in \mathbb{R}^n,\ t>0,\ 1<\alpha\le 2,\ 0<\beta\le 2, \tag{1}$$

where $(-\Delta)^{\frac{\alpha}{2}}$ is the fractional Laplacian and D_t^{β} is the Caputo time-fractional derivative of order β.

The Caputo time-fractional derivative of order $\beta > 0$ is defined by the formula

$$D_t^{\beta} u(\mathrm{x},t) = \left(I_t^{n-\beta} \frac{\partial^n u}{\partial t^n} \right)(t), \quad n-1<\beta\le n,\ n\in\mathbb{N}, \tag{2}$$

where I_t^{γ} is the Riemann–Liouville fractional integral:

$$(I_t^{\gamma} u)(t) = \begin{cases} \frac{1}{\Gamma(\gamma)} \int_0^t (t-\tau)^{\gamma-1} u(\mathrm{x},\tau)\, d\tau & \text{for } \gamma>0, \\ u(\mathrm{x},t) & \text{for } \gamma = 0. \end{cases}$$

The fractional Laplacian $(-\Delta)^{\frac{\alpha}{2}}$ is defined as a pseudo-differential operator with the symbol $|\kappa|^{\alpha}$ ([12,13]):

$$\left(\mathcal{F}\,(-\Delta)^{\frac{\alpha}{2}} f\right)(\kappa) = |\kappa|^{\alpha} (\mathcal{F}\, f)(\kappa), \tag{3}$$

where $(\mathcal{F}\, f)(\kappa)$ is the Fourier transform of a function f at the point $\kappa \in \mathbb{R}^n$ defined by

$$(\mathcal{F}\, f)(\kappa) = \hat{f}(\kappa) = \int_{\mathbb{R}^n} e^{i\kappa\cdot\mathrm{x}} f(\mathrm{x})\, d\mathrm{x}. \tag{4}$$

For $0<\alpha<m$, $m\in\mathbb{N}$ and $x \in \mathbb{R}^n$, the fractional Laplacian can be also represented as a hypersingular integral ([13]):

$$(-\Delta)^{\frac{\alpha}{2}} f(\mathrm{x}) = \frac{1}{d_{n,m}(\alpha)} \int_{\mathbb{R}^n} \frac{\left(\Delta_{\mathrm{h}}^m f\right)(\mathrm{x})}{|\mathrm{h}|^{n+\alpha}}\, d\mathrm{h} \tag{5}$$

with a suitably defined finite-differences operator $\left(\Delta_{\mathrm{h}}^{m} f\right)(\mathrm{x})$ and a normalization constant $d_{n,m}(\alpha)$.

According to [13], the representation given by Equation (5) of the fractional Laplacian in the form of the hypersingular integral does not depend on m, $m \in \mathbb{N}$ provided that $\alpha < m$.

We note that in the one-dimensional case, Equation (1) is a particular case of a more general equation with the Caputo time-fractional derivative and the Riesz-Feller space-fractional derivative that was discussed in detail in [11]. For $\alpha = 2$, the fractional Laplacian $(-\Delta)^{\frac{\alpha}{2}}$ is simply $-\Delta$, and thus Equation (1) is a particular case of the time-fractional diffusion-wave equation that was considered in many publications, including, for example, [1,2,5,14–17]. For $\alpha = 2$ and $\beta = 1$, Equation (1) is reduced to the diffusion equation, and for $\alpha = 2$ and $\beta = 2$, it is the wave equation that justifies its denotation as a fractional diffusion-wave equation.

In this paper, we deal with the Cauchy problem for Equation (1) with Dirichlet initial conditions. If the order β of the time-derivative satisfies the condition $0 < \beta \leq 1$, we pose an initial condition of the form

$$u(\mathrm{x}, 0) = \varphi(\mathrm{x}), \quad \mathrm{x} \in \mathbb{R}^n. \tag{6}$$

For the orders β satisfying the condition $1 < \beta \leq 2$, the second initial condition in the form

$$\frac{\partial u}{\partial t}(\mathrm{x}, 0) = 0, \quad \mathrm{x} \in \mathbb{R}^n \tag{7}$$

is added to the Cauchy problem.

Because the initial-value problem given by Equations (1) and (6) (or (1), (6) and (7)) is linear, its solution can be represented in the form

$$u(\mathrm{x}, t) = \int_{\mathbb{R}^n} G_{\alpha,\beta,n}(\mathrm{x} - \zeta, t)\varphi(\zeta)\, d\zeta,$$

where $G_{\alpha,\beta,n}$ is the first fundamental solution to the fractional diffusion-wave Equation (1), that is, the solution to the problem given by Equations (1) and (6) with the initial condition

$$u(\mathrm{x}, 0) = \prod_{i=1}^{n} \delta(x_i), \quad \mathrm{x} = (x_1, x_2, \ldots, x_n) \in \mathbb{R}^n$$

or to the problem given by Equations (1), (6) and (7) with the initial conditions

$$u(\mathrm{x}, 0) = \prod_{i=1}^{n} \delta(x_i), \quad \mathrm{x} = (x_1, x_2, \ldots, x_n) \in \mathbb{R}^n$$

and

$$\frac{\partial u}{\partial t}(\mathrm{x}, 0) = 0, \quad \mathrm{x} \in \mathbb{R}^n$$

for $0 < \beta \leq 1$ or $1 < \beta \leq 2$, respectively, with δ being the Dirac delta function.

Thus the behavior of the solutions to the problem given by Equations (1) and (6) (or (1), (6) and (7)) is determined by the fundamental solution $G_{\alpha,\beta,n}(\mathrm{x}, t)$, and the focus of this paper is on the derivation of the new properties of the fundamental solution.

2.2. Mellin-Barnes Representations of the Fundamental Solution

A Mellin-Barnes representation of the fundamental solution to the multi-dimensional space- and time-fractional diffusion-wave Equation (1) was derived for the first time in [7] for the case $\beta = \alpha$ (see also [8]), in [10] for the case $\beta = \alpha/2$, and in [9] for the general case. For the reader's convenience, we present here a short schema of its derivation.

The application of the multi-dimensional Fourier transform (4) with respect to the spatial variable $\mathrm{x} \in \mathbb{R}^n$ to Equation (1) and to the initial conditions given by Equation (6) with $\varphi(\mathrm{x}) = \prod_{i=1}^{n} \delta(x_i)$, and

Equation (7) (the last condition is relevant only if $\beta > 1$) leads to the ordinary fractional differential equation in the Fourier domain:

$$D_t^{\beta}\hat{G}_{\alpha,\beta,n}(\kappa,t) + |\kappa|^{\alpha}\hat{G}_{\alpha,\beta,n}(\kappa,t) = 0, \tag{8}$$

along with the initial conditions

$$\hat{G}_{\alpha,\beta,n}(\kappa,0) = 1 \tag{9}$$

in the case for $0 < \beta \leq 1$, or with the initial conditions

$$\hat{G}_{\alpha,\beta,n}(\kappa,0) = 1,\ \frac{\partial}{\partial t}\hat{G}_{\alpha,\beta,n}(\kappa,0) = 0 \tag{10}$$

in the case for $1 < \beta \leq 2$.

In both cases, the unique solution of Equation (8) with the initial conditions given by Equations (9) or (9) and (10) has the following form (see, e.g., [18]):

$$\hat{G}_{\alpha,\beta,n}(\kappa,t) = E_{\beta}\left(-|\kappa|^{\alpha}t^{\beta}\right) \tag{11}$$

in terms of the Mittag-Leffler function $E_{\beta}(z)$ that is defined by a convergent series:

$$E_{\beta}(z) = \sum_{n=0}^{\infty}\frac{z^n}{\Gamma(1+\beta n)},\quad \beta > 0,\ z \in \mathbb{C}. \tag{12}$$

Because of the asymptotic formula (see, e.g., [19]):

$$E_{\beta}(-x) = -\sum_{k=1}^{m}\frac{(-x)^{-k}}{\Gamma(1-\beta k)} + O(|x|^{-1-m}),\ m \in \mathbb{N},\ x \to +\infty,\ 0 < \beta < 2 \tag{13}$$

we have the inclusion $\hat{G}_{\alpha,\beta,n} \in L_1(\mathbb{R}^n)$ under the condition $\alpha > 1$, and thus the inverse Fourier transform of Equation (11) can be represented as follows:

$$G_{\alpha,\beta,n}(\mathrm{x},t) = \frac{1}{(2\pi)^n}\int_{\mathbb{R}^n} e^{-i\kappa\cdot\mathrm{x}}E_{\beta}\left(-|\kappa|^{\alpha}t^{\beta}\right)d\kappa,\quad \mathrm{x} \in \mathbb{R}^n, t > 0. \tag{14}$$

Because $E_{\beta}\left(-|\kappa|^{\alpha}t^{\beta}\right)$ is a radial function, the known formula (see, e.g., [13])

$$\frac{1}{(2\pi)^n}\int_{\mathbb{R}^n} e^{-i\kappa\cdot\mathrm{x}}\varphi(|\kappa|)\,d\kappa = \frac{|\mathrm{x}|^{1-\frac{n}{2}}}{(2\pi)^{\frac{n}{2}}}\int_0^{\infty}\varphi(\tau)\tau^{\frac{n}{2}}J_{\frac{n}{2}-1}(\tau|\mathrm{x}|)\,d\tau \tag{15}$$

for the Fourier transform of the radial functions can be applied, where J_{ν} denotes the Bessel function with index ν (for the properties of the the Bessel function, see, e.g., [20]), and we arrive at the representation

$$G_{\alpha,\beta,n}(\mathrm{x},t) = \frac{|\mathrm{x}|^{1-\frac{n}{2}}}{(2\pi)^{\frac{n}{2}}}\int_0^{\infty}E_{\beta}\left(-\tau^{\alpha}t^{\beta}\right)\tau^{\frac{n}{2}}J_{\frac{n}{2}-1}(\tau|\mathrm{x}|)\,d\tau \tag{16}$$

whenever the integral in Equation (16) converges absolutely or at least conditionally.

The representation given by Equation (16) can be transformed to a Mellin-Barnes integral.

We start with the case $|\mathrm{x}| = 0$ ($\mathrm{x} = (0,\ldots,0)$) and obtain the formula

$$G_{\alpha,\beta,n}(0,t) = \frac{1}{(2\pi)^n}\int_{\mathbb{R}^n} E_{\beta}(-|\kappa|^{\alpha}t^{\beta})d\kappa,$$

which can be represented in the form

$$G_{\alpha,\beta,n}(0,t) = \frac{1}{(2\pi)^n}\frac{2\pi^{\frac{n}{2}}}{\Gamma(\frac{n}{2})}\int_0^\infty E_\beta(-\tau^\alpha t^\beta)\,\tau^{n-1}\,d\tau \tag{17}$$

as a result of the known formula (see, e.g., [13]):

$$\int_{\mathbb{R}^n} f(|\mathrm{x}|)d\mathrm{x} = \frac{2\pi^{\frac{n}{2}}}{\Gamma(\frac{n}{2})}\int_0^\infty \tau^{n-1}f(\tau)d\tau. \tag{18}$$

The asymptotics of the Mittag-Leffler function ensures convergence of the integral in Equation (17) under the condition $0 < n < \alpha$. Thus, for $1 < \alpha \le 2$, the fundamental solution $G_{\alpha,\beta,n}$ is finite at $|x| = 0$ only in the one-dimensional case and we obtain the formula

$$G_{\alpha,\beta,1}(0,t) = \frac{t^{-\frac{\beta}{\alpha}}}{\alpha\pi}\int_0^\infty E_\beta(-u)\,u^{\frac{1}{\alpha}-1}\,du \;=\; \frac{t^{-\frac{\beta}{\alpha}}}{\alpha\pi}\frac{\Gamma\left(\frac{1}{\alpha}\right)\Gamma\left(1-\frac{1}{\alpha}\right)}{\Gamma\left(1-\frac{\beta}{\alpha}\right)},$$

which is valid for $\alpha > 1$ if $0 < \beta < 2$ and for $\alpha > 2$ if $\beta = 2$. This formula is an easy consequence from the known Mellin integral transform of the Mittag-Leffler function (see, e.g., [21,22]):

$$\int_0^\infty E_\beta(-u)\,u^{s-1}\,du \;=\; \frac{\Gamma(s)\Gamma(1-s)}{\Gamma(1-\beta s)} \;\text{ if }\; \begin{cases} 0 < \Re(s) < 1 \text{ for } 0 < \beta < 2, \\ 0 < \Re(s) < 1/2 \text{ for } \beta = 2. \end{cases} \tag{19}$$

The Mellin integral transform plays an important role in fractional calculus in general and for the derivation of the results of this paper in particular; thus we recall the definitions of the Mellin transform and the inverse Mellin transform, respectively:

$$f^*(s) = (\mathcal{M}f(\tau))(s) = \int_0^\infty f(\tau)\tau^{s-1}\,d\tau, \quad \tau > 0, \tag{20}$$

$$f(\tau) = (\mathcal{M}^{-1}f^*(s))(\tau) = \frac{1}{2\pi i}\int_{\gamma-i\infty}^{\gamma+i\infty} f^*(s)\tau^{-s}\,ds, \gamma_1 < \Re(s) = \gamma < \gamma_2. \tag{21}$$

The Mellin integral transform exists in particular for the functions continuous on the intervals $(0,\varepsilon]$ and $[E,+\infty)$ and integrable on the interval (ϵ, E) with any ε, E, $0 < \varepsilon < E < +\infty$ that satisfy the estimates $|f(\tau)| \le M_1\tau^{-\gamma_1}$ for $0 < \tau < \varepsilon$ and $|f(\tau)| \le M_2\tau^{-\gamma_2}$ for $\tau > E$ with $\gamma_1 < \gamma_2$ and some constants M_1 and M_2. In this case, the Mellin integral transform $f^*(s)$ is analytic in the vertical strip $\gamma_1 < \Re(s) = \gamma < \gamma_2$.

If f is piecewise differentiable and $\tau^{\gamma-1}f(\tau) \in L^c(0,\infty)$, then Equation (21) holds at all points of continuity for f. The integral in Equation (21) has to be considered in the sense of the Cauchy principal value.

For the general theory of the Mellin integral transform, we refer the reader to [22]. Several applications of the Mellin integral transform in fractional calculus are discussed in [7,21].

If the dimension n of Equation (1) is greater that one, the fundamental solution $G_{\alpha,\beta,n}(x,t)$ has an integrable singularity at the point $|x| = 0$.

Now we proceed with the case $\mathrm{x} \neq 0$ and first discuss the convergence of the integral in the integral representation given by Equation (16). It follows from the asymptotic formulas for the Mittag-Leffler function and the known asymptotic behavior of the Bessel function (see, e.g., [20]) that the integral in Equation (16) converges conditionally in the case $n < 2\alpha + 1$ and absolutely in the case $n < 2\alpha - 1$. Thus for $1 < \alpha \le 2$ and $n = 1, 2, 3$, the integral in Equation (16) is at least conditionally convergent.

Now the technique of the Mellin integral transform is applied to deduce a Mellin-Barnes representation of the fundamental solution $G_{\alpha,\beta,n}(\mathrm{x},t)$. In particular, we use the convolution theorem for the Mellin integral transform that reads as

$$\int_0^{\infty} f_1(\tau) f_2\left(\frac{y}{\tau}\right) \frac{d\tau}{\tau} \overset{\mathcal{M}}{\longleftrightarrow} f_1^*(s) f_2^*(s), \tag{22}$$

where by $\overset{\mathcal{M}}{\longleftrightarrow}$ the juxtaposition of a function f with its Mellin transform f^* is denoted.

It can be easily seen that for $\mathrm{x} \neq 0$ the integral on the right-hand side of Equation (16) is the Mellin convolution of the functions

$$f_1(\tau) = E_\beta(-\tau^\alpha t^\beta) \quad \text{and} \quad f_2(\tau) = \frac{|\mathrm{x}|^{-n}}{(2\pi)^{\frac{n}{2}}} \tau^{-\frac{n}{2}-1} J_{\frac{n}{2}-1}\left(\frac{1}{\tau}\right)$$

at the point $y = \frac{1}{|\mathrm{x}|}$.

The Mellin transform of the Mittag-Leffler function (Equation (19)), the known Mellin integral transform of the Bessel function ([22]):

$$J_\nu(2\sqrt{\tau}) \overset{\mathcal{M}}{\longleftrightarrow} \frac{\Gamma(\nu/2+s)}{\Gamma(\nu/2+1-s)}, \; -\Re(\nu/2) < \Re(s) < 3/4,$$

and some elementary properties of the Mellin integral transform (see, e.g., [21,22]) lead to the Mellin transform formulas:

$$f_1^*(s) = \frac{t^{-\frac{\beta}{\alpha}s}}{\alpha} \frac{\Gamma(\frac{s}{\alpha})\Gamma(1-\frac{s}{\alpha})}{\Gamma(1-\frac{\beta}{\alpha}s)}, \quad 0 < \Re(s) < \alpha,$$

$$f_2^*(s) = \frac{|\mathrm{x}|^{-n}}{(2\pi)^{\frac{n}{2}}} \left(\frac{1}{2}\right)^{-\frac{n}{2}+s} \frac{\Gamma\left(\frac{n}{2}-\frac{s}{2}\right)}{\Gamma\left(\frac{s}{2}\right)}, \quad \frac{n}{2}-\frac{1}{2} < \Re(s) < n.$$

These two formulas, the convolution Theorem (Equation (22)) for the Mellin transform, and the inverse Mellin transform Equation (21) result in the following Mellin-Barnes integral representation of the fundamental solution $G_{\alpha,\beta,n}$:

$$G_{\alpha,\beta,n}(\mathrm{x},t) = \frac{1}{\alpha} \frac{|\mathrm{x}|^{-n}}{\pi^{\frac{n}{2}}} \frac{1}{2\pi i} \int_{\gamma-i\infty}^{\gamma+i\infty} \frac{\Gamma\left(\frac{n}{2}-\frac{s}{2}\right)\Gamma\left(\frac{s}{\alpha}\right)\Gamma\left(1-\frac{s}{\alpha}\right)}{\Gamma\left(1-\frac{\beta}{\alpha}s\right)\Gamma\left(\frac{s}{2}\right)} \left(\frac{2t^{\frac{\beta}{\alpha}}}{|\mathrm{x}|}\right)^{-s} ds, \tag{23}$$

where $\frac{n}{2}-\frac{1}{2} < \gamma < \min(\alpha, n)$. We note that the Mellin-Barnes integral given by Equation (23) can also be interpreted as a particular case of the Fox H-function. The theory of the H-function, its properties, and applications are presented in a number of textbooks and papers (see, e.g., [23–28]); thus, here we do not discuss this subject in detail and prefer to directly deduce the properties of the fundamental solution $G_{\alpha,\beta,n}$ from its Mellin-Barnes representation (Equation (23)). Starting with this representation and using simple linear variables' substitutions, we can easily derive some other forms of this representation that will be useful for further discussions. For example, the substitutions $s \to -s$ and then $s \to s-n$ in the Mellin-Barnes representation given by Equation (23) result in two other equivalent representations:

$$G_{\alpha,\beta,n}(\mathrm{x},t) = \frac{1}{\alpha} \frac{|\mathrm{x}|^{-n}}{\pi^{\frac{n}{2}}} \frac{1}{2\pi i} \int_{\gamma-i\infty}^{\gamma+i\infty} \frac{\Gamma\left(\frac{n}{2}+\frac{s}{2}\right)\Gamma\left(-\frac{s}{\alpha}\right)\Gamma\left(1+\frac{s}{\alpha}\right)}{\Gamma\left(1+\frac{\beta}{\alpha}s\right)\Gamma\left(-\frac{s}{2}\right)} \left(\frac{|\mathrm{x}|}{2t^{\frac{\beta}{\alpha}}}\right)^{-s} ds \tag{24}$$

and

$$G_{\alpha,\beta,n}(\mathrm{x},t) = \frac{1}{\alpha}\frac{t^{-\frac{\beta n}{\alpha}}}{(4\pi)^{\frac{n}{2}}}\frac{1}{2\pi i}\int_{\gamma-i\infty}^{\gamma+i\infty}\frac{\Gamma\left(\frac{s}{2}\right)\Gamma\left(\frac{n}{\alpha}-\frac{s}{\alpha}\right)\Gamma\left(1-\frac{n}{\alpha}+\frac{s}{\alpha}\right)}{\Gamma\left(1-\frac{\beta}{\alpha}n+\frac{\beta}{\alpha}s\right)\Gamma\left(\frac{n}{2}-\frac{s}{2}\right)}\left(\frac{|\mathrm{x}|}{2t^{\frac{\beta}{\alpha}}}\right)^{-s} ds \tag{25}$$

under the conditions $-\min(\alpha,n) < \gamma < \frac{1}{2}-\frac{n}{2}$ or $\max(n-\alpha,0) < \gamma < n$, respectively.

Finally, we demonstrate how these integral representations can be used, for example, for deriving some series representations of $G_{\alpha,\beta,n}(\mathrm{x},t)$ and then its representations in terms of elementary or special functions of the hypergeometric type. To this end, we consider a simple example. In the case $\beta = 1$ and $\alpha = 2$ (standard diffusion equation), the representation given by Equation (25) takes the following form (two pairs of the gamma functions in the integral on the right-hand side of Equation (25) are canceled):

$$G_{2,1,n}(\mathrm{x},t) = \frac{t^{-\frac{n}{2}}}{2\,(4\pi)^{\frac{n}{2}}}\frac{1}{2\pi i}\int_{\gamma-i\infty}^{\gamma+i\infty}\Gamma\left(\frac{s}{2}\right)\left(\frac{z}{2}\right)^{-s} ds,\ z = \frac{|x|}{\sqrt{t}}.$$

The substitution of the variables $s \to 2s$ leads to an even simpler representation:

$$G_{2,1,n}(\mathrm{x},t) = \frac{t^{-\frac{n}{2}}}{(4\pi)^{\frac{n}{2}}}\frac{1}{2\pi i}\int_{\gamma-i\infty}^{\gamma+i\infty}\Gamma(s)\left(\frac{z}{2}\right)^{-2s} ds,\ z = \frac{|x|}{\sqrt{t}}.$$

According to the Cauchy theorem, the contour of integration in the integral on the right-hand side of the previous formula can be transformed to the loop $L_{-\infty}$ starting and ending at $-\infty$ and encircling all poles $s_k = -k,\ k = 0,1,2,\ldots$ of the function $\Gamma(s)$. Taking into account the Jordan lemma, the formula

$$\mathrm{res}_{s=-k}\Gamma(s) = \frac{(-1)^k}{k!},\ k = 0,1,2,\ldots$$

and the Cauchy residue theorem lead to a series representation of $G_{2,1,n}(\mathrm{x},t)$:

$$G_{2,1,n}(\mathrm{x},t) = \frac{t^{-\frac{n}{2}}}{(4\pi)^{\frac{n}{2}}}\int_{\gamma-i\infty}^{\gamma+i\infty}\Gamma(s)\left(\frac{z}{2}\right)^{-2s} ds = \frac{t^{-\frac{n}{2}}}{(4\pi)^{\frac{n}{2}}}\sum_{k=0}^{\infty}\frac{(-1)^k}{k!}\left(\frac{z}{2}\right)^{2k},\ z = \frac{|x|}{\sqrt{t}}.$$

Thus the fundamental solution $G_{2,1,n}$ to the n-dimensional diffusion equation takes its standard form:

$$G_{2,1,n}(\mathrm{x},t) = \frac{1}{(\sqrt{4\pi t})^n}\exp\left(-\frac{|\mathrm{x}|^2}{4t}\right). \tag{26}$$

2.3. *Special Functions of the Wright Type*

The fundamental solutions to different time-, space-, or time- and space-fractional partial differential equations are closely connected to the special functions of hypergeometric type. In the general situation, some particular cases of the Fox H-function are often involved (see, e.g., [1,2]). However, for particular cases of the orders of the fractional derivatives, the H-function can sometimes be reduced to some simpler special functions, mainly of Wright type (see, e.g., [29] for the one-dimensional case of the time-fractional diffusion-wave equation). Because the Fox H-function has still not been investigated in all its details and, in particular, because no packages for its numerical calculation are available, this reduction is very welcome. In this paper, some new reduction formulas for the fundamental solution to the multi-dimensional time- and space-fractional diffusion-wave Equation (1) are derived. In this subsection, we shortly discuss the special functions of the Wright type that appear in these derivations. For more details regarding theory and applications of these special functions, we refer the reader to, for example, [30–36].

We start with the Wright function:

$$W_{a,\mu}(z) = \sum_{k=0}^{\infty}\frac{z^k}{k!\Gamma(a+\mu k)},\ \mu > -1,\ a,z \in \mathbb{C} \tag{27}$$

that was introduced for the first time in [37] for the case $\mu > 0$. In particular, in [37,38], Wright investigated some elementary properties and asymptotic behavior of this function in connection with his research on the asymptotic theory of partitions.

Because of the relation

$$J_\nu(z) = \left(\frac{z}{2}\right)^\nu W_{1+\nu,1}\left(-\frac{1}{4}z^2\right), \tag{28}$$

the Wright function can be considered as a generalization of the Bessel function $J_\nu(z)$. In turn, the Wright function is a particular case of the Fox H-function (see, e.g., [25,39]):

$$W_{a,\mu}(-z) = H^{1,0}_{0,2}\left[z \,\middle|\, \begin{matrix} - \\ (0,1),(1-a,\mu) \end{matrix}\right]. \tag{29}$$

The Wright function is an entire function for all real values of the parameter μ (both positive and negative) under the condition $-1 < \mu$, but its asymptotic behavior is different in the cases $\mu > 0$, $\mu = 0$, and $\mu < 0$ (see [40] for details).

Two particular cases of the Wright function, namely, the functions $M(z;\beta) = W_{1-\beta,-\beta}(-z)$ and $F(z;\beta) = W_{0,-\beta}(-z)$ with the parameter β between 0 and 1, have been introduced and investigated in detail in [41,42]. These functions play an important role as fundamental solutions of the Cauchy and signaling problems for the one-dimensional time-fractional diffusion-wave equation ([29]).

In this paper, a four-parameter Wright function in the form

$$W_{(a,\mu),(b,\nu)}(z) := \sum_{k=0}^{\infty} \frac{z^k}{\Gamma(a+\mu k)\Gamma(b+\nu k)}, \quad \mu,\nu \in \mathbb{R},\ a,\ b,\ z \in \mathbb{C} \tag{30}$$

is also used. Wright himself investigated this function in [43] for the case $\mu > 0,\ \nu > 0$. For $a = \mu = 1$ or $b = \nu = 1$, the four-parameter Wright function is reduced to the Wright function (Equation (27)). In [44], Luchko and Gorenflo investigated the four-parameter Wright function for the first time in the case for which one of the parameters μ or ν is negative. In particular, they proved that the function $W_{(a,\mu),(b,\nu)}(z)$ is an entire function provided that $0 < \mu+\nu,\ a, b \in \mathbb{C}$.

It is important to emphasize that the function $W_{(a,\mu),(b,\nu)}(z)$ can have an algebraic asymptotic expansion on the positive real semi-axis in the case of suitably restricted parameters (see [44] for details):

$$W_{(a,\mu),(b,\nu)}(x) = \sum_{l=0}^{L-1} \frac{x^{(a-1-l)/(-\mu)}}{(-\mu)\Gamma(l+1)\Gamma(b+\nu(a-l-1)/(-\mu))} \tag{31}$$

$$-\sum_{k=1}^{P} \frac{x^{-k}}{\Gamma(b-\nu k)\Gamma(a-\mu k)} + O(x^{(a-1-L)/(-\mu)}) + O(x^{-1-P}),\ x \to +\infty$$

when $0 < \nu/3 < -\mu < \nu \le 2,\ L, P \in \mathbb{N}$.

In the important case $\mu + \nu = 0$, the four-parameter Wright function is no longer an entire function. Indeed, in this case, the convergence radius of the series from Equation (30) is equal to 1 rather than to infinity, as can be seen from the asymptotics of the series terms as $k \to \infty$:

$$\left|\frac{1}{\Gamma(a-\nu k)\Gamma(b+\nu k)}\right| = \left|\frac{\sin(\pi(a-\nu k))}{\pi}\frac{\Gamma(1-a+\nu k)}{\Gamma(b+\nu k)}\right| =$$

$$\left|\frac{\cosh(\pi\Im(a))}{\pi}(\nu k)^{1-a-b}\left[1+O(k^{-1})\right]\right|,\ k \to +\infty.$$

In the chain of the equalities above, the following known formulas for the gamma function were employed:

$$\frac{\Gamma(z)}{\Gamma(1-z)} = \frac{\pi}{\sin(\pi z)},$$

$$\frac{\Gamma(s+a)}{\Gamma(s+b)} = s^{a-b}\left[1 + O(s^{-1})\right], \ |s| \to +\infty, \ |\arg(s)| < \pi.$$

Finally, we mention here the generalized Wright function that is defined by the following series (in the case of its convergence):

$$ {}_p\Psi_q \left[\begin{matrix} (a_1, A_1), \ldots, (a_p, A_p) \\ (b_1, B_1) \ldots (b_q, B_q) \end{matrix} ; z \right] = \sum_{k=0}^{\infty} \frac{\prod_{i=1}^{p} \Gamma(a_i + A_i k)}{\prod_{i=1}^{q} \Gamma(b_i + B_i k)} \frac{z^k}{k!}. \tag{32}$$

This function was introduced and investigated by Wright in [43]. For details regarding the generalized Wright function, we refer the readers to the recent book [45].

3. New Integral Representations of the Fundamental Solution

In the previous section, we derived the following integral representation of the fundamental solution:

$$G_{\alpha,\beta,n}(x,t) = \frac{|x|^{1-\frac{n}{2}}}{(2\pi)^{\frac{n}{2}}} \int_0^\infty E_\beta\left(-\tau^\alpha t^\beta\right) \tau^{\frac{n}{2}} J_{\frac{n}{2}-1}(\tau|x|)\, d\tau. \tag{33}$$

In this section, we demonstrate how the Mellin-Barnes representations of the fundamental solution can be employed to obtain other integral representations of the same type. The idea is very simple. For example, we start with the Mellin-Barnes representation given by Equation (25) and consider the kernel function:

$$L_{\alpha,\beta,n}(s) = \frac{\Gamma\left(\frac{s}{2}\right)\Gamma\left(\frac{n}{\alpha} - \frac{s}{\alpha}\right)\Gamma\left(1 - \frac{n}{\alpha} + \frac{s}{\alpha}\right)}{\Gamma\left(1 - \frac{\beta}{\alpha}n + \frac{\beta}{\alpha}s\right)\Gamma\left(\frac{n}{2} - \frac{s}{2}\right)}. \tag{34}$$

When the kernel function is represented as a product of two factors, the convolution theorem for the Mellin integral transform can be applied, and we obtain an integral representation of $G_{\alpha,\beta,n}$ of the type given by Equation (33). For example, we obtained the integral representation given by Equation (33) by employing the Mellin integral transform formulas for the Mittag-Leffler function and for the Bessel function, that is, by representing the kernel function $L_{\alpha,\beta,n}(s)$ as the following product:

$$L_{\alpha,\beta,n}(s) = \frac{\Gamma\left(\frac{n}{\alpha} - \frac{s}{\alpha}\right)\Gamma\left(1 - \frac{n}{\alpha} + \frac{s}{\alpha}\right)}{\Gamma\left(1 - \frac{\beta}{\alpha}n + \frac{\beta}{\alpha}s\right)} \times \frac{\Gamma\left(\frac{s}{2}\right)}{\Gamma\left(\frac{n}{2} - \frac{s}{2}\right)}. \tag{35}$$

In what follows, we consider other possibilities of the representation of the kernel function $L_{\alpha,\beta,n}(s)$ as a product of two factors. Of course, these factors should be chosen in a way that makes it possible to easily obtain the inverse Mellin integral transform of these factors in terms of the known elementary or special functions. In the following theorem, two possible representations are given.

Theorem 1. *Let the inequalities* $1 < \alpha \leq 2$, $0 < \beta \leq 2$ *hold true. Then the first fundamental solution* $G_{\alpha,\beta,n}$ *of the multi-dimensional space- and time-fractional diffusion-wave Equation* (1) *has the following integral representations of the Mellin convolution type:*

$$G_{\alpha,\beta,n}(x,t) = \frac{1}{(\sqrt{\pi}|x|)^n} \int_0^\infty e^{-\tau} \tau^{\frac{n}{2}-1} W_{(1,\beta),(0,-\alpha/2)}\left(-\frac{\tau^{\alpha/2} t^\beta}{(|x|/2)^\alpha}\right) d\tau \quad \text{if } \beta > \alpha/2, \tag{36}$$

$$G_{\alpha,\beta,n}(x,t) = \frac{1}{(\sqrt{\pi}|x|)^n} \int_0^\infty W_{\frac{\alpha}{2},\frac{\alpha}{2}}(-\tau)\, {}_1\Psi_1 \left[\begin{matrix} (\frac{n}{2}, \frac{\alpha}{2}) \\ (1, \beta) \end{matrix} ; -\frac{\tau t^\beta}{(|x|/2)^\alpha} \right] d\tau. \tag{37}$$

Proof. To make calculations easier, we first perform the variables' substitution $s \to 2s$ in the integral representation given by Equation (25). We obtain

$$G_{\alpha,\beta,n}(\mathbf{x},t) = \frac{2}{\alpha}\frac{t^{-\frac{\beta n}{\alpha}}}{(4\pi)^{\frac{n}{2}}}\frac{1}{2\pi i}\int_{\gamma-i\infty}^{\gamma+i\infty}\frac{\Gamma(s)\,\Gamma\left(\frac{n}{\alpha}-\frac{2}{\alpha}s\right)\Gamma\left(1-\frac{n}{\alpha}+\frac{2}{\alpha}s\right)}{\Gamma\left(1-\frac{\beta}{\alpha}n+\frac{2\beta}{\alpha}s\right)\Gamma\left(\frac{n}{2}-s\right)}\left(z^2\right)^{-s}ds\,,\ z=\frac{|\mathbf{x}|}{2t^{\frac{\beta}{\alpha}}}. \tag{38}$$

Now we represent the kernel function of the previous integral as follows:

$$L_{\alpha,\beta,n}(s) = \Gamma(s) \times \frac{\Gamma\left(\frac{n}{\alpha}-\frac{2}{\alpha}s\right)\Gamma\left(1-\frac{n}{\alpha}+\frac{2}{\alpha}s\right)}{\Gamma\left(1-\frac{\beta}{\alpha}n+\frac{2\beta}{\alpha}s\right)\Gamma\left(\frac{n}{2}-s\right)}. \tag{39}$$

The inverse Mellin integral transform of $\Gamma(s)$ is simply the exponential function $exp(-\tau)$ ([22]):

$$f_1(\tau) = \frac{1}{2\pi i}\int_{\gamma-i\infty}^{\gamma+i\infty}\Gamma(s)\,\tau^{-s}\,ds = e^{-\tau}. \tag{40}$$

To calculate the inverse Mellin transform of the second factor, the variables' substitution $s \to \frac{\alpha}{2}s$ is first applied. We then obtain the formula

$$f_2(\tau) = \frac{\alpha}{2}\frac{1}{2\pi i}\int_{\gamma-i\infty}^{\gamma+i\infty}\frac{\Gamma\left(\frac{n}{\alpha}-s\right)\Gamma\left(1-\frac{n}{\alpha}+s\right)}{\Gamma\left(1-\frac{\beta}{\alpha}n+\beta s\right)\Gamma\left(\frac{n}{2}-\frac{\alpha}{2}s\right)}\left(\tau^{\frac{\alpha}{2}}\right)^{-s}ds. \tag{41}$$

To obtain a series representation of the function f_2, we employ the standard technique for the Mellin-Barnes integrals. According to the Cauchy theorem, the contour of integration in the integral on the right-hand side of the previous formula can be transformed to the loop $L_{+\infty}$ starting and ending at $+\infty$ and encircling all poles $s_k = k+\frac{n}{\alpha},\ k=0,1,2,\ldots$ of the function $\Gamma\left(\frac{n}{\alpha}-s\right)$. Taking into account the Jordan lemma and the formula for the residual of the gamma function, the Cauchy residue theorem leads to a series representation of f_2:

$$f_2(\tau) = \frac{\alpha}{2}\sum_{k=0}^{\infty}\frac{(-1)^k}{k!}\frac{\Gamma(k+1)}{\Gamma(1+\beta k)\,\Gamma\left(-\frac{\alpha}{2}k\right)}\left(\tau^{\frac{\alpha}{2}}\right)^{-k-\frac{n}{\alpha}}. \tag{42}$$

We thus have obtained a representation of f_2 in terms of the four-parameter Wright function (Equation (30)):

$$f_2(\tau) = \frac{\alpha}{2}\tau^{-n/2}\,W_{(1,\beta),(0,-\alpha/2)}\left(-\tau^{-\alpha/2}\right), \tag{43}$$

which is valid under the condition $\beta > \alpha/2$.

Now we take into consideration the Mellin-Barnes integral given by Equation (38), Equations (40) and (43), and the Mellin transform convolution theorem, and thus we obtain the integral representation given by Equation (36).

The same procedure can be applied for other representations of the kernel function $L_{\alpha,\beta,n}(s)$ as a product of two factors. We again start with the Mellin-Barnes integral given by Equation (25) and perform the variables' substitution $s \to \alpha s$. Then we obtain the representation

$$G_{\alpha,\beta,n}(\mathbf{x},t) = \frac{t^{-\frac{\beta n}{\alpha}}}{(4\pi)^{\frac{n}{2}}}\frac{1}{2\pi i}\int_{\gamma-i\infty}^{\gamma+i\infty}\frac{\Gamma\left(\frac{\alpha}{2}s\right)\Gamma\left(\frac{n}{\alpha}-s\right)\Gamma\left(1-\frac{n}{\alpha}+s\right)}{\Gamma\left(1-\frac{\beta}{\alpha}n+\beta s\right)\Gamma\left(\frac{n}{2}-\frac{\alpha}{2}s\right)}\left(z^{\alpha}\right)^{-s}ds\,,\ z=\frac{|\mathbf{x}|}{2t^{\frac{\beta}{\alpha}}}. \tag{44}$$

The next step is a representation of the kernel function of the previous integral as a product of two factors:

$$L_{\alpha,\beta,n}(s) = \frac{\Gamma\left(1 - \frac{n}{\alpha} + s\right)}{\Gamma\left(\frac{n}{2} - \frac{\alpha}{2}s\right)} \times \frac{\Gamma\left(\frac{\alpha}{2}s\right)\Gamma\left(\frac{n}{\alpha} - s\right)}{\Gamma\left(1 - \frac{\beta}{\alpha}n + \beta s\right)}. \tag{45}$$

Now we calculate the inverse Mellin integral transforms of the factors. For the first factor, we employ the same technique as above and obtain the series representation

$$f_1(\tau) = \frac{1}{2\pi i}\int_{\gamma - i\infty}^{\gamma + i\infty} \frac{\Gamma\left(1 - \frac{n}{\alpha} + s\right)}{\Gamma\left(\frac{n}{2} - \frac{\alpha}{2}s\right)} \tau^{-s}\, ds = \sum_{k=0}^{\infty} \frac{(-1)^k}{k!} \frac{1}{\Gamma\left(\frac{\alpha}{2} + \frac{\alpha}{2}k\right)} \tau^{k+1-\frac{n}{\alpha}}. \tag{46}$$

Thus the function f_1 can be represented in terms of the Wright function (Equation (27)):

$$f_1(\tau) = \tau^{1-\frac{n}{\alpha}} W_{\frac{\alpha}{2},\frac{\alpha}{2}}(-\tau). \tag{47}$$

As for the second factor, we first obtain the series representation:

$$f_2(\tau) = \frac{1}{2\pi i}\int_{\gamma - i\infty}^{\gamma + i\infty} \frac{\Gamma\left(\frac{\alpha}{2}s\right)\Gamma\left(\frac{n}{\alpha} - s\right)}{\Gamma\left(1 - \frac{\beta}{\alpha}n + \beta s\right)}\, ds = \sum_{k=0}^{\infty} \frac{(-1)^k}{k!} \frac{\Gamma\left(\frac{n}{2} + \frac{\alpha}{2}k\right)}{\Gamma(1 + \beta k)} \tau^{-k-\frac{n}{\alpha}} \tag{48}$$

and then its representation in terms of the generalized Wright function (Equation (32)):

$$f_2(\tau) = \tau^{-\frac{n}{\alpha}}\, {}_1\Psi_1\left[\begin{matrix}\left(\frac{n}{2}, \frac{\alpha}{2}\right) \\ (1, \beta)\end{matrix}; -\frac{1}{\tau}\right]. \tag{49}$$

Combining Equations (44), (47), and (49) together and applying the Mellin convolution theorem, we finally arrive at the integral representation given by Equation (37) of the fundamental solution in terms of the Wright function and the generalized Wright function. □

4. New Closed-Form Formulas for Particular Cases of the Fundamental Solution

In the paper [9], the Mellin-Barnes representations of the fundamental solution to the multi-dimensional time- and space-fractional diffusion-wave equation were employed to derive some new particular cases of the solution in terms of the elementary functions and the special functions of the Wright type. In particular, the closed-form formulas for the fundamental solution to the neutral-fractional diffusion equation ($\beta = \alpha$ in Equation (1)) in terms of elementary functions were deduced for the odd-dimensional case ($n = 1, 3, \ldots$). In this section, we derive among other things a representation of the fundamental solution to the neutral-fractional diffusion equation in the two-dimensional case in terms of the four-parameter Wright function (Equation (30)).

Theorem 2. *The first fundamental solution to the multi-dimensional space- and time-fractional diffusion Equation (1) can be represented in terms of the Wright-type functions:*

(a) For $\beta = \alpha$ and $n = 2$ under the condition $1 < \alpha \le 2$:

$$G_{\alpha,\alpha,2}(x,t) = \begin{cases} \frac{|x|^{\alpha-2}}{\sqrt{\pi} t^{\alpha}} W_{\left(\frac{1}{2} - \frac{\alpha}{2}, -\frac{\alpha}{2}\right), \left(\frac{\alpha}{2}, \frac{\alpha}{2}\right)}\left(-\left(\frac{|x|}{t}\right)^{\alpha}\right) & \text{if } |x| < t, \\ \frac{|x|^{-2}}{\sqrt{\pi}} W_{\left(0, -\frac{\alpha}{2}\right), \left(\frac{1}{2}, \frac{\alpha}{2}\right)}\left(-\left(\frac{t}{|x|}\right)^{\alpha}\right) & \text{if } |x| > t. \end{cases} \tag{50}$$

(b) For $\beta = \frac{3}{2}\alpha$ and $n = 2$ under the condition $1 < \alpha \le \frac{4}{3}$:

$$G_{\alpha,\frac{3}{2}\alpha,2}(x,t) = \frac{\sqrt{3}}{2\pi^2|x|^2}\, {}_1\Psi_3\left[\begin{matrix}(1,1) \\ \left(\frac{1}{3}, \frac{\alpha}{2}\right), \left(\frac{2}{3}, \frac{\alpha}{2}\right), \left(0, -\frac{\alpha}{2}\right)\end{matrix}; -\left(\frac{|x|}{2(3t)^{\frac{3}{2}}}\right)^{\alpha}\right]. \tag{51}$$

Proof. Once again we start with the Mellin-Barnes integral representation given by Equation (25), which for $\beta = \alpha$ and $n = 2$, takes the following form:

$$G_{\alpha,\alpha,2}(x,t) = \frac{1}{\alpha}\frac{t^{-2}}{4\pi}\frac{1}{2\pi i}\int_{\gamma-i\infty}^{\gamma+i\infty}\frac{\Gamma\left(\frac{s}{2}\right)\Gamma\left(\frac{2}{\alpha}-\frac{s}{\alpha}\right)\Gamma\left(1-\frac{2}{\alpha}+\frac{s}{\alpha}\right)}{\Gamma(-1+s)\Gamma\left(1-\frac{s}{2}\right)}\left(\frac{|x|}{2t}\right)^{-s} ds\,. \tag{52}$$

The general theory of Mellin-Barnes integrals (see, e.g., [22]) states that for $|x| \le 2t$, a series representation of Equation (52) can be obtained by transforming the contour of integration in the integral on the right-hand side of Equation (52) to the loop $L_{-\infty}$ starting and ending at $-\infty$ and encircling all poles of the functions $\Gamma\left(\frac{s}{2}\right)$ and $\Gamma\left(1-\frac{2}{\alpha}+\frac{s}{\alpha}\right)$. The problem now is that we have to take into consideration the cases in which some of the poles of $\Gamma\left(\frac{s}{2}\right)$ coincide with the poles $\Gamma\left(1-\frac{2}{\alpha}+\frac{s}{\alpha}\right)$, making the series representation become very complicated.

To avoid this problem, we aim to "eliminate" one of these gamma functions. The application of the duplication formula for the gamma function:

$$\Gamma(2s) = \frac{2^{2s-1}}{\sqrt{\pi}}\Gamma(s)\Gamma\left(s+\frac{1}{2}\right)$$

to the function $\Gamma(-1+s)$ (one of the Gamma-functions in the denominator of the kernel function from the integral in Equation (52)) results in the following representation:

$$\Gamma(1-s) = \Gamma\left(2\left(-\frac{1}{2}+\frac{s}{2}\right)\right) = \frac{2^{s-2}}{\sqrt{\pi}}\Gamma\left(-\frac{1}{2}+\frac{s}{2}\right)\Gamma\left(\frac{s}{2}\right).$$

Now we substitute the previous formula into the integral in Equation (52) and obtain another Mellin-Barnes representation:

$$G_{\alpha,\alpha,2}(x,t) = \frac{1}{\alpha}\frac{t^{-2}}{\sqrt{\pi}}\frac{1}{2\pi i}\int_{\gamma-i\infty}^{\gamma+i\infty}\frac{\Gamma\left(\frac{2}{\alpha}-\frac{s}{\alpha}\right)\Gamma\left(1-\frac{2}{\alpha}+\frac{s}{\alpha}\right)}{\Gamma\left(-\frac{1}{2}+\frac{s}{2}\right)\Gamma\left(1-\frac{s}{2}\right)}\left(\frac{|x|}{t}\right)^{-s} ds\,. \tag{53}$$

In contrast to the representation given by Equation (52), the numerator of the kernel function in Equation (53) has just one gamma function with the poles tending to $-\infty$ and one gamma function with the poles tending to $+\infty$, and thus this representation is very suitable for the derivation of a series representation of $G_{\alpha,\alpha,2}$.

To proceed, the variables' substitution $s \to \alpha s$ is first employed in the integral from Equation (53). We then obtain the representation

$$G_{\alpha,\alpha,2}(x,t) = \frac{t^{-2}}{\sqrt{\pi}}\frac{1}{2\pi i}\int_{\gamma-i\infty}^{\gamma+i\infty}\frac{\Gamma\left(\frac{2}{\alpha}-s\right)\Gamma\left(1-\frac{2}{\alpha}+s\right)}{\Gamma\left(-\frac{1}{2}+\frac{\alpha}{2}s\right)\Gamma\left(1-\frac{\alpha}{2}s\right)}\left(\left(\frac{|x|}{t}\right)^{\alpha}\right)^{-s} ds\,. \tag{54}$$

To obtain the series representation of the Mellin-Barnes integral (Equation (54)), we have to consider two cases:

(i) $|x| < t$;

(ii) $|x| > t$.

In the first case, the contour of integration in the integral on the right-hand side of Equation (54) can be transformed to the loop $L_{-\infty}$ starting and ending at $-\infty$ and encircling all poles of the function

$\Gamma\left(1-\frac{2}{\alpha}+s\right)$. Taking into account the Jordan lemma and the formula for the residuals of the gamma function, the Cauchy residue theorem leads to the following series representation of $G_{\alpha,\alpha,2}$:

$$G_{\alpha,\alpha,2}(x,t) = \frac{t^{-2}}{\sqrt{\pi}} \sum_{k=0}^{\infty} \frac{(-1)^k}{k!} \frac{k! \left(\left(\frac{|x|}{t}\right)^{\alpha}\right)^{1+k-\frac{2}{\alpha}}}{\Gamma\left(\frac{1}{2}-\frac{\alpha}{2}-\frac{\alpha}{2}k\right)\Gamma\left(\frac{\alpha}{2}+\frac{\alpha}{2}k\right)}. \tag{55}$$

We thus arrive at the closed-form formula:

$$G_{\alpha,\alpha,2}(x,t) = \frac{|x|^{\alpha-2}}{\sqrt{\pi}t^{\alpha}} W_{\left(\frac{1}{2}-\frac{\alpha}{2},-\frac{\alpha}{2}\right),\left(\frac{\alpha}{2},\frac{\alpha}{2}\right)}\left(-\left(\frac{|x|}{t}\right)^{\alpha}\right) \tag{56}$$

in terms of the four-parameter Wright function (Equation (30)) that is valid for $|x| < t$.

In the case $|x| > t$, the contour of integration in the integral on the right-hand side of Equation (54) can be transformed to the loop $L_{+\infty}$ starting and ending at $+\infty$ and encircling all poles of the function $\Gamma\left(\frac{2}{\alpha}-s\right)$. Proceeding as in case i), we first obtain a series representation of $G_{\alpha,\alpha,2}$ in the form

$$G_{\alpha,\alpha,2}(x,t) = \frac{t^{-2}}{\sqrt{\pi}} \sum_{k=0}^{\infty} \frac{(-1)^k}{k!} \frac{k! \left(\left(\frac{|x|}{t}\right)^{\alpha}\right)^{-k-\frac{2}{\alpha}}}{\Gamma\left(-\frac{\alpha}{2}k\right)\Gamma\left(\frac{1}{2}+\frac{\alpha}{2}k\right)} \tag{57}$$

and then obtain the closed-form formula:

$$G_{\alpha,\alpha,2}(x,t) = \frac{|x|^{-2}}{\sqrt{\pi}} W_{\left(0,-\frac{\alpha}{2}\right),\left(\frac{1}{2},\frac{\alpha}{2}\right)}\left(-\left(\frac{t}{|x|}\right)^{\alpha}\right) \tag{58}$$

in terms of the four-parameter Wright function that is valid for $|x| > t$.

Combining Equations (56) and (58), the obtain the representation given by Equation (50) of the fundamental solution $G_{\alpha,\alpha,2}$ in terms of the four-parameter Wright function.

In the case $|x| = t$, both series are divergent, and the problem of determining a series representation of $G_{\alpha,\alpha,2}$ is more complicated; it will be considered elsewhere.

The method described above can be used for the derivation of other closed-form formulas for particular cases of the fundamental solution $G_{\alpha,\beta,n}$ in terms of the Wright-type functions. For example, we consider the case $\beta = \frac{3}{2}\alpha$ and $n = 2$ (because of the condition $\beta \le 2$, in this case, the inequalities $1 < \alpha \le \frac{4}{3}$ have to be satisfied). The Mellin-Barnes representation of $G_{\alpha,\frac{3}{2}\alpha,2}$ is as follows:

$$G_{\alpha,\frac{3}{2}\alpha,2}(x,t) = \frac{1}{\alpha}\frac{t^{-3}}{4\pi}\frac{1}{2\pi i}\int_{\gamma-i\infty}^{\gamma+i\infty} \frac{\Gamma\left(\frac{s}{2}\right)\Gamma\left(\frac{2}{\alpha}-\frac{s}{\alpha}\right)\Gamma\left(1-\frac{2}{\alpha}+\frac{s}{\alpha}\right)}{\Gamma\left(-2+\frac{3}{2}s\right)\Gamma\left(1-\frac{s}{2}\right)} \left(\frac{|x|}{2t^{\frac{3}{2}}}\right)^{-s} ds\,. \tag{59}$$

To proceed, we apply the multiplication formula for the gamma function:

$$\Gamma(ms) = m^{ms-\frac{1}{2}}(2\pi)^{\frac{1-m}{2}} \prod_{k=0}^{m-1} \Gamma\left(s+\frac{k}{m}\right),\ m = 2,3,4,\ldots$$

with $m = 3$ to the gamma function $\Gamma\left(-2+\frac{3}{2}s\right)$ from the denominator of the kernel function from the Mellin-Barnes representation given by Equation (59). We thus obtain the representation

$$\Gamma\left(-2+\frac{3}{2}s\right) = \Gamma\left(3\left(-\frac{2}{3}+\frac{1}{2}s\right)\right) = 3^{-\frac{5}{2}+\frac{3}{2}s}(2\pi)^{-1}\Gamma\left(-\frac{2}{3}+\frac{1}{2}s\right)\Gamma\left(-\frac{1}{3}+\frac{1}{2}s\right)\Gamma\left(\frac{1}{2}s\right).$$

By applying this formula to Equation (59) and by the variables' substitution $s \to \alpha s$, we arrive at the following Mellin-Barnes representation:

$$G_{\alpha,\frac{3}{2}\alpha,2}(\mathbf{x},t) = \frac{t^{-3}}{4\pi}\frac{3^{-\frac{5}{2}}}{2\pi}\frac{1}{2\pi i}\int_{\gamma-i\infty}^{\gamma+i\infty} \frac{\Gamma\left(\frac{2}{\alpha}-s\right)\Gamma\left(1-\frac{2}{\alpha}+s\right)}{\Gamma\left(-\frac{2}{3}+\frac{\alpha}{2}s\right)\Gamma\left(-\frac{1}{3}+\frac{\alpha}{2}s\right)\Gamma\left(1-\frac{\alpha}{2}s\right)}\left(\left(\frac{|\mathbf{x}|}{2(3t)^{\frac{3}{2}}}\right)^{\alpha}\right)^{-s} ds. \quad (60)$$

Using the technique presented above, the representation given by Equation (60) leads first to a series representation of $G_{\alpha,\frac{3}{2}\alpha,2}$ in the following form:

$$G_{\alpha,\frac{3}{2}\alpha,2}(\mathbf{x},t) = \frac{\sqrt{3}}{2\pi^2|\mathbf{x}|^2}\sum_{k=0}^{\infty}\frac{\left(-\left(\frac{|\mathbf{x}|}{2(3t)^{\frac{3}{2}}}\right)^{\alpha}\right)^{k}}{\Gamma\left(\frac{1}{3}+\frac{\alpha}{2}k\right)\Gamma\left(\frac{2}{3}+\frac{\alpha}{2}k\right)\Gamma\left(-\frac{\alpha}{2}k\right)},$$

which can be represented as a particular case of the generalized Wright function (Equation (51)). □

5. Discussion

This paper is devoted to some applications of the Mellin-Barnes integral representations of the fundamental solution to the multi-dimensional space- and time-fractional diffusion-wave equation for the analysis of its properties. In particular, this representation is used to obtain two new representations of the fundamental solution in the form of the Mellin convolution of the special functions of Wright type and for the derivation of some new closed-form formulas for particular cases of the fundamental solution. Among other things, the open problem of the representation of the fundamental solution to the two-dimensional neutral-fractional diffusion-wave equation in terms of the known special functions is solved. The potential of the Mellin-Barnes integral representation of the fundamental solution to the multi-dimensional space- and time-fractional diffusion-wave equation is of course not yet ladled. It can be used among other things for the derivation of the new closed-form formulas for its particular cases, for asymptotic formulas for the fundamental solution, and for relationships between the fundamental solutions for different values of the derivative orders α and β. These problems will be considered elsewhere in further publications.

Conflicts of Interest: The author declares no conflict of interest.

References

1. Kochubei, A.N. Fractional-order diffusion. *Differ. Equ.* **1990**, *26*, 485–492.
2. Schneider, W.R.; Wyss, W. Fractional diffusion and wave equations. *J. Math. Phys.* **1989**, *30*, 134–144.
3. Hanyga, A. Multidimensional solutions of space-fractional diffusion equations. *Proc. R. Soc. Lond. A* **2001**, *457*, 2993–3005.
4. Hanyga, A. Multi-dimensional solutions of space-time-fractional diffusion equations. *Proc. R. Soc. Lond. A* **2002**, *458*, 429–450.
5. Hanyga, A. Multidimensional solutions of time-fractional diffusion-wave equations. *Proc. R. Soc. Lond. A* **2002**, *458*, 933–957.
6. Luchko, Y. Fractional wave equation and damped waves. *J. Math. Phys.* **2013**, *54*, 031505.
7. Luchko, Y. Multi-dimensional fractional wave equation and some properties of its fundamental solution. *Commun. Appl. Ind. Math.* **2014**, *6*, e485.
8. Luchko, Y. Wave-diffusion dualism of the neutral-fractional processes. *J. Comput. Phys.* **2015**, *293*, 40–52.
9. Boyadjiev, L.; Luchko, Y. Mellin integral transform approach to analyze the multidimensional diffusion-wave equations. *Chaos Solitons Fractals* **2017**, *102*, 127–134.
10. Boyadjiev, L.; Luchko, Y. Multi-dimensional α-fractional diffusion-wave equation and some properties of its fundamental solution. *Comput. Math. Appl.* **2017**, *73*, 2561–2572.

11. Mainardi, F.; Luchko, Y.; Pagnini, G. The fundamental solution of the space-time fractional diffusion equation. *Fract. Calc. Appl. Anal.* **2001**, *4*, 153–192.
12. Saichev, A.; Zaslavsky, G. Fractional kinetic equations: Solutions and applications. *Chaos* **1997**, *7*, 753–764.
13. Samko, S.G.; Kilbas, A.A.; Marichev, O.I. *Fractional Integrals and Derivatives: Theory and Applications*; Gordon and Breach: New York, NY, USA, 1993.
14. Ferreira, M.; Vieira, N. Fundamental solutions of the time fractional diffusion-wave and parabolic Dirac operators. *J. Math. Anal. Appl.* **2016**, *447*, 329–353.
15. Kochubei, A.N. Cauchy problem for fractional diffusion-wave equations with variable coefficients. *Appl. Anal.* **2014**, *93*, 2211–2242.
16. Luchko, Y. Some uniqueness and existence results for the initial-boundary-value problems for the generalized time-fractional diffusion equation. *Comput. Math. Appl.* **2010**, *59*, 1766–1772.
17. Eidelman, S.D.; Kochubei, A.N. Cauchy problem for fractional diffusion equations. *J. Differ. Equ.* **2004**, *199*, 211–255.
18. Luchko, Y. Operational method in fractional calculus. *Fract. Calc. Appl. Anal.* **1999**, *2*, 463–489.
19. Erdélyi, A. *Higher Transcendental Functions, Volume 3*; McGraw-Hill: New York, NY, USA, 1955.
20. Erdélyi, A. *Higher Transcendental Functions, Volume 2*; McGraw-Hill: New York, NY, USA, 1953.
21. Luchko, Y.; Kiryakova, V. The Mellin integral transform in fractional calculus. *Fract. Calc. Appl. Anal.* **2013**, *16*, 405–430.
22. Marichev, O.I. *Handbook of Integral Transforms of Higher Transcendental Functions, Theory and Algorithmic Tables*; Ellis Horwood: Chichester, UK, 1983.
23. Fox, C. The *G*- and *H*-functions as symmetrical Fourier kernels. *Trans. Am. Math. Soc.* **1961**, *98*, 395–429.
24. Kilbas, A.A.; Saigo, M. *H-Transform. Theory and Applications*; Chapman and Hall: Boca Raton, FL, USA, 2004.
25. Kiryakova, V. *Generalized Fractional Calculus and Applications*; Longman: Harlow, UK, 1994.
26. Mainardi, F.; Pagnini, G. Salvatore Pincherle: The pioneer of the Mellin-Barnes integrals. *J. Comput. Appl. Math.* **2003**, *153*, 331–342.
27. Mathai, A.M.; Saxena, R.K. *The H-Functions with Applications in Statistics and Other Disciplines*; John Wiley: New York, NY, USA, 1978.
28. Yakubovich, S.; Luchko, Y. *The Hypergeometric Approach to Integral Transforms and Convolutions*; Kluwer Academic Publishers: Dordrecht, The Netherlands, 1994.
29. Mainardi, F. Fractional relaxation-oscillation and fractional diffusion-wave phenomena. *Chaos Solitons Fractals* **1996**, *7*, 1461–1477.
30. Gorenflo, R.; Luchko, Y.; Mainardi, F. Analytical properties and applications of the Wright function. *Fract. Calc. Appl. Anal.* **1999**, *2*, 383–414.
31. Gorenflo, R.; Luchko, Y.; Mainardi, F. Wright functions as scale-invariant solutions of the diffusion-wave equation. *J. Comput. Appl. Math.* **2000**, *118*, 175–191.
32. Luchko, Y. Algorithms for evaluation of the Wright function for the real arguments' values. *Fract. Calc. Appl. Anal.* **2008**, *11*, 57–75.
33. Luchko, Y.; Mainardi, F. Cauchy and signaling problems for the time-fractional diffusion-wave equation. *ASME J. Vib. Acoust.* **2014**, *135*, doi:10.1115/1.4026892.
34. Mainardi, F. *Fractional Calculus and Waves in Linear Viscoelasticity*; Imperial College Press: London, UK, 2010.
35. Pagnini, G. The M-Wright function as a generalization of the Gaussian density for fractional diffusion processes. *Fract. Calc. Appl. Anal.* **2013**, *16*, 436–453.
36. Stanković, B. On the function of E.M. Wright. *Publ. l'Inst. Math. Beogr. Nouv. Sèr.* **1970**, *10*, 113–124.
37. Wright, E.M. On the coefficients of power series having exponential singularities. *J. Lond. Math. Soc.* **1933**, *8*, 71–79.
38. Wright, E.M. The asymptotic expansion of the generalized Bessel function. *Proc. Lond. Math. Soc.* **1935**, *38*, 257–270.
39. Gorenflo, R.; Mainardi, F.; Srivastava, H.M. Special functions in fractional relaxation-oscillation and fractional diffusion-wave phenomena. In *Proceedings VIII International Colloquium on Differential Equations*; Bainov, D., Ed.; VSP: Utrecht, The Netherlands, 1998; pp. 195–202.
40. Wright, E.M. The generalized Bessel function of order greater than one. *Quart. J. Math.* **1940**, *11*, 36-48.

41. Mainardi, F. Fractional calculus: some basic problems in continuum and statistical mechanics. In *Fractals and Fractional Calculus in Continuum Mechanics*; Carpinteri, A., Mainardi, F., Eds.; Springer: Wien, Austria, 1997; pp. 291–348.
42. Mainardi, F.; Tomirotti, M. On a special function arising in the time fractional diffusion-wave equation. In *Transform Methods and Special Functions*; Rusev, P., Dimovski, I., Kiryakova, V., Eds; Science Culture Technology: Singapore, 1995; pp. 171–183.
43. Wright, E.M. The asymptotic expansion of the generalized hypergeometric function. *J. Lond. Math. Soc.* **1935**, *10*, 287–293.
44. Luchko, Y.; Gorenflo, R. Scale-invariant solutions of a partial differential equation of fractional order. *Fract. Calc. Appl. Anal.* **1998**, *1*, 63–78.
45. Gorenflo, R.; Kilbas, A.A.; Mainardi, F.; Rogosin, S.V. *Mittag-Leffler Functions, Related Topics and Applications*; Springer: Berlin, Germany, 2014.

mathematics

MDPI

Article

An Iterative Method for Solving a Class of Fractional Functional Differential Equations with "Maxima"

Khadidja Nisse [1,2,*] **and Lamine Nisse** [1,2]

1 Department of Mathematics, Faculty of Exact Sciences, Echahide Hamma Lakhdar University, B.P. 789, El Oued 39000, Algeria

2 Laboratory of Applied Mathematics, Badji Mokhtar University, B.P. 12, Annaba 23000, Algeria; laminisse@gmail.com

* Correspondence: khadidjanisse@gmail.com; Tel.: +213-549-641-422

Received: 14 November 2017; Accepted: 19 December 2017; Published: 22 December 2017

Abstract: In the present work, we deal with nonlinear fractional differential equations with "maxima" and deviating arguments. The nonlinear part of the problem under consideration depends on the maximum values of the unknown function taken in time-dependent intervals. Proceeding by an iterative approach, we obtain the existence and uniqueness of the solution, in a context that does not fit within the framework of fixed point theory methods for the self-mappings, frequently used in the study of such problems. An example illustrating our main result is also given.

Keywords: functional differential equations; fractional calculus; iterative procedures

1. Introduction

One of the most interesting kinds of nonlinear functional differential equations is the case when the nonlinear part depends on the maximum values of the unknown function. These equations, called functional differential equations with "maxima", arise in many technological processes. For instance, in the automatic control theory of various technical systems, it occurs that the law of regulation depends on the maximal deviation of the regulated quantity (see [1,2]). Such problems are often modeled by differential equations that contain the maximum values of the unknown function (see [3–5]). Recently, ordinary differential equations with "maxima" have received wide attention and have been investigated in diverse directions (see, for example, [4,6–11] and the references therein). As far as we know, in the fractional case, these equations are not yet sufficiently discussed in the existing literature, and thus form a natural subject for further investigation. Motivated by the previous fact and inspired by [11], in this work, we focus on the existence and uniqueness of the solution for similar systems in a fractional context, and in more general terms. We consider the following nonlinear fractional differential equation with "maxima" and deviating arguments:

$$ {}^{C}D^{\alpha}u(t) = f\left(t, \max_{\sigma\in[a(t),b(t)]} u(\sigma), u(t-\tau_1(t)), \ldots, u(t-\tau_N(t))\right), \quad t>0, \tag{1} $$

with the initial condition function

$$ u(t) = \phi(t), \quad t \le 0, \tag{2} $$

where ${}^{C}D^{\alpha}$ denotes the Caputo fractional derivative operator of order $\alpha \in [0,1]$, N is a positive integer, a, b and τ_i (with $1 \le i \le N$) are real continuous functions defined on $\mathbb{R}_+ = [0,+\infty]$ subject to conditions that will be specified later, $\phi : [-\infty, 0] \longrightarrow \mathbb{R}$ is a continuous function such that $\phi(0) = \phi_0 > 0$, and $f : \mathbb{R}_+ \times \mathbb{R}^{1+N} \longrightarrow \mathbb{R}$ is a nonlinear continuous function.

Our aim is to give sufficient assumptions leading to an iterative process that converges to the unique continuous solutions of Equations (1) and (2). These being under weaker conditions compared to the usual contractions (see Remark 3), and in a setting for which the standard process of Picard's iterations fails to be well defined.

It should be pointed out here, that the maximums in Equation (1) are taken on time-dependent intervals and not on a fixed one as is the case of the example given in [11].

Moreover, the Equation (1) will be supposedly of mixed type, namely with both retarded and advanced deviations τ_i, while, in [11], only the delays are considered. It is also important to note that, in the Lipschitz condition of the nonlinear function f, we take into account the direction of maximums too, which is not the case of the corresponding assumption in [11].

Due to all of these generalizations, our work attempts to extend the application of [11] (Theorem 3) to the fractional case by a constructive approach.

To our knowledge, the studies devoted to the question of the existence and uniqueness of the solutions for fractional differential equations are based on different variants from the fixed point theory for self-mappings, or on the upper and lower solutions method (see, e.g., [12–18] and the references therein). We emphasize here that our result answers this question for a class of problems of the forms Equations (1) and (2), even when the previous versions of the theory fail to do so directly. That is, when the integral operator associated with Equations (1) and (2) is allowed to be a non self-mapping (see Remark 1).

The rest of the paper is organized as follows. In the next section, we introduce some basic definitions from the fractional calculus as well as preliminary lemmas. In Section 3, under some sufficient conditions allowing the integral operator associated with Equations (1) and (2) to be non self, we prove an existence–uniqueness result by means of an iterative process. The applicability of our theoretical result is illustrated in Section 4.

2. Preliminaries

We start by recalling the definitions of the Riemann–Liouville fractional integrals and the Caputo fractional derivatives on the half real axis. For further details on the historical account and essential properties about the fractional calculus, we refer to [19–22].

Definition 1. *The Riemann–Liouville fractional integral of a function* $u : \mathbb{R}_+ \longrightarrow \mathbb{R}$ *of order* $\alpha \in \mathbb{R}_+$ *is defined by*

$$I^{\alpha} u(t) := \frac{1}{\Gamma(\alpha)} \int_0^t (t-s)^{\alpha-1} u(s)\, ds, \quad t > 0,$$

where $\Gamma(.)$ *is the Gamma function, provided that the right side is pointwise defined on* $[0, \infty]$.

In the following definition, n denotes the positive integer such that $n-1 < \alpha \leq n$ and d^n/dt^n is the classical derivative operator of order n. For simplicity, we set $du/dt = u'(t)$.

Definition 2. *The Caputo fractional derivative of a function* $u : \mathbb{R}_+ \longrightarrow \mathbb{R}$ *of order* $\alpha \in \mathbb{R}_+$ *is defined by*

$${}^{C}D^{\alpha} u(t) := I^{n-\alpha} \frac{d^n}{dt^n} u(t) := \frac{1}{\Gamma(n-\alpha)} \int_0^t (t-s)^{n-\alpha-1} \frac{d^n}{ds^n} u(s)\, ds, \quad t > 0,$$

provided that the right-hand side exists pointwise on $[0, \infty]$.

In particular, when $0 < \alpha < 1$,

$${}^{C}D^{\alpha} u(t) := I^{1-\alpha} u'(t) := \frac{1}{\Gamma(1-\alpha)} \int_0^t \frac{u'(s)}{(t-s)^{\alpha}} ds, \quad t > 0. \tag{3}$$

Let us denote by $\mathcal{C}(\mathbb{R})$ the set of all real continuous functions on $\mathbb{R}$. Applying the Riemann–Liouville fractional integral operator I^{α} of order α to both sides of Equation (1) and using its properties (see [19,21]), together with the initial condition Equation (2), we easily get the following lemma.

Lemma 1. *If f, a, b and τ_i (with $1 \le i \le N$) are continuous functions, then $u \in \{v \in \mathcal{C}(\mathbb{R})$ s.t. $v(t) = \phi(t)$ for $t \le 0\}$ is a solution of Equations (1) and (2) if and only if $u(t) = Fu(t)$, where*

$$Fu(t) = \phi_0 + \int_0^t \frac{(t-s)^{\alpha-1}}{\Gamma(\alpha)} f\left(s, \max_{\sigma\in[a(s),b(s)]} u(\sigma), u(s-\tau_1(s)), ..., u(s-\tau_N(s))\right) ds,\ t > 0, \tag{4}$$

$$Fu(t) = \phi(t), \quad t \le 0. \tag{5}$$

Proof. Let $u \in \mathcal{C}(\mathbb{R})$. The functions a and b are continuous, so, according to the Remark in [7] (page 8) , see also [4] (Remark 3.1.1, page 62), $\max_{\sigma\in[a(t),b(t)]} u(\sigma)$ is continuous too. Moreover, since τ_i are continuous, then $f\left(t, \max_{\sigma\in[a(t),b(t)]} u(\sigma), u(t-\tau_1(t)), ..., u(t-\tau_N(t))\right)$ as a composition of continuous functions, it is also continuous. Now, we are able to follow the usual approach to show this type of result (see [15,19,21,23,24]). Note first that the Caputo fractional derivative of order $\alpha \in [0,1]$ can be expressed by means of the Riemann–Liouville fractional derivative denoted by D^{α}, as follows (see [21] (2.4.4) or [19] (Definition 3.2)):

$${}^{C}D^{\alpha}u(t) = D^{\alpha}[u(t) - u(0)] := \frac{d}{dt} I^{1-\alpha}[u(t) - u(0)]. \tag{6}$$

Let now $u \in \{v \in \mathcal{C}(\mathbb{R})$ s.t. $v(t) = \phi(t)$ for $t \le 0\}$ be a solution of Equations (1) and (2). Thus, in view of the first equality in Equation (6), Equation (1) can be rewritten as

$$D^{\alpha}[u(t) - u(0)] = f\left(t, \max_{\sigma\in[a(t),b(t)]} u(\sigma), u(t-\tau_1(t)), ..., u(t-\tau_N(t))\right), \quad t > 0. \tag{7}$$

Since the right-hand side of Equation (7) is continuous, then according to the definition of the Riemann–Liouville fractional derivative given by the second equality in Equation (6), we have

$$I^{1-\alpha}[u(t) - u(0)] \in \mathcal{C}^1(\mathbb{R}_+). \tag{8}$$

Thus, using [21] (Lemma 2.9, (d) with $\gamma = 0$), we have

$$I^{\alpha}D^{\alpha}[u(t) - u(0)] = [u(t) - u(0)] - \frac{1}{\Gamma(\alpha)} I^{1-\alpha}U(0)\, t^{\alpha-1}, \tag{9}$$

where

$$U(t) := [u(t) - u(0)]. \tag{10}$$

Since U is continuous, for every $T > 0$, there exists $L > 0$ such that $|U(t)| \le L$: for all $t \in [0,T]$. Thus, the following inequality holds true for every $t > 0$, sufficiently small

$$\left|I^{1-\alpha}U(t)\right| \le \frac{1}{\Gamma(1-\alpha)} \int_0^t \frac{|U(s)|}{(t-s)^{\alpha}} ds \le \frac{L}{\Gamma(2-\alpha)} t^{1-\alpha}.$$

Hence, the fact that $1 - \alpha > 0$, together with the continuity of $I^{1-\alpha}U$ resulting from Equation (8), imply that $I^{1-\alpha}U(0) = 0$. Consequently, Equation (9) becomes

$$I^{\alpha}D^{\alpha}\left[u\left(t\right)-u\left(0\right)\right]=\left[u\left(t\right)-u\left(0\right)\right]. \tag{11}$$

Now, returning to Equation (7), applying the Riemann–Liouville fractional integral to both sides, and then using Equation (11) together with Equation (2), we obtain Equation (4).

Suppose now that $u\in\{v\in\mathcal{C}(\mathbb{R})\text{ s.t. } v(t)=\phi(t)\text{ for } t\leq 0\}$ is a solution of Equations (4) and (5). Then, in view of Definition 1, we can rewrite Equation (4) as

$$u\left(t\right)=\phi_0+I^{\alpha}f\left(t,\max_{\sigma\in[a(t),b(t)]}u\left(\sigma\right),u\left(t-\tau_1\left(t\right)\right),...,u\left(t-\tau_N\left(t\right)\right)\right).$$

Since u is continuous, then the right-hand side above is continuous too. By applying the Caputo fractional derivative operator ${}^{C}D^{\alpha}$ to both sides, then using its linearity (see [19] (Theorem 3.16)), as well as the fact that the derivative of a constant (in the sense of Caputo) is equal to zero [21] (Property 2.16), together with [21] (Lemma 2.21), we get Equation (1). □

In the present work, the state space will be regarded as a complete Hausdorff locally convex space. For further details on these spaces, we refer to [25]. In the sequel of this paper, we make use of the following lemma, which can be found in [26] ([Lemma 2).

Lemma 2. *Let X be a complete Hausdorff locally convex space, $\mathbf{E}$ a closed subset of X and $u,v\in X$. If $u\in\mathbf{E}$ and $v\notin\mathbf{E}$, then there exists $\beta\in[0,1]$ such that $w_\beta:=(1-\beta)u+\beta v\in\partial\mathbf{E}$, where $\partial\mathbf{E}$ denotes the boundary of $\mathbf{E}$. Furthermore, if $u\notin\partial\mathbf{E}$, then $\beta\in[0,1]$.*

3. The Main Results

In this section, we not only prove the existence–uniqueness result for Equations (1) and (2), but we also give this solution as a limit of an iterative process.

First, let us set the following hypotheses:

(H_1) $\forall t\geq 0: 0\leq a_*\leq a(t)\leq b(t)\leq b^*$, with $a_*:=\inf_{t\in[0,+\infty[}a(t)$, and $b^*:=\sup_{t\in[0,+\infty[}b(t)$. Furthermore, for all $t\in[0,b^*]$, we assume that $a(t)=a_*$ and $b(t)=b^*$. In other words, the functions a and b are constant on the interval $[0,b^*]$.

(H_2) $\exists\tau>0$, such that, for $i=1,...,N:\tau_i\left(t\right)>t-\tau$, $\forall t>0$.

(H_3) For $i=1,...,N, \exists t_i>0:\tau_i\left(t\right)\geq t$, $\forall t\in[0,t_i]$, and $\tau_i\left(t\right)<t$, $\forall t\in[t_i,+\infty]$.

(H_4) There exist positive constants l_1 and l_2, such that f satisfies the Lipschitz condition

$$\left|f\left(t,\xi,x_1,...,x_N\right)-f\left(t,\eta,y_1,...,y_N\right)\right|\leq l_1\left|\xi-\eta\right|+l_2\sum_{i=1}^{N}\left|x_i-y_i\right|.$$

(H_5) There exists a positive constant $M>\phi_0$ such that

$$\frac{1}{\Gamma\left(\alpha\right)\alpha}f\left(t,M,x_1,...,x_N\right)\leq\frac{M-\phi_0}{b^{*\alpha}},\quad\forall\left(t,x_1,...,x_N\right)\in\left[0,b^*\right]\times\mathbb{R}^N.$$

(H_6) f is a non negative function, and, moreover, $\exists h\in[\phi_0,M]$ such that $\forall t\in[0,b^*]$

$$\frac{1}{\Gamma\left(\alpha\right)\alpha}f\left(t,h,x_1,...,x_N\right)>\frac{M-\phi_0}{\left|b^*-\max_{1\leq i\leq N}t_i\right|^{\alpha}},\ \forall\left(x_1,...,x_N\right)\in\left(\left[\phi_0,\phi_0+\tfrac{h-\phi_0}{b^*}\tau\right]\right)^N.$$

Let $X=\mathcal{C}\left(\mathbb{R}\right)$ be the locally convex sequentially complete Hausdorff space of all real valued continuous functions defined on $\mathbb{R}$, and $\{P_K : K\in\mathcal{K}\}$ be the saturated family of semi-norms, generating the topology of X, defined by

$$P_K\left(u\right)=\sup_{t\in K}\left\{e^{-\lambda t}\left|u\left(t\right)\right|\right\}, \tag{12}$$

where K runs over the set of all compact subsets of $\mathbb{R}$ denoted by $\mathcal{K}$, and λ is a positive real number to be specified later.

We denote by $\mathbf{E}_{\phi,M}$, the subset of X defined by

$$\mathbf{E}_{\phi,M} = \{u \in X : u(t) = \phi(t) \text{ for } t \leq 0, \text{ and } u(t) \leq M \text{ for } t \in [a_*, b^*]\},$$

where a_*, b^* and M are the constants given by (H_1) and (H_5). It can be easily seen that $\mathbf{E}_{\phi,M}$ is a closed subset of X and its boundary is

$$\partial\mathbf{E}_{\phi,M} = \left\{u \in X : u(t) = \phi(t) \text{ for } t \leq 0 \text{ and } \max_{t\in[a_*,b^*]} u(t) = M\right\}.$$

Throughout the remaining of this paper, F denotes the operator defined on $\mathbf{E}_{\phi,M}$ by Equations (4) and (5). Thus, according to Lemma 1, F maps $\mathbf{E}_{\phi,M}$ into X and the fixed points of F are continuous solutions of problems Equations (1) and (2).

Remark 1. *It should be pointed out that under hypotheses (H_1)–(H_3), (H_6) with the additional condition $\max_{1\leq i\leq N} t_i < b^*$, F is a non-self mapping on $\mathbf{E}_{\phi,M}$. Indeed, as is noted in the proof of [11] (Theorem 3), for any function $u \in \mathbf{E}_{\phi,M}$ defined by $u(t) = \phi_0 + (h - \phi_0)t/b^*$, where $t \in [0, b^*]$ and h is the constant given by (H_6), it can be easily seen that $Fu \notin \mathbf{E}_{\phi,M}$. This will be checked by the example of the last section.*

The introduction of a self-mapping of the index set in uniform spaces is motivated by applications in the theory of neutral functional differential equations [11,27,28]. Following this idea, let us define a map $j : \mathcal{K} \longrightarrow \mathcal{K}$ by

$$j(K) := \begin{cases} K, & \text{if } K_+ = \emptyset, \\ [0, \max\{K_m, \tau, b^*\}], & \text{if } K_+ \neq \emptyset, \end{cases} \tag{13}$$

where $K_+ := K \cap [0, +\infty]$, $K_m = \sup K$, τ and b^* are the positive constants given in (H_1)–(H_2). For $n \in \mathbb{N}^*$, $j^n(K)$ is the compact set defined inductively by $j^n(K) = j(j^{n-1}(K))$ and $j^0(K) = K$.

Remark 2. *Note that, for every $K \in \mathcal{K}$ and every integer n greater than 1, we have $j^n(K) = j(K)$.*

In the next proposition, we show that F satisfies Equation (14), which is a weakened version of the usual contraction when $L_\lambda < 1$ (see Remark 3).

Proposition 1. *Under hypotheses (H_1)–(H_4), the operator $F : \mathbf{E}_{\phi,M} \to X$ satisfies for each $u, v \in \mathbf{E}_{\phi,M}$ and every $K \in \mathcal{K}$*

$$P_K(Fx - Fy) \leq L_\lambda P_{j(K)}(x - y) \tag{14}$$

with

$$L_\lambda = \frac{l_1}{\lambda^\alpha \Gamma(\alpha)} \Gamma\left(\alpha^2\right)^{\frac{1}{1+\alpha}} \left(\frac{\alpha}{1+\alpha}\right)^{\frac{\alpha}{1+\alpha}} e^{\lambda b^*} + \frac{N l_2 e^{\lambda\tau}}{\lambda^\alpha}. \tag{15}$$

Proof. Note that it suffices to consider $K_+ \neq \emptyset$, since otherwise $P_K(Fu - Fv) = 0$. Letting $t \in K_+$, we obtain by means of hypotheses (H_3) and (H_4)

$$|Fu(t) - Fv(t)| \leq \int_0^t \frac{(t-s)^{\alpha-1}}{\Gamma(\alpha)} l_1 \left| \max_{\sigma\in[a(s),b(s)]} u(\sigma) - \max_{\sigma\in[a(s),b(s)]} v(\sigma) \right| ds$$

$$+\int_0^t \frac{(t-s)^{\alpha-1}}{\Gamma(\alpha)} + l_2 \sum_{i=1}^{N} |u(s-\tau_i(s)) - v(s-\tau_i(s))| ds$$

$$\leq l_1 \int_0^t \frac{(t-s)^{\alpha-1}}{\Gamma(\alpha)} \max_{\sigma\in[a(s),b(s)]} |u(\sigma)-v(\sigma)| ds + l_2 \sum_{i=1}^{N} \int_0^{\min\{t_i,t\}} \frac{(t-s)^{\alpha-1}}{\Gamma(\alpha)} |\phi(r_i(s))-\phi(r_i(s))| ds$$

$$+ l_2 \sum_{i\in\{1,\ldots,N:t_i\leq t\}} \int_{t_i}^t \frac{(t-s)^{\alpha-1}}{\Gamma(\alpha)} |u(r_i(s)) - v(r_i(s))| ds$$

$$= l_1 \int_0^t \frac{(t-s)^{\alpha-1}}{\Gamma(\alpha)} \max_{\sigma\in[a(s),b(s)]} |u(\sigma)-v(\sigma)| ds + l_2 \sum_{i\in\{1,\ldots,N:t_i\leq t\}} \int_{t_i}^t \frac{(t-s)^{\alpha-1}}{\Gamma(\alpha)} |u(r_i(s))-v(r_i(s))| ds$$

$$\leq l_1 \int_0^t \frac{(t-s)^{\alpha-1}}{\Gamma(\alpha)} e^{\lambda b(s)} \max_{\sigma\in[a(s),b(s)]} e^{-\lambda\sigma} |u(\sigma) - v(\sigma)| ds$$

$$+ l_2 \sum_{i\in\{1,\ldots,N:t_i\leq t\}} \int_{t_i}^t \frac{(t-s)^{\alpha-1}}{\Gamma(\alpha)} e^{\lambda r_i(s)} e^{-\lambda r_i(s)} |u(r_i(s)) - v(r_i(s))| ds$$

$$\leq l_1 \max_{\sigma\in[a_*,b^*]} e^{-\lambda\sigma} |u(\sigma)-v(\sigma)| \int_0^t \frac{(t-s)^{\alpha-1}}{\Gamma(\alpha)} e^{\lambda b(s)} ds$$

$$+ l_2 \sum_{i\in\{1,\ldots,N:t_i\leq t\}} \int_{t_i}^t \frac{(t-s)^{\alpha-1}}{\Gamma(\alpha)} e^{\lambda r_i(s)} e^{-\lambda r_i(s)} |u(r_i(s)) - v(r_i(s))| ds,$$

where $r_i(s) = s - \tau_i(s)$. Note that, due to the definition Equation (13) and under hypothesis (H_1), it is clear that, for every $K \in \mathcal{K}$ with $K_+ \neq \emptyset$, we have $[a_*, b^*] \subset j(K)$ and further (H_2)–(H_3) lead to $r_i(s) \in j(K)$ when $t_i \leq s \leq t$. Hence,

$$|Fx(t) - Fy(t)| \leq l_1 P_{j(K)}(u-v) \int_0^t \frac{(t-s)^{\alpha-1}}{\Gamma(\alpha)} e^{\lambda b(s)} ds$$

$$+ l_2 \sum_{i\in\{1,\ldots,N:t_i\leq t\}} \int_{t_i}^t \frac{(t-s)^{\alpha-1}}{\Gamma(\alpha)} e^{\lambda r_i(s)} \max_{\xi\in j(K)} e^{-\lambda\xi} |u(\xi)-v(\xi)| ds$$

$$= l_1 P_{j(K)}(u-v) \int_0^t \frac{(t-s)^{\alpha-1}}{\Gamma(\alpha)} e^{\lambda b(s)} ds + l_2 P_{j(K)}(u-v) \sum_{i\in\{1,\ldots,N:t_i\leq t\}} \int_{t_i}^t \frac{(t-s)^{\alpha-1}}{\Gamma(\alpha)} e^{\lambda r_i(s)} ds.$$

Now, multiplying the both sides of the above inequality by $e^{-\lambda t}$, then performing the change of variable $u = \lambda(t-s)$, we get

$$e^{-\lambda t} |Fu(t) - Fv(t)| \leq l_1 P_{j(K)}(u-v) \int_0^t \frac{(t-s)^{\alpha-1}}{\Gamma(\alpha)} e^{-\lambda(t-b(s))} ds$$

$$+ l_2 P_{j(K)}(u-v) \sum_{i\in\{1,\ldots,N:t_i\leq t\}} \int_{t_i}^t \frac{(t-s)^{\alpha-1}}{\Gamma(\alpha)} e^{-\lambda(t-r_i(s))} ds$$

$$= \frac{l_1}{\lambda^\alpha \Gamma(\alpha)} p_{j(K)}(u-v) \int_0^{\lambda t} x^{\alpha-1} e^{-\lambda(t-b(t-\frac{x}{\lambda}))} dx$$

$$+\frac{l_2}{\lambda^\alpha}P_{j(K)}(u-v)\sum_{i\in\{1,\ldots,N:t_i\le t\}}\int_0^{\lambda(t-t_i)}\frac{x^{\alpha-1}}{\Gamma(\alpha)}e^{-x}e^{-\lambda\tau_i\left(t-\frac{x}{\lambda}\right)}dx$$

$$\le\frac{l_1}{\lambda^\alpha\Gamma(\alpha)}P_{j(K)}(u-v)\int_0^{\lambda t}x^{\alpha-1}e^{-x}e^{-\lambda\left(\left(t-\frac{x}{\lambda}\right)-b\left(t-\frac{x}{\lambda}\right)\right)}dx$$

$$+\frac{l_2}{\lambda^\alpha}P_{j(K)}(u-v)\sum_{i\in\{1,\ldots,N:t_i\le t\}}\int_0^{\lambda(t-t_i)}\frac{x^{\alpha-1}}{\Gamma(\alpha)}e^{-x}e^{-\lambda\tau_i\left(t-\frac{x}{\lambda}\right)}dx.$$

Let $\mu:=1+\alpha$ and $\nu:=1+1/\alpha$. Taking into account (H_3), Hölder's inequality gives

$$e^{-\lambda t}\left|Fu(t)-Fv(t)\right|\le,$$

$$\left\{\frac{l_1}{\lambda^\alpha\Gamma(\alpha)}\left(\int_0^{\lambda t}x^{\mu(\alpha-1)}e^{-\mu x}dx\right)^{\frac{1}{\mu}}\left(\int_0^{\lambda t}e^{-\nu\lambda\left(\left(t-\frac{x}{\lambda}\right)-b\left(t-\frac{x}{\lambda}\right)\right)}dx\right)^{\frac{1}{\nu}}+\frac{Nl_2e^{\lambda\tau}}{\lambda^\alpha}\right\}P_{j(K)}(u-v)$$

$$=\left\{\frac{l_1}{\lambda^\alpha\Gamma(\alpha)}\left(\int_0^{\lambda t}x^{\mu(\alpha-1)}e^{-\mu x}dx\right)^{\frac{1}{\mu}}\left(\lambda\int_0^{t}e^{-\nu\lambda(s-b(s))}ds\right)^{\frac{1}{\nu}}+\frac{Nl_2e^{\lambda\tau}}{\lambda^\alpha}\right\}P_{j(K)}(u-v)$$

$$\le\left\{\frac{l_1}{\lambda^\alpha\Gamma(\alpha)}\left(\int_0^{\lambda t}x^{\mu(\alpha-1)}e^{-\mu x}dx\right)^{\frac{1}{\mu}}\left(\lambda\int_0^{t}e^{-\nu\lambda(s-b^*)}ds\right)^{\frac{1}{\nu}}+\frac{Nl_2e^{\lambda\tau}}{\lambda^\alpha}\right\}P_{j(K)}(u-v)$$

$$\le\left\{\frac{l_1}{\lambda^\alpha\Gamma(\alpha)}\Gamma\left(\alpha^2\right)^{\frac{1}{\mu}}\frac{1}{\nu}^{\frac{1}{\nu}}e^{\lambda b^*}+\frac{Nl_2e^{\lambda\tau}}{\lambda^\alpha}\right\}P_{j(K)}(u-v).$$

Thus, the result is obtained by taking the supremum on K. □

Remark 3. *Since $K_+\subset j(K)$, if $P_K(Fu-Fv)\le L_\lambda P_K(u-v)$ is satisfied, then Equation (14) holds true. Therefore, due to the choice of j, in the present context, the usual contraction is a particular case of Equation (14) when $L_\lambda<1$.*

To reach our aim, we proceed by adapting the proof of [11], [Theorem 1] with some completeness, for the construction of an iterative process converging to the unique continuous solutions of Equations (1) and (2).

According to Remark 1, the standard process of Picard's iterations fails to be well defined. To overcome this fact, we make use of Lemma 2 to construct a sequence of elements of $\mathbf{E}_{\phi,M}$ as follows: starting from an arbitrary point $u_0\in\mathbf{E}_{\phi,M}$, we define the terms of a sequence $\{u_n\}_{n\in\mathbb{N}^*}$ in $\mathbf{E}_{\phi,M}$ iteratively as follows:

$$\begin{cases}u_n=Fu_{n-1}, & \text{if } Fu_{n-1}\in\mathbf{E}_{\phi,M},\\ u_n=(1-\beta_n)u_{n-1}+\beta_nFu_{n-1}\in\partial\mathbf{E}_{\phi,M}\text{ with }\beta_n\in[0,1[, & \text{if } Fu_{n-1}\notin\mathbf{E}_{\phi,M}.\end{cases}\tag{16}$$

Note that the terms of the sequence $\{u_n\}_{n\in\mathbb{N}^*}$ belong to $\mathbf{A}\cup\mathbf{B}\subset\mathbf{E}_{\phi,M}$, with $\mathbf{B}\subset\partial\mathbf{E}_{\phi,M}$, where

$$\mathbf{A}:=\{u_i\in\{u_n\}_{n\in\mathbb{N}^*}:u_i=Fu_{i-1}\}\text{ and }\mathbf{B}:=\{u_i\in\{u_n\}_{n\in\mathbb{N}^*}:u_i\ne Fu_{i-1}\}.$$

Furthermore, if $u_n\in\mathbf{B}$, a straightforward computation leads to

$$P_K(u_{n-1}-u_n)+P_K(u_n-Fu_{n-1})=P_K(u_{n-1}-Fu_{n-1})\ \forall K\in\mathcal{K}.\tag{17}$$

Proposition 2. *Let $u_0 \in \mathbf{E}_{\phi,M}$, and $\{u_n\}_{n\in\mathbb{N}^*}$ be the sequence defined iteratively by Equation (16). Then, under hypotheses (H_1)–(H_5), for each $K \in \mathcal{K}$ and every integer m greater than or equal to 1, the following estimation holds true:*

$$\max\{P_K(u_{2m}-u_{2m+1}),\, P_K(u_{2m+1}-u_{2m+2})\} \leq 2^{2m-1}L_\lambda^m C_K, \tag{18}$$

where L_λ is given by Equation (15) and

$$C_K = \max\{P_{j(K)}(u_0-u_1),\, P_{j(K)}(u_1-u_2),\, P_{j(K)}(u_2-u_3)\}.$$

Proof. If $\max\limits_{t\in[a_*,b^*]} u(t) = M$, then, for $a_* \leq t \leq b^*$, hypothesis (H_5) gives

$$\begin{aligned} Fu(t) &= \phi_0 + \frac{1}{\Gamma(\alpha)}\int_0^t (t-s)^{\alpha-1} f(s, M, u(s-\tau_1(s)), ..., u(s-\tau_N(s)))\,ds \\ &\leq \phi_0 + \tfrac{\alpha(M-\phi_0)}{b^{*\alpha}} \int_0^t (t-s)^{\alpha-1}\,ds \leq M, \end{aligned}$$

which means that $F(\partial\mathbf{E}_{\phi,M}) \subset \mathbf{E}_{\phi,M}$. Consequently, two consecutive terms of the sequence $\{u_n\}_{n\in\mathbb{N}^*}$ can not belong to $\mathbf{B}$ (recall that $\mathbf{B} \subset \partial\mathbf{E}_{\phi,M}$). Thus, it suffices to consider the three cases below.

Case 1. $u_n,\, u_{n+1} \in \mathbf{A}$. From Equation (14), we have

$$P_K(u_n-u_{n+1}) = P_K(Fu_{n-1}-Fu_n) \leq L_\lambda \cdot P_{j(K)}(u_{n-1}-u_n).$$

Case 2. $u_n \in \mathbf{A}$, $u_{n+1} \in \mathbf{B}$. From the condition Equation (14) together with Equation (17) (for u_{n+1} instead of u_n), we get

$$\begin{aligned} P_K(u_n-u_{n+1}) &= P_K(u_n-Fu_n) - P_K(u_{n+1}-Fu_n) \\ &\leq P_K(Fu_{n-1}-Fu_n) \leq L_\lambda \cdot P_{j(K)}(u_{n-1}-u_n). \end{aligned}$$

Case 3. $u_n \in \mathbf{B}$, $u_{n+1} \in \mathbf{A}$. Then, $\exists\beta_n \in [0,1] : u_n = (1-\beta_n)u_{n-1} + \beta_n Fu_{n-1}$, which implies that

$$P_K(u_n-u_{n+1}) \leq \max\{P_K(u_{n-1}-u_{n+1}),\, P_K(Fu_{n-1}-u_{n+1})\}.$$

Thus, by Equation (14), for every integer number $n \geq 2$, we obtain either

$$P_K(u_n-u_{n+1}) \leq L_\lambda \cdot p_{j(K)}(u_{n-1}-u_n), \text{ or } P_K(u_n-u_{n+1}) \leq L_\lambda \cdot P_{j(K)}(u_{n-2}-u_n).$$

Moreover,

$$\begin{aligned} P_{j(K)}(u_{n-2}-u_n) &\leq P_{j(K)}(u_{n-2}-u_{n-1}) + P_{j(K)}(u_{n-1}-u_n) \\ &\leq 2\max\{P_{j(K)}(u_{n-2}-u_{n-1}),\, P_{j(K)}(u_{n-1}-u_n)\}. \end{aligned}$$

In summary, the following inequality is true in all cases

$$P_K(u_n-u_{n+1}) \leq 2L_\lambda \max\{P_{j(K)}(u_{n-2}-u_{n-1}),\, P_{j(K)}(u_{n-1}-u_n)\}. \tag{19}$$

We now prove Equation (18) by induction. Using Equation (19), we have either

$$P_K(u_2-u_3) \leq 2L_\lambda \cdot P_{j(K)}(u_0-u_1) \leq 2L_\lambda \cdot C_K,$$

or

$$P_K(u_2-u_3) \leq 2L_\lambda \cdot P_{j(K)}(u_1-u_2) \leq 2L_\lambda \cdot C_K,$$

and similarly we obtain

$$P_K(u_3 - u_4) \leq 2L_\lambda \cdot C_K.$$

Consequently, Equation (18) is satisfied for $m = 1$. Assume now that Equation (18) holds true for some $m > 1$. Using Equation (19), we get either

$$P_K(u_{2m+2} - u_{2m+3}) \leq 2L_\lambda \cdot P_{j(K)}(u_{2m} - u_{2m+1}),$$

or

$$P_K(u_{2m+2} - u_{2m+3}) \leq 2L_\lambda \cdot P_{j(K)}(u_{2m+1} - u_{2m+2}).$$

Thus, the fact that $C_{j(K)} = C_K$, which follows from Remark 2, leads to

$$P_K(u_{2m+2} - u_{2m+3}) \leq 2L_\lambda 2^{2m-1} {L_\lambda}^m C_{j(K)} = 2^{2m} {L_\lambda}^{m+1} C_K \leq 2^{2(m+1)-1} {L_\lambda}^{m+1} C_K.$$

In the same way, we get

$$P_K(u_{2m+3} - u_{2m+4}) \leq 2^{2(m+1)-1} {L_\lambda}^{m+1} C_K,$$

which means that Equation (18) holds for $m+1$, and this completes the proof. □

We are now ready to prove our main result.

Theorem 1. *Let $u_0 \in \mathbf{E}_{\phi,M}$, then under hypotheses (H_1)–(H_6) with*

$$\max_{1 \leq i \leq N} t_i < b^*, \tag{20}$$

the sequence $\{u_n\}_{n \in \mathbb{N}^*}$ defined iteratively by Equation (16), converges in $\mathbf{E}_{\phi,M}$ to the unique continuous solution of Equations (1) and (2) provided that

$$\left\{ \frac{l_1}{\Gamma(\alpha)} \Gamma\left(\alpha^2\right)^{\frac{1}{1+\alpha}} \left(\frac{\alpha}{1+\alpha}\right)^{\frac{\alpha}{1+\alpha}} + Nl_2 \right\} e \max\{\tau, b^*\}^\alpha < \frac{1}{4}. \tag{21}$$

Proof. Let us put $\lambda = 1/\max\{\tau, b^*\}$ in Equation (12). Thus, according to Proposition 1, for every $K \in \mathcal{K}$ and $u, v \in \mathbf{E}_{\phi,M}$, Equation (14) holds true with $L_\lambda < 1/4$. Therefore, for an arbitrary fixed $K \in \mathcal{K}$, and, for each $\varepsilon > 0$, there exists a positive integer s satisfying

$$\sum_{m=s}^{\infty} 2^{2m} L_\lambda^m < \frac{\varepsilon}{C_K}. \tag{22}$$

Hence, for $n \geq 2s$, $q \geq 1$ and a sufficiently large l, we get, by means of Equations (18) and (22),

$$\begin{aligned} P_K(u_n - u_{n+q}) &\leq P_K(u_n - u_{n+1}) + P_K(u_{n+1} - u_{n+2}) + \cdots + P_K(u_{n+q-1} - u_{n+q}) \\ &\leq \sum_{m=s}^{l} \{P_K(u_{2m} - u_{2m+1}) + P_K(u_{2m+1} - u_{2m+2})\} \\ &\leq \sum_{m=s}^{l} 2^{2m} L_\lambda^m \cdot C_K \leq C_K \cdot \sum_{m=s}^{\infty} 2^{2m} L_\lambda^m < \varepsilon. \end{aligned}$$

Consequently, $\{u_n\}_{n \in \mathbb{N}^*}$ is a Cauchy sequence in the closed subset $\mathbf{E}_{\phi,M}$ of the complete locally convex space X, and so it converges to a point $u \in \mathbf{E}_{\phi,M}$. Let $\{u_{n_k}\}_{k \geq 1}$ be a sub-sequence of $\{u_n\}_{n \geq 1}$ in $\mathbf{A}$, which is $u_{n_k+1} = F u_{n_k}$ for every positive integer k. Then, for each compact $K \in \mathcal{K}$, we have

$$\begin{aligned} P_K(u - Fu) &\leq P_K(u - u_{n_k}) + P_K(u_{n_k} - Fu) = P_K(u - u_{n_k}) + P_K(Fu_{n_k - 1} - Fu) \\ &\leq P_K(u - u_{n_k}) + L_\lambda \cdot P_{j(K)}(u_{n_k - 1} - u) \xrightarrow[k \to +\infty]{} 0. \end{aligned}$$

Therefore, $u = Fu$ and so, according to Lemma 1, u is a solution of Equations (1) and (2). For the uniqueness, assume that there exists another solution $v \in \mathbf{E}_{\phi,M}$ such that $u \neq v$. Since X is Hausdorff, then $P_{K_0}(u - v) \neq 0$ for some compact $K_0 \in \mathcal{K}$. Using Equation (14) and Remark 2, we get for every positive integer n

$$\begin{aligned} 0 < P_{K_0}(u - v) &= P_{K_0}(Fu - Fv) \leq L_\lambda \cdot P_{j(K_0)}(u - v) = L_\lambda \cdot P_{j(K_0)}(Fu - Fv) \\ &\leq L_\lambda^2 \cdot P_{j(K_0)}(u - v) \leq \ldots \leq L_\lambda^n P_{j(K_0)}(u - v), \end{aligned}$$

which contradicts the fact that $L_\lambda < \frac{1}{4}$. This completes the proof. □

4. Example

The following example illustrates the applicability of our theoretical result. Let us consider the following equation

$$^{C}D^{0.5}u\,(t) = \frac{1.132}{2.45 + 10^{-2}|\max_{t \in [10^{-1}, 2]} u(t)| + 10^{-4}|u(\frac{0.994}{1+t}t - 0.004)|}, \quad t > 0, \tag{23}$$

subject to the initial condition function

$$u\,(t) = 2t^2 + 1.168, \quad t \leq 0, \tag{24}$$

Problems in Equations (23) and (24) are identified to Equations (1) and (2) with $\alpha = 0.5$, $N = 1$, $a(t) = 10^{-1}$, $b(t) = 2$, $\tau_1\,(t) = t - \dfrac{0.994}{1+t}t + 0.004$, $\phi\,(t) = 2t^2 + 1.168$ and

$$f\,(t, \xi, \eta) = \frac{1.132}{2.45 + 10^{-2}\,|\xi| + 10^{-4}\,|\eta|}.$$

It can be easily seen that hypotheses (H_1)–(H_4) are satisfied with $a_* = 10^{-1}$, $b^* = 2$, $\tau = 0.994$, $t_1 = 4 \times 10^{-3}/(1 - 10^{-2})$, $l_1 = 1132 \times 10^{-5}$ and $l_2 = 1132 \times 10^{-7}$.

In addition, there exists $M = 1.9$ ($M > \phi_0 = 1.168$), such that, for every $\eta \in \mathbb{R}$, we have

$$f\,(t, M, \eta) \leq \frac{1.132}{2.45 + 10^{-2} \times M} \simeq 0.45845216686918,$$

and

$$0.5\Gamma\,(0.5)\,\frac{M - \phi_0}{b^{*0.5}} \simeq 0.458712974257473.$$

Thus, (H_5) holds true. To check (H_6), let $h = 1.168001$ ($\phi_0 < h < M$), and then we have for every $\eta \in [\phi_0, \phi_0 + \frac{h - \phi_0}{b^*}\tau[=]1.168,\, 1.168 + (9.97 \times 10^{-5})]$

$$f\,(t, h, \eta) > \frac{1.132}{2.45 + h \times 10^{-2} + \left(\phi_0 + \frac{h - \phi_0}{b^*}\tau\right) \times 10^{-4}} \simeq 0.461798645149363,$$

and

$$0.5\Gamma\,(0.5)\,\frac{M - \phi_0}{(b^* - t_1)^{0.5}} \simeq 0.459177023920154.$$

Since, moreover, f is clearly non negative, hypothesis (H_6) is satisfied too.

Furthermore, we have $t_1 < b^*$ and

$$\left\{ \frac{l_1}{\Gamma(0.5)} \Gamma(0.25)^{\frac{1}{1.5}} \left(\frac{0.5}{1.5} \right)^{\frac{0.5}{1.5}} + Nl_2 \right\} e \max\{\tau, b^*\}^{0.5} \simeq 0.041231690678434,$$

which is all conditions of Theorem 1 being fulfilled. Now, let u_0 be the function defined by

$$u_0(t) = \begin{cases} 1.168 + (5 \times 10^{-7})t : & t \in [0, b^*], \\ 1.168001 : & t \geq b^*, \\ 2t^2 + 1.168 : & t \leq 0. \end{cases}$$

It is clear that $u_0 \in \mathbf{E}_{\phi,1.9}$, where

$$\mathbf{E}_{\phi,1.9} = \left\{ u \in C(\mathbb{R}) : u(t) = 2t^2 + 1.168 \text{ for } t \leq 0 \text{ and } u(t) \leq 1.9 \text{ for } t \in \left[10^{-1}, 2\right] \right\}.$$

Note that, for $0 < t - \tau_1(t) < \tau$, we have

$$\phi_0 = 1.168 < u_0(t - \tau_1(t)) < 1.168 + (5 \times 10^{-7})\tau = \phi_0 + \frac{h - \phi_0}{b^*}\tau.$$

Then, hypothesis (H_6) yields to

$$Fu_0(2) = Fu_0(b^*) = \phi_0 + \int_0^{b^*} \frac{(b^* - s)^{\alpha - 1}}{\Gamma(\alpha)} f(s, h, u_0(s - \tau_1(s)))\, ds$$

.

$$\geq \phi_0 + \int_{t_1}^{b^*} \frac{(b^* - s)^{\alpha - 1}}{\Gamma(\alpha)} f(s, h, u_0(s - \tau_1(s)))\, ds$$

$$> \phi_0 + \alpha \frac{M - \phi_0}{(b^* - t_1)^{\alpha}} \int_{t_1}^{b^*} (b^* - s)^{\alpha - 1}\, ds = M = 1.9,$$

which means that $Fu_0 \notin \mathbf{E}_{\phi,1.9}$. Thus, in this framework, the iterative processes usually used in the self-mapping context can not be applied, while, according to Theorem 1, the process defined by Equation (16), converges in $\mathbf{E}_{\phi,1.9}$, to the unique continuous solutions of Equations (23) and (24). The first term is approximately given by

$$u_1(t) \simeq \begin{cases} 1.168 + (12115 \times 10^{-13})\, t + 0.5176017180\sqrt{t} : \ 0 < t \leq t_1, \\ 1.168 + (12115 \times 10^{-13})\, t + (509761847 \times 10^{-18})\sqrt{27225\, t - 110} + 0.5176017180\sqrt{t} : \\ \ \ t_1 < t \leq b^*, \\ 1.168000002 + (509761847 \times 10^{-18})\sqrt{27225\, t - 110} + 0.517601718\sqrt{t} : \ \ t \geq b^*, \\ 2t^2 + 1.168 : \ \ t \leq 0. \end{cases}$$

For $t > 0$, the other terms can be computed using the following formulas:

$$u_2(t) = \phi_0 + \frac{1.132}{\Gamma(\alpha)} \int_0^t \frac{(t - s)^{\alpha - 1}}{2.45 + 10^{-2} \left| \max\limits_{t \in [10^{-1}, 2]} u_1(s) \right| + 10^{-4} \left| u_1\left(\frac{0.994}{1+s} s - 0.004 \right) \right|} ds.$$

By successive iterations up to the order $n - 1$, the term u_n is given by

$$u_n(t) = \int_0^t \frac{(t - s)^{\alpha - 1}}{2.45 + 10^{-2} \left| \max\limits_{t \in [10^{-1}, 2]} u_{n-1}(s) \right| + 10^{-4} \left| u_{n-1}\left(\frac{0.994}{1+s} s - 0.004 \right) \right|} ds,$$

if the right-hand side belongs to $\mathbf{E}_{\phi,1.9}$. If not, we have

$$u_n(t) = (1-\beta_n)u_{n-1} + \beta_n \left(\phi_0 + \frac{1.132}{\Gamma(\alpha)} \int_0^t \frac{(t-s)^{\alpha-1}}{2.45 + 10^{-2} \left| \max\limits_{t\in[10^{-1},2]} u_{n-1}(s) \right| + 10^{-4} \left| u_{n-1}\left(\frac{0.994}{1+s}s - 0.004\right) \right|} ds \right),$$

with $\beta_n \in [0,1]$, such that the right-hand side belongs to $\partial\mathbf{E}_{\phi,1.9}$.

5. Conclusions

In this contribution, the investigated question concerns the existence and uniqueness of the solution for a class of nonlinear functional differential equations of fractional order. The considered problems in Equations (1) and (2) are distinguished by the fact that the nonlinear part depends on maximum values of the unknown function, which is not frequently discussed in the existing literature. These maximums are taken on time-dependent intervals and, moreover, the equation is of mixed type, i.e., with both retarded and advanced deviations. It should be noted that, if the hypotheses (H_6) and Equation (20) are omitted, the operator F can be a self mapping, and thus, by the usual contraction methods, it can be shown that the result of Theorem 1 remains valid with the bound in Equation (21) weakened to 1. When additional conditions are necessary to meet the physical or mechanical requirements of the phenomenon governed by Equations (1) and (2), we leave the previous usual framework of study. In this case, our main result of Theorem 1 shows that the condition in Equation (21) is sufficient for the existence and uniqueness of the solution.

Acknowledgments: The authors thank the anonymous reviewers for their relevant comments and valuable suggestions, which significantly contributed to improving the quality of the paper. This research was supported by the Ministry of Higher Education and Scientific Research of Algeria, under the program of National Commission for the Evaluation of University Research Projects, Grant Agreement No. B01120140060.

Author Contributions: K. Nisse and L. Nisse contributed equally in this work. All authors read and approved the final manuscript.

Conflicts of Interest: The authors declare that they have no competing interests.

References

1. Popov, E.P. *Automatic Regulation and Control*; Nauka: Moscow, Russia, 1966.
2. Popov, E.P. *The Dynamics of Automatic Control Systems*; Pergamon Press: Oxford, UK, 1962.
3. Agarwal, R.P.; Hristova, S. Strict stability in terms of tow measures for impulsive differential equations with supremum. *Appl. Anal.* **2012**, *91*, 1379–1392.
4. Bainov, D.D.; Hristova, S.G. *Differential Equations with Maxima*; Chapman and Hall/CRC: Boca Raton, FL, USA, 2011.
5. Stepanov, E. On solvability of some boundary value problems for differential equations with maxima. *Topol. Methods Nonlinear Anal.* **1996**, *8*, 315–326.
6. Agarwal, R.P.; Hristova, S. Quasilinearization for initial value problems involving differential equations with "maxima". *Math. Comput. Model.* **2012**, *55*, 2096–2105.
7. Angelov, V.G.; Bainov, D.D. On functional differential equations with Maximums. *Period. Math. Hung.* **1987**, *18*, 7–15.
8. Bohner, M.J.; Georgieva, A.T.; Hristova, S.G. Nonlinear differential equations with maxima: Parametric stability in terms of two measures. *Appl. Math. Inf. Sci.* **2013**, *7*, 41–48.
9. Golev, A.; Hristova, S.; Rahnev, A. An algorithm for approximate solving of differential equations with "maxima". *Comput. Math. Appl.* **2010**, *60*, 2771–2778.
10. Otrocol, D. Systems of functional differential equations with maxima of mixed type, Volume 191 of Graduate Texts in Mathematics. *Electron. J. Qual. Theory Differ. Equ.* **2014**, *5*, 1–9.
11. Tsachev, T.; Angelov, V.G. Fixed points of nonself-mappings and applications. *Nonlinear Anal. Theory Methods Appl.* **1993**, *21*, 1–12.

12. Araci, A.; Sen, E.; Acikgoz, M.; Srivastava, H.M. Existence and uniqueness of positive and nondecreasing solutions for a class of fractional boundary value problems involving the p-Laplacian operator. *Adv. Differ. Equ.* **2015**, *40*, doi:10.186/s13662-015-0375-0.
13. Jankowski, T. Fractional equations of Volterra type involving a Riemann-Liouville derivative. *Appl. Math. Lett.* **2013**, *26* , 344–350.
14. Mishra, L.N.; Srivastava, H.M.; Sen, M. Existence results for some nonlinear functional-integral equations in Banach algebra with applications. *Int. J. Anal. Appl.* **2016**, *11*, 1–10.
15. Nisse, L.; Bouaziz, A. Existence and stability of the solutions for systems of nonlinear fractional differential equations with deviating arguments. *Adv. Differ. Equ.* **2014**, *275*, doi:10.1186/1687-1847-2014-275.
16. Wang, G.; Agarwal, R.P.; Cabada, A. Existence results and the monotone iterative technique for systems of nonlinear fractional differential equations. *Appl. Math. Lett.* **2012**, *25*, 1019–1024.
17. Ahmad, B.; Ntouyas, S.K. Initial value problems of fractional order Hadamard-type functional differential equations. *Electron. J. Differ. Equ.* **2015**, *77*, 1–9.
18. Tisdell, C.C. Basic existence and a priori bound results for solutions to systems of boundary value problems for fractional differential equations. *Electron. J. Differ. Equ.* **2016**, *84*, 1–9.
19. Diethelm, K. *The Analysis of Fractional Differential Equations*; Springer: Berlin, Germany, **2004**.
20. Gorenflo, R.; Mainardi, F. Fractional calculus: Integral and differential equations of fractional order. In *Fractional Calculus in Continuum and Statistical Mechanics (Udine, 1996)*; CISM Courses and Lectures; Springer: Viena, Austria, 1997; pp. 223–276.
21. Kilbas, A.A.; Srivastava, H.M.; Trujillo, J.J. *Theory and Applications of Fractional Differential Equations*; Elsevier: Amsterdam, The Netherlands, 2006.
22. Mainardi, F. Fractional calculus: Some basic problems in Continuum and Statistical Mechanics. In *Fractal and Fractional Calculus in Continuum Mechanics (Udine, 1996)*; CIAM Courses and Lectures; Springer: Wien, Austria; New York, NY, USA, 1997; pp. 291–348.
23. Diethelm, K. Analysis of Fractional Differential Equations. *J. Math. Anal. Appl.* **2002**, *265*, 229–248.
24. Jalilian, Y.; Jalilian, R. Existence of solution for delay fractional differential equations. *Mediterr. J. Math.* **2013**, *10*, 1731–1747.
25. Koethe, G. *Topological Vector Spaces I*; Springer: Berlin, Germany, 1969.
26. Sehgal, V.M.; Singh, S.P. On a fixed point theorem of Krasnoselskii for locally convex spaces. *Pac. J. Math.* **1976**, *62*, 561–567.
27. Angelov, V.G. A Converse to a Contraction Mapping Theorem in Uniform Spaces. *Nonlinear Anal. Theory Methods Appl.* **1988**, *12*, 989–996.
28. Angelov, V.G. Fixed point theorems in uniform spaces and applications. *Czechoslov. Math. J.* **1987**, *37*, 19–33.

mathematics

MDPI

Article

Numerical Solution of Fractional Differential Equations: A Survey and a Software Tutorial

Roberto Garrappa

Dipartimento di Matematica, Università Degli Studi di Bari, Via E. Orabona 4, 70126 Bari, Italy; roberto.garrappa@uniba.it

Received: 8 December 2017; Accepted: 14 January 2018; Published: 23 January 2018

Abstract: Solving differential equations of fractional (i.e., non-integer) order in an accurate, reliable and efficient way is much more difficult than in the standard integer-order case; moreover, the majority of the computational tools do not provide built-in functions for this kind of problem. In this paper, we review two of the most effective families of numerical methods for fractional-order problems, and we discuss some of the major computational issues such as the efficient treatment of the persistent memory term and the solution of the nonlinear systems involved in implicit methods. We present therefore a set of MATLAB routines specifically devised for solving three families of fractional-order problems: fractional differential equations (FDEs) (also for the non-scalar case), multi-order systems (MOSs) of FDEs and multi-term FDEs (also for the non-scalar case); some examples are provided to illustrate the use of the routines.

Keywords: fractional differential equations (FDEs); numerical methods; multi-order systems (MOSs); multi-term equations; product integration (PI); fractional linear multi-step methods (FLMMs); MATLAB routines

1. Introduction

The increasing interest in applications of fractional calculus has motivated the development and the investigation of numerical methods specifically devised to solve fractional differential equations (FDEs). Finding analytical solutions of FDEs is, indeed, even more difficult than solving standard ordinary differential equations (ODEs) and, in the majority of cases, it is only possible to provide a numerical approximation of the solution.

Although several computing environments (such as, for instance, Maple, Mathematica, MATLAB and Python) provide robust and easy-to-use codes for numerically solving ODEs, the solution of FDEs still seems not to have been addressed by almost all computational tools, and usually, researchers have to write codes by themselves for the numerical treatment of FDEs.

When numerically solving FDEs, one faces some non-trivial difficulties, mainly related to the presence of a persistent memory (which makes the computation extremely slow and expensive), to the low-order accuracy of the majority of the methods, to the not always straightforward computation of the coefficients of several schemes, and so on.

Writing reliable codes for FDEs can be therefore a quite difficult task for researchers and users with no particular expertise in computational mathematics, and it would be surely preferable to rely on efficient and already tested routines.

The aim of this paper is to illustrate the basic principles behind some methods for FDEs, thus to provide a short tutorial on the numerical solution of FDEs, and discuss some non-trivial issues related to the effective implementation of methods as, for instance, the treatment of the persistent memory term, the solution of equations involved by implicit methods, and so on; at the same time, we present some MATLAB routines for the solution of a wide range of FDEs.

This paper is organized as follows. In Section 2, we recall some basic definitions concerning fractional-order operators, and we present some of the most useful properties that will be used throughout the paper. Section 3 is devoted to illustrating multi-step methods for FDEs; in particular, we discuss product-integration (PI) rules and Lubich's fractional linear multi-step methods (FLMMs); we also discuss in detail the main issues and advantages related to the use of implicit methods, and we illustrate a technique based on the fast Fourier transform (FFT) algorithm to efficiently treat the persistent memory term.

In Section 4, we consider two special cases of FDEs: multi-order systems (MOSs) in which each equation has a different fractional-order and multi-term FDEs in which there is more than one fractional derivative in a single equation; in particular, we will see how standard methods studied in the previous section can be adapted to solve these particular problems.

In Section 5, we present some MATLAB routines for solving different families of FDEs and explain their use in detail; finally, Section 6 is devoted to showing the application of the routines to a selection of test problems.

2. Preliminary Material on Fractional Calculus

For the sake of clarity, we review in this section some of the most useful definitions in fractional calculus, and we recall the properties that we will use in the subsequent sections. For a more comprehensive introduction to this subject, the reader is referred to any of the available textbooks [1–5] or review papers [6,7] and, in particular, to the book by Diethelm [8] by which this introductory section is mainly inspired.

As the starting point for introducing fractional-order operators, we consider the Riemann–Liouville (RL) integral; for a function $y(t) \in L^1([t_0, T])$ (as usual, L^1 is the set of Lebesgue integrable functions), the RL fractional integral of order $\alpha > 0$ and origin at t_0 is defined as:

$$J_{t_0}^{\alpha} y(t) = \frac{1}{\Gamma(\alpha)} \int_{t_0}^{t} (t-\tau)^{\alpha-1} y(\tau) \mathrm{d}\tau. \tag{1}$$

It provides a generalization of the standard integral, which, indeed, can be considered a particular case of the RL integral (1) when $\alpha = 1$. The left inverse of $J_{t_0}^{\alpha}$ is the RL fractional derivative:

$$\widehat{D}_{t_0}^{\alpha} y(t) := D^m J_{t_0}^{m-\alpha} y(t) = \frac{1}{\Gamma(m-\alpha)} \frac{\mathrm{d}^m}{\mathrm{d}t^m} \int_{t_0}^{t} (t-\tau)^{m-\alpha-1} y(\tau) \mathrm{d}\tau, \tag{2}$$

where $m = \lceil \alpha \rceil$ is the smallest integer greater or equal to α and D^m, $y^{(m)}$ or $\mathrm{d}^m/\mathrm{d}t^m$ denotes the standard integer-order derivative.

An alternative definition of the fractional derivative, obtained after interchanging differentiation and integration in Equation (2), is the so-called Caputo derivative, which, for a sufficiently differentiable function, namely for $y \in A^m([t_0, T])$ (i.e., $y^{(m-1)}$ absolutely continuous), is given by:

$$D_{t_0}^{\alpha} y(t) := J_{t_0}^{m-\alpha} D^m y(t) = \frac{1}{\Gamma(m-\alpha)} \int_{t_0}^{t} (t-\tau)^{m-\alpha-1} y^{(m)}(\tau) \mathrm{d}\tau. \tag{3}$$

We observe that also $D_{t_0}^{\alpha} y(t)$ is a left inverse of the RL integral, namely $D_{t_0}^{\alpha} J_{t_0}^{\alpha} y = y$ [8] (Theorem 3.7), but not its right inverse, since [8] (Theorem 3.8):

$$J_{t_0}^{\alpha} D_{t_0}^{\alpha} y(t) = y(t) - T_{m-1}[y; t_0](t), \tag{4}$$

where $T_{m-1}[y; t_0](t)$ is the Taylor polynomial of degree $m-1$ for the function $y(t)$ centered at t_0, that is:

$$T_{m-1}[y; t_0](t) = \sum_{k=0}^{m-1} \frac{(t-t_0)^k}{k!} y^{(k)}(t_0).$$

More generally speaking, by combining [1] (Lemma 2.3) and [8] (Theorem 3.8), it is also possible to observe that for any $\beta > \alpha$, it holds:

$$J_{t_0}^{\beta} D_{t_0}^{\alpha} y(t) = J_{t_0}^{\beta} \widehat{D}_{t_0}^{\alpha} \left[y(t) - T_{m-1}[y;t_0](t) \right] = J_{t_0}^{\beta-\alpha} \left[y(t) - T_{m-1}[y;t_0](t) \right], \tag{5}$$

a relationship that will be useful, in a particular way, in Section 4.2 on multi-term FDEs.

The two definitions (2) and (3) are interrelated, and indeed, by deriving both sides of Equation (4) in the RL sense, it is possible to observe that:

$$D_{t_0}^{\alpha} y(t) = \widehat{D}_{t_0}^{\alpha} \left[y(t) - T_{m-1}[y;t_0](t) \right]$$

and, consequently:

$$\widehat{D}_{t_0}^{\alpha} y(t) = D_{t_0}^{\alpha} y(t) + \sum_{k=0}^{m-1} \frac{(t-t_0)^{k-\alpha}}{\Gamma(k+1-\alpha)} y^{(k)}(t_0).$$

Observe that in the special case $0 < \alpha < 1$, the above relationship becomes:

$$\widehat{D}_{t_0}^{\alpha} y(t) = D_{t_0}^{\alpha} y(t) + \frac{(t-t_0)^{-\alpha}}{\Gamma(1-\alpha)} y(t_0)$$

clearly showing how the Caputo derivative is a sort of regularization of the RL derivative at t_0. Another feature that justifies the introduction of the Caputo derivative is related to the differentiation of constant function; indeed, since:

$$\widehat{D}_{t_0}^{\alpha} 1 = \frac{1}{\Gamma(1-\alpha)} (t-t_0)^{-\alpha}, \quad D_{t_0}^{\alpha} 1 = 0,$$

in several applications, it is preferable to deal with operators for which the derivative of a constant is zero as in the case of Caputo's derivative.

One of the most important applications of Caputo's derivative is however in FDEs. Unlike FDEs with the RL derivative, which are initialized by derivatives of non-integer order, an initial value problem for an FDE (or a system of FDEs) with Caputo's derivative can be formulated as:

$$\begin{cases} D_{t_0}^{\alpha} y(t) = f(t, y(t)) \\ y(t_0) = y_0, \, y'(t_0) = y_0^{(1)}, \, \ldots, \, y^{(m-1)}(t_0) = y_0^{(m-1)} \end{cases} \tag{6}$$

where $f(t,y)$ is assumed to be continuous and $y_0, y_0^{(1)}, \ldots, y_0^{(m-1)}$ are the assigned values of the derivatives at t_0. Clearly, initializing the FDE with assigned values of integer-order derivatives is more useful since they have a more clear physical meaning with respect to fractional-order derivatives.

The application to both sides of Equation (6) of the RL integral $J_{t_0}^{\alpha}$, together with Equation (4), leads to the reformulation of the FDE in terms of the weakly-singular Volterra integral equation (VIE):

$$y(t) = T_{m-1}[y;t_0](t) + \frac{1}{\Gamma(\alpha)} \int_{t_0}^{t} (t-\tau)^{\alpha-1} f(\tau, y(\tau)) \mathrm{d}\tau. \tag{7}$$

The integral Formulation (7) is surely useful since it allows exploiting theoretical and numerical results already available for this class of VIEs in order to study and solve FDEs.

We stress the nonlocal nature of FDEs: the presence of a real power in the kernel makes it not possible to split the solution of Equation (7) at any point t_n as the solution at some previous point $t_n - h$ plus the increment term related to the interval $[t_n - h, t_n]$, as is common with ODEs.

Furthermore, as proved by Lubich [9], the solution of the VIE (7) presents an expansion in mixed (i.e., integer and fractional) powers:

$$y(t) = T_{m-1}(t) + \sum_{i,j \in \mathbb{N}} (t-t_0)^{i+j\alpha} Y_{i,j}, \tag{8}$$

thus showing a non-smooth behavior at t_0; as is well-known, the absence of smoothness at $t = t_0$ poses some problems for the numerical computation since methods based on polynomial approximations fail to provide accurate results in the presence of some lack of smoothness.

3. Multi-Step Methods for FDEs

Most of the step-by-step methods for the numerical solution of differential equations can be roughly divided into two main families: one-step and multi-step methods.

In one-step methods, just one approximation of the solution at the previous step is used to compute the solution and, hence, they are particularly suited when it is necessary to dynamically change the step-size in order to adapt the integration process to the behavior of the solution. In multi-step methods, it is instead necessary to use more previously evaluated approximations to compute the solution.

Because of the persisting memory of fractional-order operators, multi-step methods are clearly a natural choice for FDEs; anyway, although multi-step methods for FDEs are usually derived from multi-step methods for ODEs, when applied to FDEs, the number of steps involved in the computation is not fixed, but it increases as the integration proceeds forward, and the whole history of the solution is involved in each step's computation.

Multi-step methods for the FDEs (6) are therefore convolution quadrature formulas, which can be written in the general form:

$$y_n = \varphi_n + \sum_{j=0}^{n} c_{n-j} f_j, \quad f_j = f(t_j, y_j), \tag{9}$$

where φ_n and c_n are known coefficients and $t_n = t_0 + nh$ is an assigned grid, with a constant step-size $h > 0$ just for simplicity; the way in which the coefficients are derived depends on the specific method. In particular, the following two classes of multi-step methods for FDEs are mainly studied in the literature:

- product-integration (PI) rules,
- fractional linear multi-step methods (FLMMs).

Both families of methods are based on the approximation of the RL integral in the VIE (7) and generalize, on different bases, standard multi-step methods for ODEs. They allow one to write general-purpose methods requiring just the knowledge of the vector field of the differential equation.

We must mention that several other approaches have been however discussed in the literature: see, for instance, the generalized Adams methods [10], extensions of the Runge-Kutta methods [11], generalized exponential integrators [12,13], spectral methods [14,15], spectral collocation methods [16], methods based on matrix functions [17–20], and so on. In this paper, for brevity, we focus only on PI rules and FLMMs, and we refer the reader to the existing literature for alternative approaches.

3.1. Product-Integration Rules

PI rules were introduced by Young [21] in 1954 to numerically solve second-kind weakly-singular VIEs; they hence apply in a natural way to FDEs due to their formulation in Equation (7).

Given a grid $t_n = t_0 + nh$, with constant step-size $h > 0$, in PI rules, the solution of the VIE (7) at t_n is first written in a piece-wise way:

$$y(t_n) = T_{m-1}[y; t_0](t_n) + \frac{1}{\Gamma(\alpha)} \sum_{j=0}^{n-1} \int_{t_j}^{t_{j+1}} (t_n - \tau)^{\alpha-1} f(\tau, y(\tau)) \mathrm{d}\tau$$

and $f(\tau, y(\tau))$ is approximated, in each subinterval $[t_j, t_{j+1}]$, by means of some interpolant polynomial; the resulting integrals are hence evaluated in an exact way, thus to lead to y_n. According to the way in which the approximation is made, explicit or implicit rules are obtained, and this is perhaps the most straightforward way to generalize Adams multi-step methods commonly employed for integer-order ODEs [22].

For instance, to extend to FDEs the (explicit) forward and (implicit) backward Euler methods, it is sufficient to approximate, in each interval $[t_j, t_{j+1}]$, the integrand $f(\tau, y(\tau))$ by the constant values $f(t_j, y_j)$ and $f(t_{j+1}, y_{j+1})$, respectively; the resulting methods are:

$$\text{Expl. PI Rectangular:} \quad y_n = T_{m-1}[y; t_0](t_n) + h^\alpha \sum_{j=0}^{n-1} b_{n-j-1}^{(\alpha)} f(t_j, y_j) \tag{10}$$

and:

$$\text{Impl. PI Rectangular:} \quad y_n = T_{m-1}[y; t_0](t_n) + h^\alpha \sum_{j=1}^{n} b_{n-j}^{(\alpha)} f(t_j, y_j) \tag{11}$$

with $b_n^{(\alpha)} = ((n+1)^\alpha - n^\alpha)/\Gamma(\alpha+1)$; the term rectangular comes after the underlying quadrature rules used for the integration. In a similar way, when $f(\tau, y(\tau))$ is approximated by the first order interpolant polynomial:

$$f(\tau, y(\tau)) \approx f(t_{j+1}, y_{j+1}) + \frac{\tau - t_{j+1}}{h} \big(f(t_{j+1}, y_{j+1}) - f(t_j, y_j) \big), \quad \tau \in [t_j, t_{j+1}],$$

one obtains a generalization (of implicit type) of the standard trapezoidal rule:

$$\text{Impl. PI Trap.:} \quad y_n = T_{m-1}[y; t_0](t_n) + h^\alpha \left(\tilde{a}_n^{(\alpha)} f_0 + \sum_{j=1}^{n} a_{n-j}^{(\alpha)} f(t_j, y_j) \right) \tag{12}$$

with:

$$\tilde{a}_n^{(\alpha)} = \frac{(n-1)^{\alpha+1} - n^\alpha (n - \alpha - 1)}{\Gamma(\alpha+2)}, \quad a_n^{(\alpha)} = \begin{cases} \dfrac{1}{\Gamma(\alpha+2)} & n = 0 \\ \dfrac{(n-1)^{\alpha+1} - 2n^{\alpha+1} + (n+1)^{\alpha+1}}{\Gamma(\alpha+2)} & n = 1, 2, \ldots \end{cases}$$

An explicit version of the trapezoidal PI rule (12) is also possible, but it is not frequently encountered in the literature.

Unlike what one would expect, using interpolant polynomials of higher degree does not necessarily improve the accuracy of the obtained approximation. This phenomenon, already studied in [23], is related to the behavior of the solution of FDEs, which (with few exceptions [24]) have a non-smooth behavior also in the presence of a smooth given function $f(t, y)$; some of the derivatives of $y(t)$, and consequently of $f(t, y(t))$, are indeed unbounded at t_0 and hence not properly approximated by polynomials.

Thus, methods (10) and (11), as expected, converge with order one with respect to h, that is the error between the exact solution $y(t_n)$ and the approximation y_n is:

$$|y(t_n) - y_n| = \mathcal{O}(h), \quad h \to 0.$$

Differently, the convergence order of the trapezoidal PI rule (12) usually drops to $1 + \alpha$ when $0 < \alpha < 1$, and the expected order two is obtained only when $\alpha > 1$ or just for well-selected problems with a sufficiently smooth solution (see, for instance, [23–26]). Actually, as one of the most general results, the error of the trapezoidal PI rule (12) is:

$$|y(t_n) - y_n| = \mathcal{O}(h^{\min\{1+\alpha, 2\}}), \quad h \to 0,$$

although other special cases could be encountered. For this reason, PI rules of (just virtual) higher order, based on polynomial interpolation of degree two or more, are seldom considered in practice since in the majority of cases, they do not actually lead to any improvement of accuracy and convergence order.

To avoid the solution of the nonlinear equations in Equation (12) for the evaluation of y_n, a predictor-corrector (PC) approach is sometimes preferred, in which a first approximation of y_n is predicted by means of the explicit PI rectangular rule (10) and hence corrected by the implicit PI trapezoidal rule (12) according to:

$$\begin{aligned} y_n^P &= T_{m-1}[y;t_0](t_n) + h^\alpha \sum_{j=0}^{n-1} b_{n-j-1}^{(\alpha)} f(t_j, y_j), \\ y_n &= T_{m-1}[y;t_0](t_n) + h^\alpha \left(\tilde{a}_n^{(\alpha)} f_0 + \sum_{j=1}^{n-1} a_{n-j}^{(\alpha)} f(t_j, y_j) + a_0^{(\alpha)} f(t_n, y_n^P) \right). \end{aligned} \tag{13}$$

The PC method for FDEs has been extensively investigated (see, for instance, [25,27–29]). With the aim of improving the approximation, a multiple number, say μ, of corrector iterations can be applied:

$$\begin{aligned} y_n^{[0]} &= T_{m-1}[y;t_0](t_n) + h^\alpha \sum_{j=0}^{n-1} b_{n-j-1}^{(\alpha)} f(t_j, y_j) \\ y_n^{[\mu]} &= T_{m-1}[y;t_0](t_n) + h^\alpha \left(\tilde{a}_n^{(\alpha)} f_0 + \sum_{j=1}^{n-1} a_{n-j}^{(\alpha)} f(t_j, y_j) + a_0^{(\alpha)} f(t_n, y_n^{[\mu-1]}) \right), \quad \mu = 1, 2, \ldots. \end{aligned} \tag{14}$$

Each iteration is expected to increase the order of convergence of a fraction α from the first order of convergence of the predictor method, until the order of convergence of the corrector method is achieved: thus, one or very few corrector iterations are usually necessary. The explicit PI rectangular rule (10) is obtained when $\mu = 0$; the standard predictor-corrector method (13) clearly requires $\mu = 1$.

3.2. Fractional Linear Multi-Step Methods

FLMMs were introduced and extensively studied by Lubich in [30] (it is, however, also useful to refer the reader to the papers [31–33], where these methods are studied under a more general perspective in connection with wider classes of convolution integrals).

The main feature of FLMMs is that they generalize, in a robust and elegant way, quadrature rules obtained from standard linear multi-step methods (LMMs). Thus, they are one of the most powerful methods for FDEs.

Given the initial value problem:

$$y'(t) = f(t), \quad y(t_0) = y_0, \tag{15}$$

its solution can be approximated by means of an LMM given by:

$$\sum_{j=0}^{k} \rho_j y_{n-j} = h \sum_{j=0}^{k} \sigma_j f(t_{n-j}),$$

where $\rho(z) = \rho_0 z^k + \rho_1 z^{k-1} + \cdots + \rho_k$ and $\sigma(z) = \sigma_0 z^k + \sigma_1 z^{k-1} + \cdots + \sigma_k$ are the first and second characteristic polynomial of the LMM. Problem (15) can be rewritten in the integral form:

$$y(t) = y_0 + \int_{t_0}^{t} f(\tau) \mathrm{d}\tau$$

and as investigated by Wolkenfelt [34,35] and also explained in the textbook [36], the solution $y(t)$ can be approximated by using LMMs reformulated in terms of convolution quadrature formulas:

$$y_n = h \sum_{j=0}^{n} \omega_{n-j} f(t_j), \quad n \geq k$$

where the weights ω_n depend on the characteristic polynomials $\rho(z)$ and $\sigma(z)$, but not on h. The computation of the weights ω_n is usually not easy, but interestingly, it is possible to represent them as the coefficients of the formal power series (FPS) of the generating function of the LMM [37], namely:

$$\delta(\xi) = \sum_{n=0}^{\infty} \omega_n \xi^n, \quad \delta(\xi) = \frac{\rho(1/\xi)}{\sigma(1/\xi)}.$$

The idea underlying FLMMs, supported by a rigorous theoretical reasoning, is to derive convolution quadratures for the RL integral (1) with convolution weights given by the coefficients of the FPS of the function:

$$F\left(\frac{\delta(\xi)}{h}\right) = \left(\frac{\delta(\xi)}{h}\right)^{-\alpha} = h^{\alpha} \left(\frac{\rho(1/\xi)}{\sigma(1/\xi)}\right)^{\alpha}, \tag{16}$$

being $F(s) = s^{-\alpha}$ the Laplace transform of the kernel $t^{\alpha-1}/\Gamma(\alpha)$ in (1). The assumptions that make possible this generalization of LMMs are that the generating function $\delta(\xi)$ has no zeros in the closed unit disc $|\xi| \leq 1$, except for $\xi = 1$, and $|\arg \delta(\xi)| < \pi$ for $|\xi| < 1$. LMMs satisfying these assumptions are, for instance, the backward differentiation formulas (BDFs) and the trapezoidal rule, which are reported in Table 1.

Table 1. Some linear multi-step methods (LMMs) with corresponding polynomials $\rho(z)$ and $\sigma(z)$ and generating function $\delta(\xi)$.

Name	Formula	$\rho(z)$	$\sigma(z)$	$\delta(\xi)$
BDF1	$y_n = y_{n-1} + hf_n$	$z-1$	z	$1-\xi$
BDF2	$y_n - \frac{4}{3}y_{n-1} + \frac{1}{3}y_{n-2} = \frac{2}{3}hf_n$	$z^2 - \frac{4}{3}z + \frac{1}{3}$	$\frac{2}{3}z^2$	$\frac{3}{2} - 2\xi + \frac{1}{2}\xi^2$
BDF3	$y_n - \frac{18}{11}y_{n-1} + \frac{9}{11}y_{n-2} - \frac{2}{11}y_{n-3} = \frac{6}{11}hf_n$	$z^3 - \frac{18}{11}z^2 + \frac{9}{11}z - \frac{2}{11}$	$\frac{6}{11}z^3$	$\frac{11}{6} - 3\xi + \frac{3}{2}\xi^2 - \frac{1}{3}\xi^3$
Trapez.	$y_n = y_{n-1} + \frac{h}{2}(f_{n-1} + f_n)$	$z-1$	$\frac{1}{2}z + \frac{1}{2}$	$2\frac{1-\xi}{1+\xi}$

When an LMM is generalized to Equation (1) in the above Lubich sense, the resulting FLMM reads as:

$${}_hJ_{t_0}^{\alpha} f(t_n) = h^{\alpha} \sum_{j=0}^{n} \omega_{n-j}^{(\alpha)} f(t_j) \tag{17}$$

where the convolution quadrature weights $\omega_n^{(\alpha)}$ are obtained from:

$$\sum_{n=0}^{\infty} \omega_n^{(\alpha)} \xi^n = \omega^{(\alpha)}(\xi), \quad \omega^{(\alpha)}(\xi) = (\delta(\xi))^{-\alpha}.$$

When $f(t)$ is sufficiently smooth and the LMM has order p of convergence, the approximation provided by Equation (17) satisfies [33] (Theorem 2.1):

$$\left| J_{t_0}^{\alpha} f(t_n) - {}_hJ_{t_0}^{\alpha} f(t_n) \right| \leq C(t_n - t_0)^{\alpha-1-p} h^p$$

for come constant C, which does not depend on h. Anyway, when $f(t)$ lacks smoothness, for instance at t_0, it is no longer possible to preserve the order p of convergence, and for $f(t) = (t - t_0)^{\gamma}$ the following result holds (see [31] (Theorem 5.1) and [33] (Theorem 2.2)):

$$\left| J_{t_0}^{\alpha} (t_n - t_0)^{\gamma} - {}_hJ_{t_0}^{\alpha} (t_n - t_0)^{\gamma} \right| \leq \begin{cases} C(t_n - t_0)^{\alpha-p} h^{\gamma+1} & -1 < \gamma \leq p-1, \\ C(t_n - t_0)^{\alpha+\gamma-p} h^p & \gamma \geq p-1. \end{cases} \tag{18}$$

Thus, to handle non-smooth functions (as happens in the solution of fractional-order problems), it is necessary to introduce a correction term:

$$ {}_hJ^{\alpha}_{t_0} f(t_n) = h^{\alpha} \sum_{j=0}^{\nu} w_{n,j} f(t_j) + h^{\alpha} \sum_{j=0}^{n} \omega^{(\alpha)}_{n-j} f(t_j), \tag{19} $$

where the starting quadrature weights $w_{n,j}$ are suitably selected in order to eliminate low order terms in the error bounds (18) and obtain the same convergence of order p of the underlying LMM.

From the application of the discretized convolution quadrature rule (19) to integral Equation (7), we are able to derive FLMMs for the approximation of the solution of FDEs:

$$ y_n = T_{m-1}[y; t_0](t_n) + h^{\alpha} \sum_{j=0}^{s} w_{n,j} f(t_j, y_j) + h^{\alpha} \sum_{j=0}^{n} \omega^{(\alpha)}_{n-j} f(t_j, y_j) \tag{20} $$

with the starting weights $w_{n,j}$ selected in order to cope with the non-smooth behavior of $y(t)$ highlighted by Equation (8). Thus, to achieve the same order of convergence of the underlying LMM, the starting weights $w_{n,j}$ are chosen by imposing that the quadrature rule (19) is exact when applied to $f(t) = t^{\gamma}$, with γ assuming all the possible fractional values expected in the expansion of the true solution and, hence, by solving at each step the algebraic linear system:

$$ \sum_{j=0}^{s} w_{n,j} j^{\gamma} = - \sum_{j=0}^{n} \omega_{n-j} j^{\gamma} + \frac{\Gamma(\gamma+1)}{\Gamma(1+\gamma+\alpha)} n^{\gamma+\alpha}, \quad \nu \in \mathcal{A}_p, \tag{21} $$

with $\mathcal{A}_p = \{\gamma \in \mathbb{R} \mid \gamma = i + j\alpha,\ i,j \in \mathbb{N},\ \gamma < p-1\}$ and $s+1$ the cardinality of $\mathcal{A}_p$.

One of the simplest FLMMs is obtained from the implicit Euler method (or BDF1). No starting weights are necessary in this case, and since the generating function is $\delta(\xi) = 1 - \xi$ (see Table 1), we see that $\omega^{(\alpha)}_n$, $n = 0, 1, \ldots$, are the coefficients of the generalized binomial series $(1-\xi)^{-\alpha}$, namely:

$$ \omega^{(\alpha)}_n = (-1)^n \binom{-\alpha}{n} = (-1)^n \frac{\Gamma(1-\alpha)}{\Gamma(n+1)\Gamma(-\alpha-n+1)} $$

which can be also evaluated by the recurrence $\omega^{(\alpha)}_n = (1 - (1-\alpha)/n)\,\omega^{(\alpha)}_{n-1}$, with $\omega^{(\alpha)}_0 = 1$. The corresponding method:

$$ y_n = T_{m-1}[y; t_0](t_n) + h^{\alpha} \sum_{j=0}^{n} (-1)^{n-j} \binom{-\alpha}{n-j} f(t_j, y_j) $$

is commonly referred to as the Grünwald–Letnikov scheme [5].

It is possible to derive several FLMMs with second order of convergence, which mainly differ for stability properties; for this purpose, we refer the reader to the paper [26], where the MATLAB code `FLMM2.m` implementing three different FLMMs is also presented.

The regularization operated by the starting weights is one of the most attractive features of FLMMs since it makes it possible to substantially achieve the same order of the underlying LMM. Unlike PI for which, in general, it is difficult to obtain a convergence order equal to or greater than two, FLMMs make the development of high order methods possible. However, round-off errors may accumulate when solving the ill-conditioned linear systems (21) [38], and hence, it is advisable to avoid very high order methods.

3.3. *Implicit vs. Explicit Methods*

Numerical methods for solving differential equations can be of an explicit or implicit nature. In explicit methods, such as method (10) or (27), the evaluation of each y_n does not present any particular difficulty once the previous values $y_0, y_1, \ldots, y_{n-1}$ have already been evaluated. In implicit methods, such as method (11), (12), (28) or (29), the approximation of y_n is expressed by means of a functional relationship of the kind:

$$y_n = \Psi_n + c_0 f(t_n, y_n), \tag{22}$$

where with Ψ_n we denote the term collecting all the explicitly known information.

Implicit methods possess better stability properties, but they need some numerical procedure to solve the nonlinear equation, or system of nonlinear equation (22).

One of the most powerful methods is the iterative Newton–Raphson method; given an initial approximation $y_n^{(0)}$ for y_n, the Newton–Raphson method when applied to solve Equation (22) evaluates successive approximations of y_n by means of the relationship:

$$y_n^{(k+1)} = y_n^{(k)} - \left[I - c_0 \mathbf{J}_f(t_n, y_n^{(k)}) \right]^{-1} \left(y_n^{(k)} - \Psi_n - c_0 f(t_n, y_n^{(k)}) \right)$$

where $\mathbf{J}_f(t, y)$ is the Jacobian of $f(t, y)$ with respect to y and I the identity matrix of compatible size (in the scalar case, a simple derivative replaces the Jacobian matrix).

The Newton–Raphson method converges in a fast way (it indeed has second-order convergence properties, i.e., $\|y_n - y_n^{(k+1)}\| \approx \|y_n - y_n^{(k)}\|^2$), but its convergence is local, i.e., it is necessary to start sufficiently close to the solution. In general, there is no available information to localize the solution of Equation (22), and a usually satisfactory strategy is to start from the last evaluated approximation, namely $y_n^{(0)} = y_{n-1}$; since, under standard assumptions, it is reasonable to assume that at least $y_n = y_{n-1} + \mathcal{O}(h)$, a sufficiently small step-size h will in general assure the convergence of Newton–Raphson iterations unless y or f change very rapidly.

An alternative approach, which is used to reduce the computational cost, consists of evaluating the Jacobian just once and reusing it for all the following approximations, namely:

$$y_n^{(k+1)} = y_n^{(k)} - \left[I - c_0 \mathbf{J}_f(t_n, y_n^{(0)}) \right]^{-1} \left(y_n^{(k)} - \Psi_n - c_0 f(t_n, y_n^{(k)}) \right).$$

This approach, usually known as the modified Newton–Raphson method, not only allows one to save the cost of computing a new Jacobian matrix at each step, but also reduces the computational cost related to the solution of the linear algebraic system since it is possible to evaluate an LU decomposition of the matrix $\left[I - c_0 \mathbf{J}_f(t_n, y_n^{(0)}) \right]$ and solve all the systems in the iterative process by using the same decomposition.

Although the derivative (or the Jacobian in the non-scalar case) could be numerically approximated, it is not advisable to introduce a further source of error; moreover, the numerical approximation of derivatives is usually an ill-conditioned problem. Therefore, to avoid a loss of accuracy, all the codes for implicit methods presented in this paper require that derivatives or Jacobian matrices are explicitly provided. As for $f(t, y)$, some parameters could be optionally specified also for $\mathbf{J}_f(t, y)$, thus to allow solving general problems depending on user-supplied parameters.

3.4. *Efficient Treatment of the Persistent Memory*

The numerical solution of FDEs demands for a large amount of computation, which, if not suitably organized, represents a serious issue. Most of the methods for FDEs, such as PI rules or FLMMs, are indeed discrete convolution quadrature rules of the form (9), and since the computation of each approximation y_n requires a number of floating-point operations proportional to n, the whole evaluation of the solution on a grid $t_0, t_1, \ldots, t_N$ involves a number of operations proportional to:

$$\sum_{n=0}^{N} n = \frac{N(N+1)}{2} \approx N^2.$$

When the interval of integration $[t_0, T]$ is large or stability reasons demand for a very small step-size h, the required number $N = \lceil (T - t_0)/h \rceil$ of grid points can be too high to perform the computation in a reasonable time.

This is one of the most serious consequences of the persistent memory of non-local operators such as fractional integrals and derivatives. Although it is possible to apply short memory procedures relying on a truncation of the memory tail (e.g., see [39]) or on some more sophisticated approaches [40,41], their use introduces a further source of errors and/or increases the computational complexity.

For general-purpose codes, it is, instead, preferable to adopt techniques that are easy to implement and do not affect the accuracy of the solution.

An extremely powerful approach, exploiting general properties of convolution quadratures and based on the FFT algorithm, has been proposed by Hairer, Lubich and Schlichte [42,43].

This approach evaluates only the first r steps directly by means of the discrete convolution (9), namely:

$$y_n = \varphi_n + \sum_{j=0}^{n} c_{n-j} f_j, \quad n = 0, 1, \ldots, r-1.$$

with r denoting a moderately small integer value selected, for convenience, as a power of two.

To determine the following r approximations, after writing:

$$y_n = \varphi_n + \sum_{j=0}^{r-1} c_{n-j} f_j + \sum_{j=r}^{n} c_{n-j} f_j, \quad n \in \{r, r+1, \ldots, 2r-1\}$$

we observe that the set of partial sums each of length r:

$$S_r(n, 0, r-1) := \sum_{j=0}^{r-1} c_{n-j} f_j, \quad n \in \{r, r+1, \ldots, 2r-1\}$$

can be evaluated by the FFT algorithm (described, for instance, in [44]), requiring only $\mathcal{O}(2r \log_2 2r)$ floating-point operations instead of $\mathcal{O}(r^2)$, as with standard computation. The same process can be recursively repeated by doubling the time-interval; thus, for the successive $2r$ approximations y_n, for $n \in \{2r, 2r+1, \ldots, 4r-1\}$, after writing:

$$\begin{cases} y_n = \varphi_n + \sum_{j=0}^{2r-1} c_{n-j} f_j + \sum_{j=2r}^{n} c_{n-j} f_j & n \in \{2r, 2r+1, \ldots, 3r-1\} \\ y_n = \varphi_n + \sum_{j=0}^{2r-1} c_{n-j} f_j + \sum_{j=2r}^{3r-1} c_{n-j} f_j + \sum_{j=3r}^{n} c_{n-j} f_j & n \in \{3r, 3r+1, \ldots, 4r-1\} \end{cases}$$

again, it is possible to use the FFT algorithm to evaluate the two sets of partial sums:

$$S_{2r}(n, 0, 2r-1) := \sum_{j=0}^{2r-1} c_{n-j} f_j, \quad S_r(n, 2r, 3r-1) := \sum_{j=2r}^{3r-1} c_{n-j} f_j,$$

of length $2r$ and r, respectively, with a computational cost proportional to $\mathcal{O}(4r \log_2 4r)$ and $\mathcal{O}(2r \log_2 2r)$. The whole process can be iteratively repeated; for instance, to evaluate the $4r$ approximations y_n in the interval $n \in \{4r, \ldots, 8r-1\}$, we have:

$$\begin{cases} y_n = \varphi_n + \sum_{j=0}^{4r-1} c_{n-j}f_j + \sum_{j=4r}^{n} c_{n-j}f_j & n \in \{4r, 4r+1, \dots, 5r-1\} \\ y_n = \varphi_n + \sum_{j=0}^{4r-1} c_{n-j}f_j + \sum_{j=4r}^{5r-1} c_{n-j}f_j + \sum_{j=5r}^{n} c_{n-j}f_j & n \in \{5r, 3r+1, \dots, 5r-1\} \\ y_n = \varphi_n + \sum_{j=0}^{4r-1} c_{n-j}f_j + \sum_{j=4r}^{6r-1} c_{n-j}f_j + \sum_{j=6r}^{n} c_{n-j}f_j & n \in \{6r, 6r+1, \dots, 7r-1\} \\ y_n = \varphi_n + \sum_{j=0}^{4r-1} c_{n-j}f_j + \sum_{j=4r}^{6r-1} c_{n-j}f_j + \sum_{j=6r}^{7r-1} c_{n-j}f_j + \sum_{j=7r}^{n} c_{n-j}f_j & n \in \{7r, 7r+1, \dots, 8r-1\} \end{cases}$$

and the sets of partial sums:

$$S_{4r}(n, 0, 4r-1) := \sum_{j=0}^{4r-1} c_{n-j}f_j, \quad S_{2r}(n, 4r, 6r-1) := \sum_{j=4r}^{6r-1} c_{n-j}f_j, \quad S_r(n, 6r, 7r-1) := \sum_{j=6r}^{7r-1} c_{n-j}f_j$$

are evaluated in $\mathcal{O}(8r \log_2 8r)$, $\mathcal{O}(4r \log_2 4r)$ and $\mathcal{O}(2r \log_2 2r)$ floating-point operations, respectively. To better understand how this process works, Figure 1 can be of some help, where each square represents the computation of a set of partial sums (by means of the FFT algorithm), and each triangle represents the standard computation of each final convolution term:

$$T_r(p, n) = \sum_{j=p}^{n} c_{n-j}f_j, \quad p = \ell r, \quad n \in \{\ell r, \ell r+1, \dots, (\ell+1)r-1\}, \quad \ell = 0, 1, 2, \dots.$$

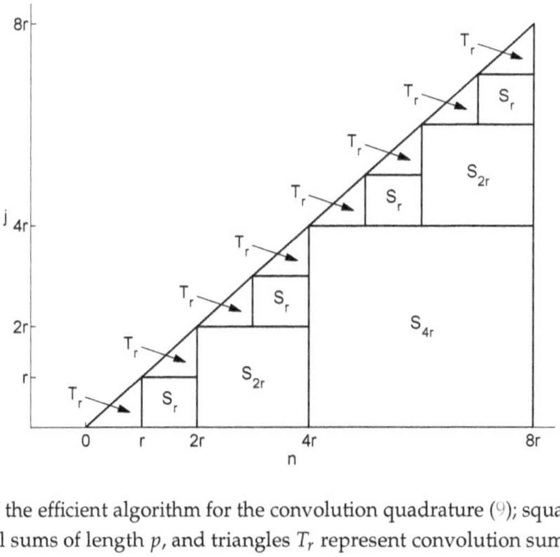

Figure 1. Scheme of the efficient algorithm for the convolution quadrature (9); squares S_p represent the evaluation of partial sums of length p, and triangles T_r represent convolution sums of fixed length r.

To determine the whole computational cost, we assume, for simplicity, that the total number N of grid points is a power of two. Hence, the computation of one partial sum of length $N/2$ (the square S_{4r} in Figure 1) is requested, involving $\mathcal{O}(N \log_2 N)$ operations, two partial sums of length $N/4$ (the squares S_{2r}) each involving $\mathcal{O}(\frac{N}{2} \log_2 \frac{N}{2})$ operations, four partial sums of length $N/8$ (the squares S_r) each involving $\mathcal{O}(\frac{N}{4} \log_2 \frac{N}{4})$ operations, and so on. Additionally, N/r convolution sums of length r (the triangles T_r) are requested each with a computational cost proportional to $r(r+1)/2$. Thus, the total amount of computation is proportional to:

$$N\log_2 N + 2\frac{N}{2}\log_2\frac{N}{2} + 4\frac{N}{4}\log_2\frac{N}{4} + \cdots + p\frac{N}{p}\log_2\frac{N}{p} + \frac{N}{r}\frac{r(r+1)}{2}, \quad p = \frac{N}{2r},$$

and by means of some simple manipulations, we are able to estimate a computational cost proportional to:

$$\sum_{j=0}^{\log_2 p} N\log_2\frac{N}{2^j} + N\frac{r+1}{2} = \mathcal{O}(N(\log_2 N)^2)$$

which, for sufficiently large N, is clearly smaller than N^2, i.e., the number of operations required by a computation performed in the standard way.

4. Some Particular Families of FDEs and Systems of FDEs

Equation (6) is perhaps the most standard example of FDEs. The numerical methods presented in the previous sections can be applied both to scalar equations and to systems of FDEs of any size.

There are however other types of fractional-order problems whose treatment necessitates a particular discussion.

4.1. Multi-Order Systems of FDEs

As a special case of non-linear systems of FDEs, we consider systems in which each equation has its own order, which can differ from the order of the other equations. The general form of a MOS of FDEs is:

$$\begin{cases} D_{t_0}^{\alpha_1} y_1(t) = f_1(t, y(t)) \\ D_{t_0}^{\alpha_2} y_2(t) = f_2(t, y(t)) \\ \quad\vdots \\ D_{t_0}^{\alpha_Q} y_Q(t) = f_Q(t, y(t)) \end{cases} \tag{23}$$

with $y(t) = (y_1(t), \ldots, y_Q(t))$; it must be coupled with the initial conditions:

$$y(t_0) = y_0, \quad \frac{\mathrm{d}}{\mathrm{d}t}y(t_0) = y_0^{(1)}, \quad \ldots, \frac{\mathrm{d}^{m-1}}{\mathrm{d}t^{m-1}}y(t_0) = y_0^{(m-1)},$$

where their number is $m = \max\{m_1, m_2, \ldots, m_Q\}$, with $m_i = \lceil \alpha_i \rceil$, $i = 1, 2, \ldots, Q$. Also in this case, it is possible to reformulate each equation of the system (23) in terms of the VIEs:

$$\begin{cases} y_1(t_n) = T_{m_1-1}[y_1; t_0](t_n) + \dfrac{1}{\Gamma(\alpha_1)}\displaystyle\int_{t_0}^{t_n} (t_n - s)^{\alpha_1-1} f_1(s, y(s))\mathrm{d}s \\ y_2(t_n) = T_{m_2-1}[y_2; t_0](t_n) + \dfrac{1}{\Gamma(\alpha_2)}\displaystyle\int_{t_0}^{t_n} (t_n - s)^{\alpha_2-1} f_2(s, y(s))\mathrm{d}s \\ \quad\vdots \\ y_Q(t_n) = T_{m_Q-1}[y_Q; t_0](t_n) + \dfrac{1}{\Gamma(\alpha_Q)}\displaystyle\int_{t_0}^{t_n} (t_n - s)^{\alpha_Q-1} f_Q(s, y(s))\mathrm{d}s \end{cases} \tag{24}$$

and apply one of the methods described in Section 3.

From the theoretical point of view, there are no particular differences with respect to the solution of a system of FDEs (6) in which all the equations have the same order. The computation is however more expensive due to the need for computing different sequences of weights and evaluating more than one discrete convolution quadrature. It is therefore necessary to optimize the codes to exploit the possible presence of equations having the same order, thus to avoid unnecessary computations; the codes presented in the next section provide this kind of optimization.

4.2. Linear Multi-Term FDEs

A further special case of FDE is when more than one fractional derivative appears in a single equation. Equations of this kind are called multi-term equations, and in the linear case, they are described as:

$$\lambda_Q D_{t_0}^{\alpha_Q} y(t) + \lambda_{Q-1} D_{t_0}^{\alpha_{Q-1}} y(t) + \cdots + \lambda_2 D_{t_0}^{\alpha_2} y(t) + \lambda_1 D_{t_0}^{\alpha_1} y(t) = f(t, y(t)), \tag{25}$$

where $\lambda_1, \lambda_2, \ldots, \lambda_{Q-1}, \lambda_Q$ are some (usually real) coefficients and the orders $\alpha_1, \alpha_2, \ldots, \alpha_{Q-1}, \alpha_Q$ of the fractional derivatives are assumed (just for convenience) to be sorted in an ascending order, i.e., $0 < \alpha_1 < \alpha_2 < \cdots < \alpha_{Q-1} < \alpha_Q$, with $\lambda_Q \neq 0$. Note that here we focus on multi-term FDEs, which are linear with respect to the fractional derivatives, but with a (possible) nonlinearity of $f(t, y)$.

The number of initial conditions is given by m_Q, where as usual, $m_i = \lceil \alpha_i \rceil$, $i = 1, \ldots, Q$ (and clearly $m_Q = \max m_i$), and they are expressed in the usual way as derivatives of the solution at the starting point t_0:

$$y(t_0) = y_0, \quad \frac{\mathrm{d}}{\mathrm{d}t} y(t_0) = y_0^{(1)}, \quad \ldots, \frac{\mathrm{d}^{m_Q-1}}{\mathrm{d}t^{m_Q-1}} y(t_0) = y_0^{(m_Q-1)}.$$

Multi-term FDEs are more difficult to solve than standard FDEs. Anyway, as proposed in [45,46], it is always possible to recast Equation (25) in such a way that some of the methods for FDEs can be easily adapted. Indeed, thanks to Equations (4) and (5) and by applying $J_{t_0}^{\alpha_Q}$ to Equation (25), we can reformulate the multi-term FDE as:

$$y(t) = T_{m_Q-1}[y; t_0](t) - \sum_{i=1}^{Q-1} \frac{\lambda_i}{\lambda_Q} J_{t_0}^{\alpha_Q - \alpha_i} \left[y(t) - T_{m_i-1}[y; t_0](t) \right] + \frac{1}{\lambda_Q} J_{t_0}^{\alpha_Q} f(t, y(t)). \tag{26}$$

Hence, numerical methods are straightforwardly devised by applying any method for the discretization of RL integrals (PIs or FLMMs).

As illustrative examples, we consider the generalization of the explicit and implicit first-order PI rules (10) and (11). For this purpose, we first observe that:

$$J_{t_0}^{\alpha} T_{m-1}[y; t_0](t) = \sum_{k=0}^{m-1} y^{(k)}(t_0) J_{t_0}^{\alpha} \frac{(t - t_0)^k}{k!} = \sum_{k=0}^{m-1} \frac{(t - t_0)^{k+\alpha}}{\Gamma(k + \alpha)} y^{(k)}(t_0)$$

and hence, after denoting:

$$\tilde{T}(t) : \ = T_{m_Q-1}[y; t_0](t) + \sum_{i=1}^{Q-1} \frac{\lambda_i}{\lambda_Q} \sum_{k=0}^{m_i-1} \frac{(t - t_0)^{k+\alpha_Q-\alpha_i}}{\Gamma(k + \alpha_Q - \alpha_i + 1)} y^{(k)}(t_0),$$

the corresponding methods for the multiterm FDE (25) are respectively:

$$\text{Expl. PI 1}: \quad y_n = \tilde{T}(t) - \sum_{i=1}^{Q-1} \frac{\lambda_i}{\lambda_Q} h^{\alpha_Q - \alpha_i} \sum_{j=0}^{n-1} b_{n-j-1}^{(\alpha_Q - \alpha_i)} y_j + \frac{1}{\lambda_Q} h^{\alpha_Q} \sum_{j=0}^{n-1} b_{n-j-1}^{(\alpha_Q)} f(t_j, y_j) \tag{27}$$

and:

$$\text{Impl. PI 1}: \quad y_n = \tilde{T}(t) - \sum_{i=1}^{Q-1} \frac{\lambda_i}{\lambda_Q} h^{\alpha_Q - \alpha_i} \sum_{j=1}^{n} b_{n-j}^{(\alpha_Q - \alpha_i)} y_j + \frac{1}{\lambda_Q} h^{\alpha_Q} \sum_{j=1}^{n} b_{n-j}^{(\alpha_Q)} f(t_j, y_j). \tag{28}$$

In a similar way, it is possible to extend to multi-term FDEs also the implicit trapezoidal rule (12):

$$\text{Impl. PI 2}: \quad y_n = \tilde{T}(t) - \sum_{i=1}^{Q-1} \frac{\lambda_i}{\lambda_Q} h^{\alpha_Q-\alpha_i} \left(\tilde{a}_n^{(\alpha_Q-\alpha_i)} y_0 + \sum_{j=1}^{n} a_{n-j}^{(\alpha_Q-\alpha_i)} y_j \right) + \frac{1}{\lambda_Q} h^{\alpha_Q} \left(\tilde{a}_n^{(\alpha_Q)} f(t_0, y_0) + \sum_{j=1}^{n} a_{n-j}^{(\alpha_Q)} f(t_j, y_j) \right), \tag{29}$$

which usually assures a higher accuracy; alternatively, in order to avoid the solution, at each step of the nonlinear equations to determine y_n (see the discussion in Section 3.3), also in this case, a predictor-corrector strategy can be of some practical help. Clearly, all these methods apply in a straightforward way to non-scalar problems, as well.

Although several other approaches have been proposed to solve linear multi-term FDEs, we think that the one discussed in this section could be privileged due to its simplicity. The application of FLMMs to Equation (26) is surely possible, but some problems must be solved to properly identify the starting weights on the basis of the behavior analysis of the exact solution; we do not address this problem in this paper.

The computational cost is proportional to Q times the cost of standard PI rules and can be kept under control by applying to each discrete convolution the technique discussed in Section 3.4.

5. MATLAB Routines for Fractional-Order Problems

In this section, we present some MATLAB routines specifically devised to solve fractional-order problems by means of the methods illustrated in this paper. The routines are listed in Table 2 with the indicated kind of problem that is aimed to be solved and the specific implemented method.

All the MATLAB routines are available on the software section of the web-page of the author, at the address: https://www.dm.uniba.it/Members/garrappa/Software.

Table 2. MATLAB routines for some fractional-order problems. FDEs: fractional differential equations; PI: product-integration.

Name	Problem	Method
`FDE_PI1_Ex.m`	System of FDEs (6) or (23)	Explicit PI rectangular rule (10)
`FDE_PI1_Im.m`	System of FDEs (6) or (23)	Implicit PI rectangular rule (11)
`FDE_PI2_Im.m`	System of FDEs (6) or (23)	Implicit PI trapezoidal rule (12)
`FDE_PI12_PC.m`	System of FDEs (6) or (23)	Predictor-corrector PI rules (13)
`MT_FDE_PI1_Ex.m`	Multi-term FDE (25)	Explicit PI rectangular rule (27)
`MT_FDE_PI1_Im.m`	Multi-term FDE (25)	Implicit PI rectangular rule (28)
`MT_FDE_PI2_Im.m`	Multi-term FDE (25)	Implicit PI trapezoidal rule (29)
`MT_FDE_PI12_PC.m`	Multi-term FDE (25)	Predictor-corrector PI rules

The number 1 in the name of routines based on the PI rectangular rule refers to the convergence order of the underlying formula, while the number 2 stands for the maximum obtainable order (under suitable smoothness assumptions) of the PI trapezoidal rule. For this reason, routines based on PC, which use both PI rectangular rules (as predictor) and PI trapezoidal rules (as corrector), have 1 and 2 in the name. These names have been selected in analogy with the names of some built-in MATLAB functions for ODEs.

The way in which the different routines can be used is quite similar. One of the main differences is in implicit methods, which, in addition to the right-hand side $f(t,y)$, denoted by the function handle `f_fun`, require also the Jacobian $\mathbf{J}_f(t,y)$ of the right-hand side, namely the function handle `J_fun`; this is necessary to solve the inner nonlinear equation by means of Newton–Raphson iterations as described in Section 3.3.

The routines for solving a standard system of FDEs (6) or an MOS (23) are used by means of the following instructions:

```
[t, y] = FDE_PI1_Ex(alpha,f_fun,t0,T,y0,h,param)
[t, y] = FDE_PI1_Im(alpha,f_fun,J_fun,t0,T,y0,h,param,tol,itmax)
[t, y] = FDE_PI2_Im(alpha,f_fun,J_fun,t0,T,y0,h,param,tol,itmax)
[t, y] = FDE_PI12_PC(alpha,f_fun,t0,T,y0,h,param,mu,mu_tol)
```

Clearly, in case of an MOS, the parameter `alpha` must be a vector of the same size of the problem, while with a standard system (6), `alpha` is a scalar value.

Codes for linear multi-term FDEs (25) are used in a slightly different way since they additionally require providing the parameters λ_i, according to:

```
[t, y] = MT_FDE_PI1_Ex(alpha,lambda,f_fun,t0,T,y0,h,param)
[t, y] = MT_FDE_PI1_Im(alpha,lambda,f_fun,J_fun,t0,T,y0,h,param,tol,itmax)
[t, y] = MT_FDE_PI2_Im(alpha,lambda,f_fun,J_fun,t0,T,y0,h,param,tol,itmax)
[t, y] = MT_FDE_PI12_PC(alpha,lambda,f_fun,t0,T,y0,h,param,mu,mu_tol)
```

The meaning of each parameter is explained as follows (note that some of the parameters are optional and can be therefore omitted):

- `alpha`: order of the fractional derivative; when solving standard systems (6), `alpha` must be a single scalar value, while with MOSs (23) and linear multi-term FDEs (25), `alpha` must be a vector;
- `lambda`: only for linear multi-term FDEs (23), `lambda` is the vector of the coefficients λ_i of each derivative in Equation (23);
- `f_fun`: function defining the right-hand side $f(t,y)$ or $f(t,y,param)$ of the equation; `param` denotes a possible optional parameter (or a set of parameters collected in a single vector);
- `J_fun`: Jacobian matrix (or derivative in the scalar case) of the right-hand side $f(t,y)$ of the equation (only for implicit methods) with respect to the second variable y; also, the Jacobian $\mathbf{J}_f(t,y)$ can have some parameters, namely $\mathbf{J}_f(t,y,param)$;
- `t0` and `T`: initial and final endpoints of the integration interval;
- `y0`: matrix of the initial conditions with the number of rows equal to the size of the problem and the number of columns equal to the smallest integer greater than $\max\{\alpha_1,\ldots,\alpha_Q\}$;
- `h`: step-size for integration; it must be real and positive;
- `param`: (optional) vector of possible parameters for the evaluation of the vector field $f(t,y)$ and its Jacobian (if not necessary, this vector can be omitted or an empty vector `[ ]` can be used);
- `tol`: (optional) fixed tolerance for stopping the Newton–Raphson iterations when solving the internal system of nonlinear equations (only for implicit methods); when not specified, the default value 10^{-6} is assumed;
- `itmax`: (optional) maximum number of iterations for the Newton–Raphson method; when not specified, the default value `itmax = 100` is used;
- `mu`: (optional) number of corrector iterations (only for predictor-corrector methods); when not specified, the default value `mu = 1` is used;
- `mu_tol`: (optional) tolerance for testing the convergence of corrector iterations when `mu=Inf`; when not specified, the default value `mu_tol =` 10^{-6} is used.

All the codes give two outputs:

- `t`: the vector of nodes on the interval $[t_0, T]$ in which the numerical solution is evaluated;
- `y`: the matrix whose columns are the values of the solution evaluated in the points of `t`.

6. Some Applicative Examples

In the following, we present some applications of the routines described in the previous section, also in order to show the way in which they can be used.

For problems for which the analytical solution is not known, we will use, as reference solution, the numerical approximation obtained with a tiny step h by the implicit trapezoidal PI rule, which, as we will see, usually shows an excellent accuracy. All the experiments are carried out in MATLAB Ver. 8.3.0.532 (R2014a) on a computer equipped with a CPU Intel i5-7400 at 3.00 GHz running under the operating system Windows 10.

The first test problem aims to show the superiority of implicit methods for stability reasons. For this purpose, we consider the simple linear test equation:

$$y'(t) = \lambda y(t), \quad y(t_0) = y_0, \tag{30}$$

whose exact solution is $y(t) = E_\alpha\big((t-t_0)^\alpha\lambda\big)$ with:

$$E_\alpha(z) = \sum_{k=0}^{\infty} \frac{z^k}{\Gamma(\alpha k + 1)}$$

the Mittag–Leffler function, which can be evaluated thanks to the algorithm described in [47]. This problem can be solved, in the interval $[0,\ 5]$ and for $\alpha = 0.6$ and $\lambda = -10$, by the following few MATLAB lines:

```
alpha = 0.6; lambda = -10 ;
f_fun = @(t,y,lam) lam * y;
J_fun = @(t,y,lam) lam ;
param = lambda ;
t0 = 0 ; T = 5 ; y0 = 1 ; h = 2^(-5) ;
```

and after calling one of the following routines:

```
[t, y] = FDE_PI1_Ex(alpha,f_fun,t0,T,y0,h,param) ;
[t, y] = FDE_PI1_Im(alpha,f_fun,J_fun,t0,T,y0,h,param) ;
[t, y] = FDE_PI2_Im(alpha,f_fun,J_fun,t0,T,y0,h,param) ;
[t, y] = FDE_PI12_PC(alpha,f_fun,t0,T,y0,h,param).
```

As we can see from the results in Table 3, the explicit methods (including the predictor-corrector, which actually works as an explicit method) provide wrong or inaccurate results for small step-sizes, while implicit methods are able to return reliable results even with large step-sizes (as usual, numbers as $6.38(-6)$ denote 6.38×10^{-6}). This issue is related to the bounded stability region of explicit methods as already investigated in the paper [29] (see also [48,49]).

Table 3. Errors and EOC at $T = 5.0$ for the FDE (30) with $\alpha = 0.6$ and $\lambda = -10.0$.

	PI 1 Expl		PI 1 Impl.		PI 2 Impl.		PI P.C.	
h	Error	EOC	Error	EOC	Error	EOC	Error	EOC
2^{-2}	7.52(12)		6.80(−4)		5.55(−4)		5.43(21)	
2^{-3}	3.57(17)	∗∗∗∗	3.31(−4)	1.036	1.81(−4)	1.614	2.57(27)	∗∗∗∗
2^{-4}	8.14(17)	∗∗∗∗	1.63(−4)	1.020	5.95(−5)	1.609	7.87(21)	∗∗∗∗
2^{-5}	1.57(−1)	∗∗∗∗	8.11(−5)	1.011	1.95(−5)	1.606	4.22(−4)	∗∗∗∗
2^{-6}	3.99(−5)	∗∗∗∗	4.04(−5)	1.006	6.43(−6)	1.604	3.96(−5)	∗∗∗∗
2^{-7}	2.00(−5)	0.997	2.01(−5)	1.003	2.12(−6)	1.602	8.90(−6)	2.153
2^{-8}	1.00(−5)	0.998	1.01(−5)	1.002	6.98(−7)	1.602	2.43(−6)	1.873

In all the tables, we denote with EOC the estimated order of convergence obtained as $\log_2\big(E(h)/E(h/2)\big)$, with $E(h)$ the error corresponding to the step-size h.

For the next test problem, we consider the equation proposed in [25]:

$$D^\alpha_{t_0} y(t) = \frac{40320}{\Gamma(9-\alpha)} t^{8-\alpha} - 3\frac{\Gamma(5+\alpha/2)}{\Gamma(5-\alpha/2)} t^{4-\frac{\alpha}{2}} + \frac{9}{4}\Gamma(\alpha+1) + \left(\frac{3}{2}t^{\frac{\alpha}{2}} - t^4\right)^3 - [y(t)]^{\frac{3}{2}} \tag{31}$$

with the exact solution given by $y(t) = t^8 - 3t^{4+\frac{\alpha}{2}} + \frac{9}{4}t^\alpha$. This problem is surely of interest because, unlike several other problems often proposed in the literature, it does not present an artificial smooth solution, which is indeed not realistic in most of the fractional-order applications.

The right-hand side and its Jacobian (for implicit methods), together with the main parameters of the problem, are defined by means of the MATLAB lines:

```
f_fun = @(t,y,al) 40320/gamma(9-al)*t.^(8-al) - ...
  3*gamma(5+al/2)/gamma(5-al/2)*t.^(4-al/2)+9/4*gamma(al+1) + ...
  (3/2*t.^(al/2)-t.^4).^3 - y.^(3/2) ;
J_fun = @(t,y,al) -3/2.*y.^(1/2) ;
alpha = 0.5 ;
param = alpha ;
t0 = 0 ; T = 1 ;
y0 = 0 ;
```

and the results concerning errors and EOCs are reported in Table 4.

Table 4. Errors and EOC at $T = 1.0$ for the FDE (31) with $\lambda = 0.5$.

	PI 1 Expl		PI 1 Impl.		PI 2 Impl.		PI P.C.	
h	Error	EOC	Error	EOC	Error	EOC	Error	EOC
2^{-4}	8.03(−2)		7.55(−2)		3.71(−3)		3.56(−3)	
2^{-5}	3.85(−2)	1.060	3.79(−2)	0.997	1.04(−3)	1.842	6.03(−4)	2.560
2^{-6}	1.89(−2)	1.025	1.90(−2)	0.998	2.76(−4)	1.907	2.28(−4)	1.407
2^{-7}	9.40(−3)	1.009	9.48(−3)	1.000	7.19(−5)	1.941	1.04(−4)	1.135
2^{-8}	4.69(−3)	1.002	4.74(−3)	1.001	1.85(−5)	1.961	4.50(−5)	1.204
2^{-9}	2.35(−3)	1.000	2.37(−3)	1.001	4.70(−6)	1.974	1.83(−5)	1.294
2^{-10}	1.17(−3)	0.999	1.18(−3)	1.002	1.19(−6)	1.982	7.15(−6)	1.359

With the aim of showing the application to MOSs of FDEs we first consider a classical fractional-order dynamical system consisting of the nonlinear Brusselator system:

$$\begin{cases} D_{t_0}^{\alpha_1} x(t) = A - (B+1)x(t) + x(t)^2 z(t) \\ D_{t_0}^{\alpha_2} z(t) = Bx(t) - x(t)^2 z(t) \\ x(t_0) = x_0, \; z(t_0) = z_0, \end{cases} \tag{32}$$

and we perform the computation for $(\alpha_1, \alpha_2) = (0.8, 0.7)$, $(A, B) = (1.0, 3.0)$ and $(x_0, z_0) = (1.2, 2.8)$ by means of the following MATLAB lines:

```
alpha = [0.8,0.7] ;
A = 1 ; B = 3 ;
param = [ A , B ] ;
f_fun = @(t,y,par) [ ...
  par(1) - (par(2)+1)*y(1) + y(1)^2*y(2) ; ...
  par(2)*y(1) - y(1)^2*y(2) ] ;
J_fun = @(t,y,par) [ ...
  -(par(2)+1) + 2*y(1)*y(2) , y(1)^2 ; ...
  par(2) - 2*y(1)*y(2) , -y(1)^2 ] ;
t0 = 0 ; T = 100 ;
y0 = [ 1.2 ; 2.8] ;
```

After showing in Figure 2 the behavior of the solution, the errors and the EOCs are presented in Table 5.

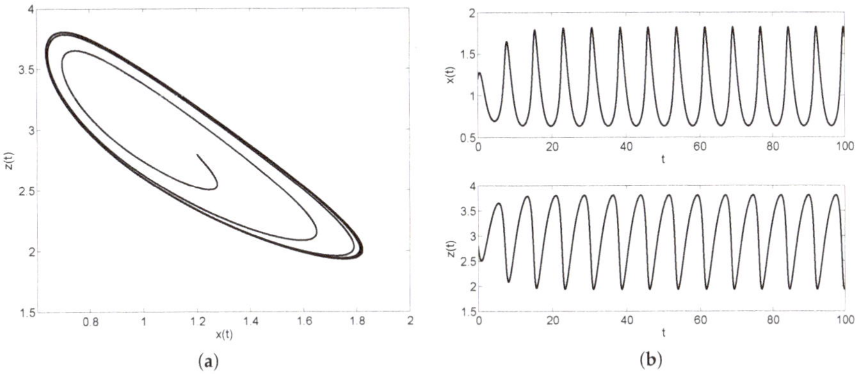

Figure 2. Behavior of the solution of the Brusselator multi-order system (MOS) (32) in the phase plane (**a**) and in the (t, x) and (t, z) planes (**b**).

Table 5. Errors and EOC at $T = 100.0$ for the Brusselator system of FDEs (32) with $(\alpha_1, \alpha_2) = (0.8, 0.7)$, $(A, B) = (1.0, 3.0)$ and $(x_0, z_0) = (1.2, 2.8)$.

	PI 1 Expl		**PI 1 Impl.**		**PI 2 Impl.**		**PI P.C.**	
h	**Error**	**EOC**	**Error**	**EOC**	**Error**	**EOC**	**Error**	**EOC**
2^{-2}	4.64(−1)		1.03(0)		4.90(−2)		1.16(0)	
2^{-3}	2.32(−1)	0.996	5.20(−1)	0.988	7.84(−3)	2.643	2.92(−1)	1.994
2^{-4}	1.22(−1)	0.926	2.25(−1)	1.211	2.85(−3)	1.460	5.80(−2)	2.333
2^{-5}	6.86(−2)	0.834	9.84(−2)	1.191	7.63(−4)	1.903	1.28(−2)	2.179
2^{-6}	3.69(−2)	0.896	4.52(−2)	1.124	1.92(−4)	1.991	3.41(−3)	1.910
2^{-7}	1.92(−2)	0.941	2.15(−2)	1.071	4.60(−5)	2.060	1.01(−3)	1.758

A more involved problem has been considered in [50] as a benchmark problem for testing software for fractional-order problems. It is defined as:

$$\begin{cases} D_{t_0}^{0.5} x(t) = \frac{1}{\sqrt{\pi}} \left(\sqrt[6]{(y(t) - 0.5)(z(t) - 0.3)} + \sqrt{t} \right) \\ D_{t_0}^{0.2} y(t) = \Gamma(2.2)(x(t) - 1) \\ D_{t_0}^{0.6} z(t) = \frac{\Gamma(2.8)}{\Gamma(2.2)} (y(t) - 0.5) \\ x(0) = 1,\ y(0) = 0.5,\ z(0) = 0.3, \end{cases} \tag{33}$$

and its analytical solution is $x(t) = t + 1$, $y(t) = t^{1.2} + 0.5$ and $z(t) = t^1.8 + 0.3$. The results of the evaluation performed on the interval $t \in [0,\ 5]$ by means of the set of MATLAB lines:

```
alpha = [0.5, 0.2, 0.6] ;
f_fun = @(t,y) [ ...
  (((y(2)-0.5).*(y(3)-0.3)).^(1/6) + sqrt(t))/sqrt(pi) ; ...
  gamma(2.2)*(y(1)-1) ; ...
  gamma(2.8)/gamma(2.2)*(y(2)-0.5) ] ;
J_fun = @(t,y) [ ...
  0 , (y(2)-0.5).^(-5/6).*(y(3)-0.3).^(1/6)/6/sqrt(pi) , ...
   (y(2)-0.5).^(1/6).*(y(3)-0.3).^(-5/6)/6/sqrt(pi) ; ...
  gamma(2.2) , 0 , 0 ; ...
  0 , gamma(2.8)/gamma(2.2) , 0 ] ;
t0 = 0 ; T = 5 ;
y0 = [ 1 ; 0.500000001 ; 0.300000001 ] ;
```

are presented in Table 6 where, as suggested in [50], relative errors are evaluated since some of the components of the system rapidly increase.

Table 6. Errors and EOC at $T = 5.0$ for the MOS of FDEs (33).

	PI 1 Expl		PI 1 Impl.		PI 2 Impl.		PI P.C.	
h	Error	EOC	Error	EOC	Error	EOC	Error	EOC
2^{-2}	2.56(−1)		1.37(−1)		7.30(−3)		7.84(−2)	
2^{-3}	1.31(−1)	0.963	7.41(−2)	0.892	3.16(−3)	1.210	3.50(−2)	1.163
2^{-4}	6.60(−2)	0.992	3.95(−2)	0.905	1.35(−3)	1.227	1.56(−2)	1.171
2^{-5}	3.29(−2)	1.005	2.09(−2)	0.918	5.72(−4)	1.238	6.89(−3)	1.175
2^{-6}	1.63(−2)	1.011	1.10(−2)	0.930	2.41(−4)	1.245	3.04(−3)	1.178
2^{-7}	8.09(−3)	1.013	5.72(−3)	0.940	1.01(−4)	1.250	1.34(−3)	1.180

The main issue with this test equation is related to the presence of a real power in the first equation, which makes the Jacobian of the given function singular at the origin, and hence, it is not possible to apply the Newton–Raphson iterative process in the same way as described in Section 3.3. There are different ways to overcome this issue; for these experiments, we have simply perturbed the initial values by a small amount $\varepsilon = 10^{-8}$; clearly, this perturbation affects the accuracy of the obtained solution, but as we can see from Table 6, where the error is evaluated with respect to the exact solution evaluated with correct initial values, the loss of accuracy is negligible.

Clearly, a comparison of the computational times with those reported in [50] is not possible due to the different features of the computers used for the experiments. Anyway, for the sake of completeness, we report here that with the step-size $h = 2^{-7}$, which provides accuracies comparable with those obtained in [50], the execution times (in seconds) of the four MATLAB routines are respectively 0.0369, 0.0923, 0.1258 and 0.0668.

We conclude this presentation with the multi-term case. As a first test equation, we consider the Bagley–Torvik equation (e.g, see [1]):

$$\begin{cases} y''(t) + aD_{t_0}^{3/2}y(t) + by(t) = f(t, y(t)), \\ y(t_0) = y_0,\ y'(t_0) = y_0^{(1)}, \end{cases} \tag{34}$$

in which, with the aim of showing the robustness of the approaches described in Section 4.2, we have replaced the standard external source $f(t)$ with a non-linear term $f(t, y(t))$ depending on the solution $y(t)$. On the interval $[0, T]$, we select the parameters $a = 2$, $b = \frac{1}{2}$, the initial conditions $y_0 = 0$ and $y_0^{(1)} = 0$ and the non-linear given function $f(t, y) = t^2 - y^{3/2}$. We first show the MATLAB code to set this problem:

```
alpha = [2 3/2 0] ;
lambda = [1 2 1/2] ;
f_fun = @(t,y) t.^2 - y.^(3/2) ;
J_fun = @(t,y) -3/2*y.^(1/2) ;
t0 = 0 ; T = 5 ;
y0 = [0 , 0 ] ;
```

and hence, the results of the numerical computation by means of the four codes:

```
[t, y] = MT_FDE_PI1_Ex(alpha,lambda,f_fun,t0,T,y0,h)
[t, y] = MT_FDE_PI1_Im(alpha,lambda,f_fun,J_fun,t0,T,y0,h)
[t, y] = MT_FDE_PI2_Im(alpha,lambda,f_fun,J_fun,t0,T,y0,h)
[t, y] = MT_FDE_PI12_PC(alpha,lambda,f_fun,t0,T,y0,h) .
```

are presented in Table 7.

Table 7. Errors and EOC at $T = 5.0$ for the multi-term FDE (34) with $(a, b) = (2.0, 0.5)$, $y(0) = 0$ and $y'(0) = 0$.

	PI 1 Expl		PI 1 Impl.		PI 2 Impl.		PI P.C.	
h	Error	EOC	Error	EOC	Error	EOC	Error	EOC
2^{-2}	3.52(−2)		8.17(−2)		2.72(−4)		8.53(−2)	
2^{-3}	2.16(−2)	0.704	3.94(−2)	1.053	7.03(−5)	1.950	2.36(−2)	1.855
2^{-4}	1.22(−2)	0.826	1.88(−2)	1.067	1.75(−5)	2.005	7.21(−3)	1.709
2^{-5}	6.58(−3)	0.888	9.00(−3)	1.063	4.30(−6)	2.026	2.36(−3)	1.609
2^{-6}	3.47(−3)	0.925	4.34(−3)	1.052	1.04(−6)	2.046	8.00(−4)	1.564
2^{-7}	1.80(−3)	0.948	2.11(−3)	1.041	2.46(−7)	2.082	2.75(−4)	1.540

Another interesting example is presented in [50] as the benchmark Problem 2 and consists of the multi-term FDE:

$$\begin{cases} y'''(t) + D_0^{5/2}y(t) + y''(t) + 4y'(t) + D_0^{1/2}y(t) + 4y(t) = 6\cos t, \\ y(0) = 1,\ y'(0) = 1,\ y''(0) = -1 \end{cases} \tag{35}$$

whose exact solution is $y(t) = \sqrt{2}\sin(t + \pi/4)$. The MATLAB lines for describing this problem on the interval [0, 100] are:

```
alpha = [3 2.5 2 1 0.5 0] ;
lambda = [1 1 1 4 1 4] ;
f_fun = @(t,y) 6*cos(t) ;
J_fun = @(t,y) 0 ;
t0 = 0 ; T = 100 ;
y0 = [1 , 1, -1 ] ;
```

and the results obtained by the four MATLAB codes for multi-term FDEs are reported in Table 8.

Table 8. Errors and EOC at $T = 100.0$ for the multi-term FDE (35).

	PI 1 Expl		PI 1 Impl.		PI 2 Impl.		PI P.C.	
h	Error	EOC	Error	EOC	Error	EOC	Error	EOC
2^{-2}	2.23(−2)		3.07(−2)		1.69(−3)		2.20(−2)	
2^{-3}	1.03(−2)	1.120	1.34(−2)	1.199	4.04(−4)	2.062	4.35(−3)	2.335
2^{-4}	4.33(−3)	1.244	6.16(−3)	1.119	9.84(−5)	2.036	1.24(−3)	1.808
2^{-5}	2.29(−3)	0.918	2.92(−3)	1.079	2.42(−5)	2.024	3.98(−4)	1.642
2^{-6}	1.20(−3)	0.934	1.40(−3)	1.055	5.97(−6)	2.018	1.34(−4)	1.575
2^{-7}	6.18(−4)	0.959	6.84(−4)	1.036	1.50(−6)	1.993	4.58(−5)	1.544

In [50], the problem of integrating Equation (35) on the very large integration interval [0, 5000] has been discussed. This challenging problem requires a remarkable computational effort, especially when high accuracy is demanded, and in Table 9, we have reported the execution times for the same step-sizes h of the previous experiments (in the second column of the table, the corresponding number N of grid-points is also indicated).

Table 9. Execution times (in seconds) for solving the multi-term FDE (35) at $T = 5000$.

h	N	PI 1 Expl	PI 1 Impl.	PI 2 Impl.	PI P.C.	$N(\log_2 N)^2$
2^{-2}	20,000	1.1224	1.8754	1.9601	2.6078	4.08×10^6
2^{-3}	40,000	2.0545	3.6506	3.8378	5.1718	9.35×10^6
2^{-4}	80,000	4.1367	7.3493	7.6927	10.4410	2.12×10^7
2^{-5}	160,000	8.3460	14.7751	15.4799	21.0344	4.78×10^7
2^{-6}	320,000	16.9107	29.6091	30.9253	42.3168	1.07×10^8
2^{-7}	640,000	34.7736	60.1003	62.8137	86.0035	2.38×10^8

We must report that integrating on $[0, 5000]$ with small step-sizes by the two methods based on the PI trapezoidal rules leads to some loss of accuracy, while the two methods based only on PI rectangular rules still continue to provide accurate results; this phenomenon, which suggests avoiding the use of PI trapezoidal rules on very large integration intervals, seems related to the accumulation of round-off errors due to the huge number of floating point operations (indeed, the same issue is not reported on smaller integration intervals); as already mentioned in Section 3.4, the number of floating-point operations is proportional to $\mathcal{O}\big(N(\log_2 N)^2\big)$ (this value is reported in the last column of Table 9), a number that becomes very high in this case.

The propagation of round-off errors for the integration of fractional-order problems on large intervals needs however to be studied in a more in-depth way; as a rule of thumb, in these cases, we just suggest to prefer PI rectangular rules to PI trapezoidal rules due to their better stability properties [29].

7. Conclusions

In this paper we have presented some of the existing methods for numerically solving systems of FDEs and we have discussed their application to multi-order systems and linear multi-term FDEs. In particular, we have focused on the efficient implementation of product integration rules and we have presented some Matlab routines by providing a tutorial guide to their use. Their application has been moreover illustrated in details by means of some examples.

Supplementary Materials: The MATLAB codes presented in the paper and listed in Table 2 are available on the software page of the author's web-site at https://www.dm.uniba.it/Members/garrappa/Software.

Acknowledgments: This work has been supported by INdAM-GNCSunder the 2017 project "Analisi Numerica per modelli descritti da operatori frazionari".

Conflicts of Interest: The author declares no conflict of interest

Abbreviations

The following abbreviations are used in this manuscript:

FDE	Fractional differential equation
ODE	Ordinary differential equation
RL	Riemann–Liouville
PI	Product integration
FLLM	Fractional linear multi-step method
LMM	Linear multi-step method
PC	Predictor-corrector
MOS	Multi-order system
FFT	Fast Fourier transform
VIE	Volterra integral equation

References

1. Kilbas, A.A.; Srivastava, H.M.; Trujillo, J.J. *Theory and Applications of Fractional Differential Equations*; North-Holland Mathematics Studies; Elsevier Science B.V.: Amsterdam, The Netherlands, 2006; Volume 204, p. 523.

2. Mainardi, F. *Fractional Calculus and Waves in Linear Viscoelasticity*; Imperial College Press: London, UK, 2010; p. 347.
3. Miller, K.S.; Ross, B. *An Introduction to the Fractional Calculus and Fractional Differential Equations*; A Wiley-Interscience Publication, John Wiley & Sons, Inc.: New York, NY, USA, 1993; p. 366.
4. Podlubny, I. *Fractional Differential Equations*; Mathematics in Science and Engineering; Academic Press Inc.: San Diego, CA, USA, 1999; Volume 198, p. 340.
5. Samko, S.G.; Kilbas, A.A.; Marichev, O.I. *Fractional Integrals and Derivatives*; Gordon and Breach Science Publishers: Yverdon, Switzerland, 1993; p. 976.
6. Gorenflo, R.; Mainardi, F. Fractional calculus: Integral and differential equations of fractional order. In *Fractals and Fractional Calculus in Continuum Mechanics (Udine, 1996)*; CISM Courses and Lectures; Springer: Vienna, Austria, 1997; Volume 378, pp. 223–276.
7. Mainardi, F.; Gorenflo, R. Time-fractional derivatives in relaxation processes: A tutorial survey. *Fract. Calc. Appl. Anal.* **2007**, *10*, 269–308.
8. Diethelm, K. *The Analysis of Fractional Differential Equations*; Lecture Notes in Mathematics; Springer: Berlin, Germany, 2010; Volume 2004, p. 247.
9. Lubich, C. Runge-Kutta theory for Volterra and Abel integral equations of the second kind. *Math. Comput.* **1983**, *41*, 87–102.
10. Aceto, L.; Magherini, C.; Novati, P. Fractional convolution quadrature based on generalized Adams methods. *Calcolo* **2014**, *51*, 441–463.
11. Garrappa, R. Stability-preserving high-order methods for multi-term fractional differential equations. *Int. J. Bifurc. Chaos Appl. Sci. Eng.* **2012**, *22*, doi:10.1142/S0218127412500733.
12. Esmaeili, S. The numerical solution of the Bagley-Torvik by exponential integrators. *Sci. Iran.* **2017**, *24*, 2941–2951.
13. Garrappa, R.; Popolizio, M. Generalized exponential time differencing methods for fractional order problems. *Comput. Math. Appl.* **2011**, *62*, 876–890.
14. Zayernouri, M.; Karniadakis, G.E. Fractional spectral collocation method. *SIAM J. Sci. Comput.* **2014**, *36*, A40–A62.
15. Zayernouri, M.; Karniadakis, G.E. Exponentially accurate spectral and spectral element methods for fractional ODEs. *J. Comput. Phys.* **2014**, *257*, 460–480.
16. Burrage, K.; Cardone, A.; D'Ambrosio, R.; Paternoster, B. Numerical solution of time fractional diffusion systems. *Appl. Numer. Math.* **2017**, *116*, 82–94.
17. Garrappa, R.; Moret, I.; Popolizio, M. On the time-fractional Schrödinger equation: Theoretical analysis and numerical solution by matrix Mittag-Leffler functions. *Comput. Math. Appl.* **2017**, *74*, 977–992.
18. Popolizio, M. A matrix approach for partial differential equations with Riesz space fractional derivatives. *Eur. Phys. J. Spec. Top.* **2013**, *222*, 1975–1985.
19. Popolizio, M. Numerical approximation of matrix functions for fractional differential equations. *Bolletino dell Unione Matematica Italiana* **2013**, *6*, 793–815.
20. Popolizio, M. Numerical Solution of Multiterm Fractional Differential Equations Using the Matrix Mittag-Leffler Functions. *Mathematics* **2018**, *6*, 7, doi:10.3390/math6010007.
21. Young, A. Approximate product-integration. *Proc. R. Soc. Lond. Ser. A* **1954**, *224*, 552–561.
22. Lambert, J.D. *Numerical Methods for Ordinary Differential Systems*; John Wiley & Sons, Ltd.: Chichester, UK, 1991; p. 293.
23. Dixon, J. On the order of the error in discretization methods for weakly singular second kind Volterra integral equations with nonsmooth solutions. *BIT* **1985**, *25*, 624–634.
24. Diethelm, K. Smoothness properties of solutions of Caputo-type fractional differential equations. *Fract. Calc. Appl. Anal.* **2007**, *10*, 151–160.
25. Diethelm, K.; Ford, N.J.; Freed, A.D. Detailed error analysis for a fractional Adams method. *Numer. Algorithms* **2004**, *36*, 31–52.
26. Garrappa, R. Trapezoidal methods for fractional differential equations: theoretical and computational aspects. *Math. Comput. Simul.* **2015**, *110*, 96–112.
27. Diethelm, K.; Ford, N.J.; Freed, A.D. A predictor-corrector approach for the numerical solution of fractional differential equations. *Nonlinear Dyn.* **2002**, *29*, 3–22.

28. Diethelm, K. Efficient solution of multi-term fractional differential equations using $P(EC)^mE$ methods. *Computing* **2003**, *71*, 305–319.
29. Garrappa, R. On linear stability of predictor-corrector algorithms for fractional differential equations. *Int. J. Comput. Math.* **2010**, *87*, 2281–2290.
30. Lubich, C. Discretized fractional calculus. *SIAM J. Math. Anal.* **1986**, *17*, 704–719.
31. Lubich, C. Convolution quadrature and discretized operational calculus. I. *Numer. Math.* **1988**, *52*, 129–145.
32. Lubich, C. Convolution quadrature and discretized operational calculus. II. *Numer. Math.* **1988**, *52*, 413–425.
33. Lubich, C. Convolution quadrature revisited. *BIT* **2004**, *44*, 503–514.
34. Wolkenfelt, P.H.M. *Linear Multistep Methods and the Construction of Quadrature Formulae for Volterra Integral and Integro-Differential Equations*; Technical Report NW 76/79; Mathematisch Centrum: Amsterdam, The Netherlands, 1979.
35. Wolkenfelt, P.H.M. The construction of reducible quadrature rules for Volterra integral and integro-differential equations. *IMA J. Numer. Anal.* **1982**, *2*, 131–152.
36. Brunner, H.; van der Houwen, P.J. *The Numerical Solution of Volterra Equations*; CWI Monographs; North-Holland Publishing Co.: Amsterdam, The Netherlands, 1986; Volume 3, p. 588.
37. Lubich, C. On the stability of linear multi-step methods for Volterra convolution equations. *IMA J. Numer. Anal.* **1983**, *3*, 439–465.
38. Diethelm, K.; Ford, J.M.; Ford, N.J.; Weilbeer, M. Pitfalls in fast numerical solvers for fractional differential equations. *J. Comput. Appl. Math.* **2006**, *186*, 482–503.
39. Deng, W. Short memory principle and a predictor-corrector approach for fractional differential equations. *J. Comput. Appl. Math.* **2007**, *206*, 174–188.
40. Schädle, A.; López-Fernández, M.A.; Lubich, C. Fast and oblivious convolution quadrature. *SIAM J. Sci. Comput.* **2006**, *28*, 421–438.
41. Aceto, L.; Magherini, C.; Novati, P. On the construction and properties of m-step methods for FDEs. *SIAM J. Sci. Comput.* **2015**, *37*, A653–A675.
42. Hairer, E.; Lubich, C.; Schlichte, M. Fast numerical solution of nonlinear Volterra convolution equations. *SIAM J. Sci. Stat. Comput.* **1985**, *6*, 532–541.
43. Hairer, E.; Lubich, C.; Schlichte, M. Fast numerical solution of weakly singular Volterra integral equations. *J. Comput. Appl. Math.* **1988**, *23*, 87–98.
44. Henrici, P. Fast Fourier methods in computational complex analysis. *SIAM Rev.* **1979**, *21*, 481–527.
45. Diethelm, K.; Luchko, Y. Numerical solution of linear multi-term initial value problems of fractional order. *J. Comput. Anal. Appl.* **2004**, *6*, 243–263.
46. Nkamnang, A. Diskretisierung von Mehrgliedrigen Abelschen Integralgleichungen und GewöHnlichen Differentialgleichungen Gebrochener Ordnung. Ph.D. Thesis, Freie Universiät Berlin, Berlin, Germany, 1998.
47. Garrappa, R. Numerical evaluation of two and three parameter Mittag-Leffler functions. *SIAM J. Numer. Anal.* **2015**, *53*, 1350–1369.
48. Garrappa, R.; Messina, E.; Vecchio, A. Effect of perturbation in the numerical solution of fractional differential equations. *Discret. Contin. Dyn. Syst. Ser. B* **2018**, doi:10.3934/dcdsb.2017188.
49. Messina, E.; Vecchio, A. Stability and boundedness of numerical approximations to Volterra integral equations. *Appl. Numer. Math.* **2017**, *116*, 230–237.
50. Xue, D.; Bai, L. Benchmark problems for Caputo fractional-order ordinary differential equations. *Fract. Calc. Appl. Anal.* **2017**, *20*, 1305–1312.

mathematics

MDPI

Article

Numerical Solution of Multiterm Fractional Differential Equations Using the Matrix Mittag–Leffler Functions

Marina Popolizio

Dipartimento di Matematica e Fisica "Ennio De Giorgi", Università del Salento, Via per Arnesano, 73100 Lecce, Italy; marina.popolizio@unisalento.it

Received: 14 December 2017; Accepted: 1 January 2018; Published: 9 January 2018

Abstract: Multiterm fractional differential equations (MTFDEs) nowadays represent a widely used tool to model many important processes, particularly for multirate systems. Their numerical solution is then a compelling subject that deserves great attention, not least because of the difficulties to apply general purpose methods for fractional differential equations (FDEs) to this case. In this paper, we first transform the MTFDEs into equivalent systems of FDEs, as done by Diethelm and Ford; in this way, the solution can be expressed in terms of Mittag–Leffler (ML) functions evaluated at matrix arguments. We then propose to compute it by resorting to the matrix approach proposed by Garrappa and Popolizio. Several numerical tests are presented that clearly show that this matrix approach is very accurate and fast, also in comparison with other numerical methods.

Keywords: fractional differential equations; multiterm differential equations; Mittag–Leffler function; matrix function; fractional calculus

1. Introduction

The use of fractional order derivatives is nowadays widespread in many fields. Indeed, the possibility to use any real order improves the modeling of several phenomena in physics, engineering and many application areas. The available literature on fractional calculus is very rich, and we cite only, among the others, [1–5]. It is a fact that the theoretical analysis of fractional differential equations (FDEs) is much more advanced than finding their numerical solution. This topic is indeed very delicate and much more difficult than finding the numerical solution of differential equations of integer order (ODEs). The introduction of effective numerical methods is recent (see, e.g., [6–13] and the books [4,14] together with the references therein). Several numerical methods for FDEs are generalizations of well-established methods for ODEs with appropriate changes. A common problem when dealing with FDEs is the loss of order with respect to the ODE case. This is mainly due to the fact that solutions of FDEs (and their fractional derivatives) are usually not smooth.

In this paper, we address the numerical solution of multiterm fractional differential equations (MTFDEs), that is, FDEs in which multiple fractional derivatives are involved. These turn out to be very helpful in many fields, particularly to model complex multirate physical processes. However, even if many numerical methods for FDEs can be extended to MTFDEs, delicate issues such as numerical stability, convergence or accuracy cannot be easily predicted in this case. Many authors have worked thoroughly on their numerical solution [15–22]. We restrict our attention to the linear case that includes important models, such as the Bagley–Torvik equation [23], the fractional oscillation equation [24] and many others.

Crucial contributions to the numerical solution of MTFDEs came from Diethelm and coauthors [4,16,17,25]. An important result therein is the equivalence between a MTFDE and a single-order system of FDEs [16]. In this paper, we propose a numerical approach that is grounded on this equivalent formulation; indeed, its solution can be expressed in terms of the Mittag–Leffler (ML) function evaluated in the coefficient matrix. The ML function is fundamental in the analysis of fractional calculus, not least because the exact solution of many FDEs can be expressed in terms of this function. However, for a long time, this has been considered only as a theoretical tool because of the lack of effective methods to numerically approximate this function. Only recently have many advances been made for the numerical evaluation of the scalar ML function [26–29]; the case of matrix arguments has since been analyzed [30,31], and finally a numerical algorithm has been accomplished, which reaches very high accuracies [32]. In this paper, we show the effectiveness of the matrix approach when solving MTFDEs, both in terms of execution time and in terms of accuracy, and also in comparison with some well-established numerical methods. The test equations we consider are well known in literature, and their exact solution is at our disposal. This is fundamental to test the reliability. However, the tests we present show an excellent accuracy, and thus the approach can be favorably applied to solve more general multiterm FDEs.

In particular, among the available numerical methods for MTFDEs, we consider the product integration (PI) rules. These nowadays represent a valuable numerical method for FDEs, although they were originally proposed for the numerical solution of Volterra integral formulas (see, e.g., [33]). Indeed, as a result of the possibility of rewriting any linear FDE as a weakly singular Volterra integral equation of second-type, a generalization of PI rules has been applied to FDEs [12,19,21,34–36]. In our numerical tests, we compare the approach we propose to the results given by methods belonging to this class.

The paper is organized as follows: In Section 2, we briefly review the main definitions of fractional calculus, and we linger on the multiterm case. We present the main theoretical tool, given by Diethelm and Ford [16,25], to transform the MTFDE into an equivalent system of FDEs. In Section 3, we address the numerical solution of this equivalent system. This essentially grounds on ML functions with matrix arguments, and we introduce the numerical method proposed by Garrappa and Popolizio [32] to compute matrix ML functions. The performance is tested in Section 4, where we give a comparison with PI methods for several tests; in the same section, a brief description of these methods is presented.

2. Fractional Differential Equations

Fractional derivatives can be introduced by means of the Riemann–Liouville (RL) definition or the Caputo definition [3]. These coincide when equipped with homogeneous initial conditions, while in the more general case, important peculiar features separate them. Both of these have been extensively analyzed and are commonly used. However, in the context of the multiterm case we discuss, the definition by Caputo is generally preferred (see the discussion in [16,17,25]). Thus for any $\alpha \in \mathbb{R}$, the fractional derivative D^α is defined as

$$D^\alpha y(t) \equiv \frac{1}{\Gamma(m-\alpha)} \int_0^t \frac{y^{(m)}(u)}{(t-u)^{\alpha+1-m}} du$$

with $\Gamma(\cdot)$ denoting Euler's gamma function and $m = \lceil \alpha \rceil$ being the smallest integer greater than or equal to α. The use of this definition allows us to use as initial conditions the values of y and its derivatives of integer order; that is, we augment the differential equation of order α with initial conditions of the form

$$y^{(k)}(0) = y_0^{(k)}, \ k = 0, 1, \ldots, m-1$$

We are interested in the numerical solution of linear MTFDEs of the form given by the equation

$$\sum_{k=0}^{n} a_k D^{\alpha_k} y(t) = f(t) \tag{1}$$

with $a_k \in \mathbb{R}$, $a_n \neq 0$ and $0 \leq \alpha_0 < \ldots < \alpha_n$. The associated initial conditions are

$$y^{(j)}(0) = y_0^{(j)}, j = 0, 1, \ldots, \lceil \alpha_n \rceil - 1$$

The MTFDE (Equation (1)) is defined as *commensurate* if the numbers $\alpha_0, \ldots, \alpha_n$ are commensurate, that is, if the quotients α_i/α_j are rational numbers for all $i, j \in \{0, \ldots, n\}$.

One of the main differences is that for MTFDEs, the RL definition would require initial conditions corresponding to each fractional derivative order that appears in the equations, while the Caputo definition merely requires the initial conditions for integer-order derivatives. Moreover, only the Caputo derivative operator has, under suitable hypotheses on continuity, the semigroup property, which is a fundamental tool to treat the multiterm case (see Theorem 1 in the following).

The analytical solution of the problem (Equation (1)) was thoroughly addressed by Gorenflo and Luchko [15], who gave its explicit expression, through ML-type functions, by using operational calculus for the Caputo fractional derivative.

The analysis of commensurate MTFDEs becomes simpler by applying an approach commonly used for ODEs. Indeed, given an ODE of order 2 or above, it can be converted to a system of first-order ODEs. Analogously, we can rewrite a commensurate MTFDE as a single-order system of FDEs, according to well-known theoretical results [16,25]. We report here the main theorem stating this equivalence (as given in [4]).

Theorem 1. *Consider the equation*

$$D^{\alpha_k} y(t) = f(t, y(t), D^{\alpha_1} y(t), D^{\alpha_2} y(t), \ldots, D^{\alpha_{k-1}} y(t)) \tag{2}$$

subject to the initial condition

$$y^{(j)}(0) = y_0^{(j)}, \quad j = 0, 1, \ldots, \lceil \alpha_k \rceil - 1$$

where $\alpha_k > \alpha_{k-1} > \ldots > \alpha_1 > 0$, $\alpha_j - \alpha_{j-1} \le 1$ for all $j = 2, 3, \ldots, k$ and $0 < \alpha_1 \le 1$. Assuming that $\alpha_j \in \mathbb{Q}$ for all $j = 1, 2, \ldots, k$, define M to be the least common multiple of the denominators of $\alpha_1, \alpha_2, \ldots, \alpha_k$, and set $\gamma = 1/M$ and $N = M\alpha_k$. Then this initial value problem is equivalent to the system of equations

$$\begin{aligned} D^\gamma y_0(t) &= y_1(t) \\ D^\gamma y_1(t) &= y_2(t) \\ &\vdots \\ D^\gamma y_{N-2}(t) &= y_{N-1}(t) \\ D^\gamma y_{N-1}(t) &= f(t, y_0(t), y_{\alpha_1/\gamma}(t), \ldots, y_{\alpha_{k-1}/\gamma}(t)) \end{aligned} \tag{3}$$

together with the initial conditions

$$y_j(0) = \begin{cases} y_0^{(j/M)} & \text{if } j/M \in \mathbb{N}_0 \\ 0 & \text{else} \end{cases}$$

in the following sense:

1. *Whenever $Y := (y_0, \ldots, y_{N-1})^T$ with $y_0 \in C^{\lceil \alpha_k \rceil}[0, b]$ for some $b > 0$ is the solution of the system given by Equation (3), the function $y := y_0$ solves the multiterm initial value problem of Equation (2).*
2. *Whenever $y \in C^{\lceil \alpha_k \rceil}[0, b]$ is a solution of the multiterm initial value problem Equation (2), the vector function $Y := (y_0, \ldots, y_{N-1})^T := (y, D^\gamma y, D^{2\gamma} y, \ldots, D^{(N-1)\gamma} y)^T$ solves the multidimensional initial value problem of Equation (3).*

The equivalence stated above can in fact also be applied to *any* multiterm equation. Indeed, Diethelm and Ford [16] showed that when the orders of the fractional derivatives are approximated by commensurate ones, the errors between the solutions of the two systems are comparable to the errors between the orders, to thus ensure the *structural stability*. In practice, because any real number can be approximated arbitrarily closely by a rational number, any MTFDE can be approximated arbitrarily closely by a commensurate one; this remark allows us to restrict our attention to the commensurate case.

3. Matrix Approach for the Solution of Linear MTFDEs

As a result of Theorem 1, the linear MTFDE (Equation (1)) can be reformulated in terms of a linear system of FDEs of the form

$$D^{\alpha}Y(t) = AY(t) + e_n f(t), \quad Y(0) = Y_0 \tag{4}$$

where $e_n = (0, 0, \ldots, 0, 1)^T \in \mathbb{R}^n$, Y_0 is composed in a suitable way on the basis of the initial values, and the coefficient matrix $A \in \mathbb{R}^{n \times n}$ is the companion matrix:

$$A = \begin{pmatrix} 0 & 1 & 0 & \ldots & 0 \\ 0 & 0 & 1 & \ldots & 0 \\ \vdots & \vdots & \vdots & \ddots & \vdots \\ 0 & 0 & 0 & \ldots & 1 \\ -\frac{a_0}{a_n} & -\frac{a_1}{a_n} & -\frac{a_2}{a_n} & \ldots & -\frac{a_{n-1}}{a_n} \end{pmatrix}$$

Once the solution $Y(t)$ of Equation (4) has been computed, we keep only its first component, which, as stated in Theorem 1, corresponds to the (scalar) solution of the MTFDE (Equation (1)).

It is well known that the exact solution $Y(t)$ of Equation (4) is

$$Y(t) = E_{\alpha,1}(t^{\alpha}A)Y_0 + \int_0^t (t-\tau)^{\alpha-1} E_{\alpha,\alpha}((t-\tau)^{\alpha}A)e_n f(\tau)d\tau \tag{5}$$

with $E_{\alpha,\beta}$ denoting the ML function that, for complex parameters α and β, with $\Re(\alpha) > 0$, is defined by means of the series

$$E_{\alpha,\beta}(z) = \sum_{j=0}^{\infty} \frac{z^j}{\Gamma(\alpha j + \beta)}, \quad z \in \mathbb{C} \tag{6}$$

$E_{\alpha,\beta}(z)$ is clearly a generalization of the exponential function to which it reduces when $\alpha = \beta = 1$, as for $j \in \mathbb{N}$, it is $\Gamma(j+1) = j!$.

The solution is even simpler when $f(t)$ can be expressed in terms of powers (possibly of non-integer order) of t, as no integral is required. Namely, if

$$f(t) = \sum_{\nu \in \mathcal{G}} c_{\nu} t^{\nu}$$

with $\mathcal{G} \subset \{\nu \in \mathbb{R}, \nu > -1\}$ being an index set and c_{ν} being some real coefficients, then, as a result of well-known theoretical results (see, e.g., [3]), the true solution can be written as

$$Y(t) = E_{\alpha,1}(t^{\alpha}A)Y_0 + \sum_{\nu \in \mathcal{G}} c_{\nu}\Gamma(\nu+1)t^{\alpha+\nu}E_{\alpha,\alpha+\nu+1}(t^{\alpha}A)e_n \tag{7}$$

Source terms of this kind are common in applications, often resulting from the approximation of given signals. We thus present the numerical solutions of test cases of this form. The more general situation described in Equation (5) requires exactly the same matrix approach we describe, combined with some quadrature methods.

It is decisive that the solution $Y(t)$ written as Equation (7) essentially relies on t and not on $[0, t]$; instead, any numerical method for FDEs has to work on the whole interval. This is a fundamental difference and a great strength of the matrix approach, particularly when integration is required for large values of t.

The solution as given in Equation (7) essentially requires the computation of the ML function with the matrix argument $t^{\alpha}A$. The numerical computation of matrix functions is an extensively studied topic that has deserved great attention during the last decades (we refer to Higham [37] for a complete treatise and a full list of references). Only recent studies have considered matrix arguments for the ML function (see, e.g., [30–32,38,39]). To be precise, even the numerical scalar case has received poor attention, and only recently has Garrappa [29] developed a powerful Matlab routine (`ml.m`, available on Matlab website) that gives very accurate results for arguments all over the complex plane. Thereafter, an effective numerical procedure for the matrix case has been proposed [32], and we apply this for our computations. Interestingly, the computation of the matrix ML function turns out to be very accurate, practically close to the machine precision. The resulting matrix approach to approximate Equation (7) thus proves to be very accurate and favorable also for its computational costs; indeed, once the matrices $E_{\alpha,\beta}(t^{\alpha}A)$ have been computed, only few additional matrix–vector products and vector sums are needed to obtain the solution.

4. Numerical Solution of FDEs

In this section, we present several numerical tests in order to show the effectiveness of the matrix approach discussed in Section 3. Moreover we compare it with PI rules. We briefly recall here the main features of these methods, while we refer to the related references listed in the introduction and to [21] for a complete treatise on their use for the numerical solution of MTFDEs.

4.1. Product Integration Rules

To introduce PI rules, we first need to recall some basics of fractional calculus (we refer to [4] for details). The RL integral is defined as

$$J^{\alpha}y(t) = \frac{1}{\Gamma(\alpha)}\int_0^t (t-u)^{\alpha-1}y(u)du \tag{8}$$

We let $T_j[y;0]$ denote the Taylor polynomial of degree j for the function $y(t)$ centered at the point 0; then the following relation holds [4]:

$$J^{\alpha}D_0^{\alpha}y(t) = y(t) - T_{m-1}[y;0](t) \quad \text{for } m = \lceil\alpha\rceil \tag{9}$$

Thus, if we compute the RL integral J^{α_n} for both terms in Equation (1), after some manipulations, we obtain

$$y(t) = T_{m_n-1}[y;0](t) - \sum_{i=1}^{n-1}\frac{a_i}{a_n}J^{\alpha_n-\alpha_i}[y(t) - T_{m_i-1}[y;0](t)] + \frac{1}{a_n}J^{\alpha_n}f(t)$$

PI rules approximate the expression above by applying suitable quadrature formulas to the involved RL integrals. More precisely, they use, on a grid with a constant step-size $h>0$, piecewise interpolant polynomials of fixed order, and the resulting integrals are evaluated exactly. For our numerical tests, we deal with PI rules using polynomials of first and second order.

4.2. Numerical Tests

The focus of our numerical tests is on both the accuracy and the computation time. For the former task, we consider MTFDEs whose exact solution is known. Moreover, as stated in Section 3, the key strength of the matrix approach arises when integration on long time intervals $[0,T]$ is required. Indeed, PI rules need to work on the whole interval, while the matrix approach computes directly at T. For this reason, we consider fairly large values of T, namely, $T = 10, 50, 100$; we stress that these values are fair, particularly when the interest is on the qualitative description of physical models.

For the matrix approach, the routine by Garrappa is used (https://www.dm.uniba.it/Members/garrappa/ml_matrix), while for PI rules, we follow the codes as described in [21].

All the experiments have been carried out in Matlab version 8.3.0.532 (R2014a) on a Intel(R) Core(TM) running at 2.50 GHz under Windows 10. The execution time results from an average of 20 runs.

The step-size h for PI methods was selected to obtain good accuracies and a reasonable execution time. The labels PI_1 and PI_2 in the following denote the first- and second-order PI methods, respectively.

4.2.1. Bagley–Torvik Equation

Fractional calculus is a common theoretical tool in the field of rheology. Here, an important model is given by Bagley and Torvik [23], who introduced the equation

$$a_1 D^2 y(t) + a_2 D^{3/2} y(t) + a_3 y(t) = f(t) \tag{10}$$

to describe the motion of a rigid plate immersed in a Newtonian fluid. The coefficients a_i, $i = 1,2,3$ are real, $a_1 \neq 0$, and homogeneous initial conditions are considered to ensure the unicity of the solution. Diethelm and Ford [25] deeply analyzed the numerical solution of Equation (10); the approach they propose starts by rewriting it as a system of four FDEs of order $1/2$ by following Theorem 1. Thus our matrix approach works with 4×4 matrices. For PI methods, we use $h = 2^{-3}$.

We consider $a_i = 1$, $i = 1,2,3$ and $f(t) = t + 1$ to let the exact solution be $y(t) = t + 1$.

From Figure 1, we may appreciate the great accuracy of the matrix approach. PI_2 is also very accurate, particularly for small values of t, while PI_1 improves as t becomes larger but never reaches accuracies comparable to other methods. In terms of execution time, Table 1 shows that PI_1 and the matrix approach are similar for $T = 10$, while the latter is more than 10 times faster for $T = 50, 100$.

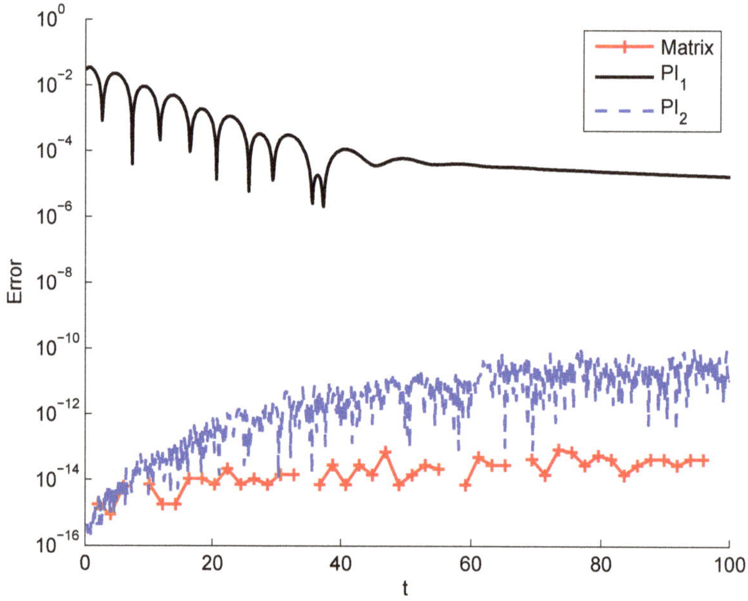

Figure 1. Error for the matrix approach and the product integration (PI) methods compared with the exact solution of the Bagley–Torvik equation. $h = 2^{-3}$ for PI.

Table 1. Comparison of execution time for the matrix approach and the product integration (PI) methods for computing the approximate solution at T.

T	Matrix	PI_1	PI_2
10	5.4×10^{-3}	$\mathbf{3.0 \times 10^{-3}}$	6.5×10^{-3}
50	$\mathbf{4.9 \times 10^{-3}}$	1.6×10^{-2}	3.1×10^{-2}
100	$\mathbf{4.0 \times 10^{-3}}$	3.2×10^{-2}	6.2×10^{-2}

4.2.2. The Basset Problem

The Basset equation is a classical model for the dynamics of a sphere immersed in an incompressible viscous fluid and subject to an elastic force. It was first considered by Basset in 1888, who interpreted the particle velocity relative to the fluid in terms of a fractional derivative of order 1/2; this term is now called the Basset force. There are many studies on the Basset equation (see, e.g., [1,2,40–42]). Moreover, a generalization of the Basset force with fractional derivatives of order $0 < \alpha < 1$ is given by Mainardi [2].

The general equation is

$$\left[\frac{d}{dt} + \delta^{1-\alpha}\frac{d^\alpha}{dt^\alpha} + 1\right] V(t) = 1, \;\; V(0^+) = 0, \;\; 0 < \alpha < 1 \tag{11}$$

The true solution of this problem is

$$V(t) = 1 - M(t;\alpha)$$

with $M(t;\alpha)$ to be determined by some inversion method. In particular, when $\alpha = p/q$, with p, q being integer numbers and $p < q$, then

$$M(t;p/q) = \sum_{k=1}^{q} C_k E_{1/q}(a_k t^{1/q})$$

where $a_1, a_2, \ldots, a_q$ are the zeros of the polynomial $P(x) = x^q + \delta^{(1-p/q)} x^p + 1$, $A_k^{-1} = \prod_{j=1}^{q}(a_k - a_j)$, $j \neq k$ and $C_k = -A_k/a_k$ [40,41].

For our numerical tests, we considered $\alpha = 1/2$, which results in the equivalent system of dimension 2×2. We considered $\delta = 9/(1+2\chi)$ with $\chi = 10$, as in [2], and $h = 2^{-4}$ for PI. For $\alpha = 0.25$ and for other values of χ, the results were very similar, and thus we omit their presentation.

Figure 2 reveals the excellent accuracy reached by the matrix approach. For this test, the method PI_2 was more accurate than PI_1, but it never caught the matrix approach. Moreover, the execution time was much longer, even 100 times greater than that of the matrix approach, as Table 2 reports.

Table 2. Comparison of execution time for the matrix approach and the product integration (PI) methods for computing the approximate solution at T for $\alpha = 0.5$.

T	Matrix	PI_1	PI_2
10	$\mathbf{1.6 \times 10^{-3}}$	6.3×10^{-3}	1.2×10^{-2}
50	$\mathbf{1.6 \times 10^{-3}}$	3.2×10^{-2}	6.2×10^{-2}
100	$\mathbf{1.1 \times 10^{-3}}$	6.1×10^{-2}	1.3×10^{-1}

4.2.3. Fractional Oscillation Equation

The fractional oscillation equation is one of the basic examples to show the generalization of standard differential equations to fractional equations [2]. Its general form is

$$D^\alpha y(t) + \omega^\alpha y(t) = 0 \tag{12}$$

for $1 < \alpha \leq 2$ and $t \geq 0$. Two initial conditions are needed to uniquely solve this equation, namely, $y(0)$ and $y'(0)$. If we assume that the latter is zero, the exact solution can be expressed in terms of the ML function:

$$y(t) = y(0)E_{\alpha,1}(-(\omega t)^{\alpha})$$

As in [43], we consider $\alpha = 1.95$, $\omega = 1$, $y(0) = 1$, and $y'(0) = 0$. With this choice, the system corresponding to Equation (12) has dimension $N = 39$. The step-size for PI is $h = 2^{-7}$.

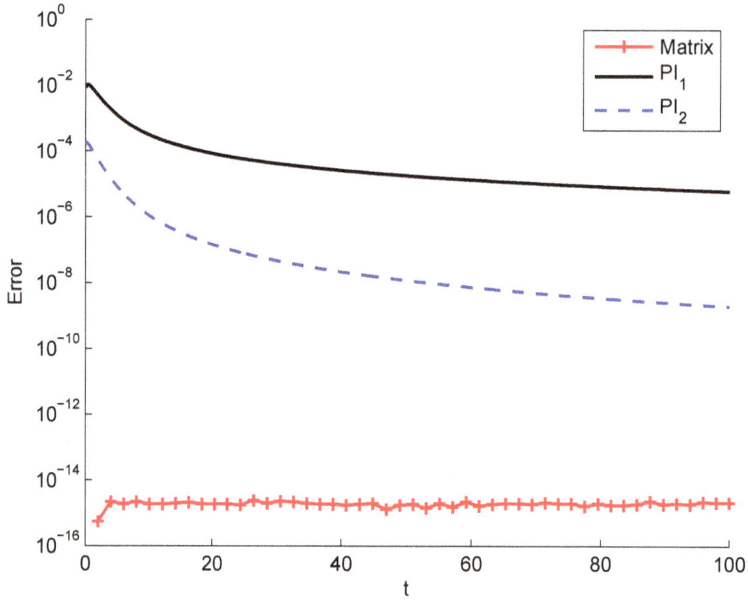

Figure 2. Error for the matrix approach and the product integration (PI) methods compared with the exact solution for $\alpha = 0.5$ for the Basset equation.

Figure 3 shows the error between the exact solution and the approximations given by the matrix approach and the PI methods. As for the previous examples, the matrix approach was definitely the most accurate and the fastest, as shown in Table 3. Indeed, for the largest T value, the execution time for the PI_2 method was more than 30 times longer than that for the matrix approach.

Table 3. Comparison of execution time for the matrix approach and the product integration (PI) methods for computing the approximate solution at T.

T	Matrix	PI_1	PI_2
10	$\mathbf{2.5 \times 10^{-2}}$	4.4×10^{-2}	8.9×10^{-2}
50	$\mathbf{2.7 \times 10^{-2}}$	2.2×10^{-1}	4.5×10^{-1}
100	$\mathbf{2.7 \times 10^{-2}}$	4.4×10^{-1}	9.0×10^{-1}

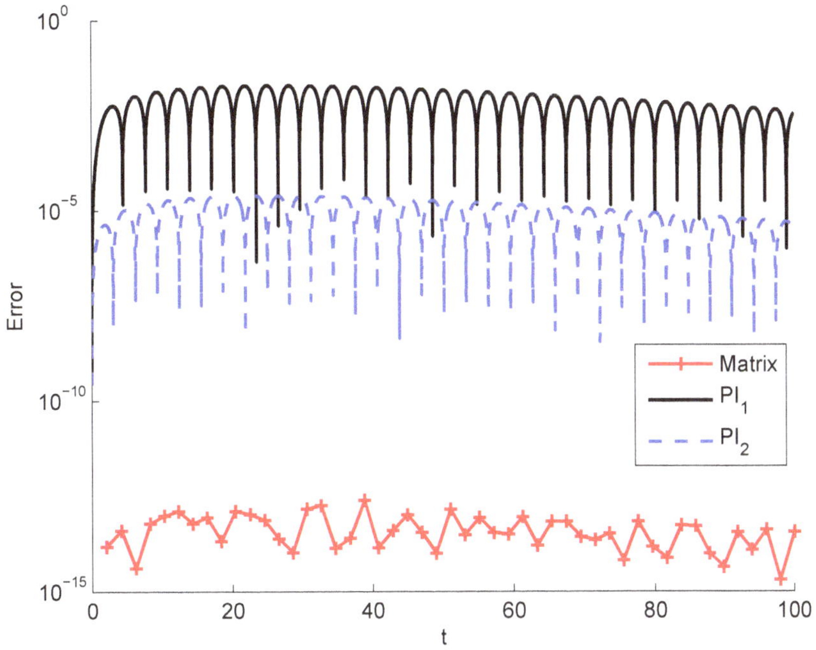

Figure 3. Error for the matrix approach and the product integration (PI) methods compared with the exact solution for the fractional oscillation equation.

4.2.4. Academic Examples

Example 1

We consider the MTFDE proposed in [44]:

$$D^2y(t) + D^{1/2}y(t) + y(t) = t^3 + 6t + \frac{3.2t^{2.5}}{\Gamma(0.5)}, \ \ y(0) = 0, \ \ y'(0) = 0 \tag{13}$$

whose exact solution is $y(t) = t^3$.

The resulting matrix has dimension 2×2. We use $h = 2^{-3}$ for PI methods.

Figure 4 shows that the matrix approach was the most accurate for this test, even if PI_2 reached very high accuracies. However the former was the cheapest in terms of execution time, as Table 4 reports.

Table 4. Comparison of execution time for the matrix approach and the product integration (PI) methods for computing the approximate solution at T of Equation (13).

T	Matrix	PI_1	PI_2
10	8.4×10^{-3}	$\mathbf{4.1 \times 10^{-3}}$	7.3×10^{-3}
50	$\mathbf{9.5 \times 10^{-3}}$	2.4×10^{-2}	5.1×10^{-2}
100	$\mathbf{5.9 \times 10^{-3}}$	3.7×10^{-2}	7.3×10^{-2}

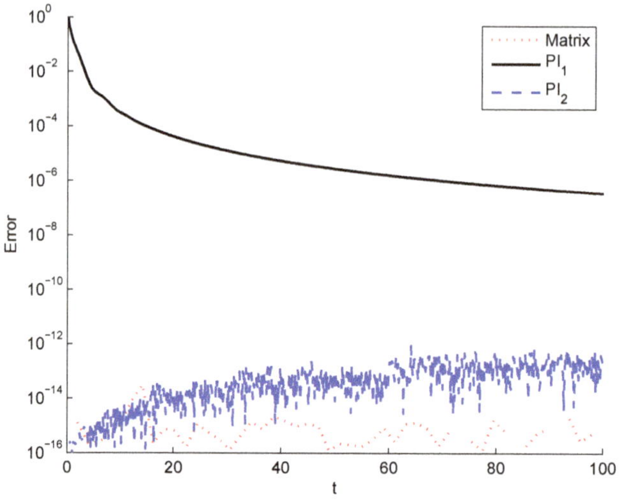

Figure 4. Relative error for the matrix approach and the product integration (PI) methods compared with the exact solution for Equation (13).

Example 2

We consider a test problem proposed in [19]. The FDE to solve is

$$D^{\alpha}y(t) + y(t) = t^m + \frac{m!}{\Gamma(m+1-\alpha)}t^{m-\alpha}, \;\; y(0) = 0, \;\; 0 < \alpha < 1, \;\; m \in \mathbb{N} \tag{14}$$

whose exact solution is $y(t) = t^m$.

We use, as in [19], the values $m = 4$ and $\alpha = 0.4$. For the matrix approach, matrices of dimension 2×2 come into play, while for the PI methods, we use $h = 2^{-7}$.

The comments for the previous tests apply also for this case: from Figure 5 we may appreciate the excellent accuracy of the matrix approach while Table 5 reveals an even greater gap in terms of execution time, with a difference of 3 orders of magnitude between the matrix approach and PI_2.

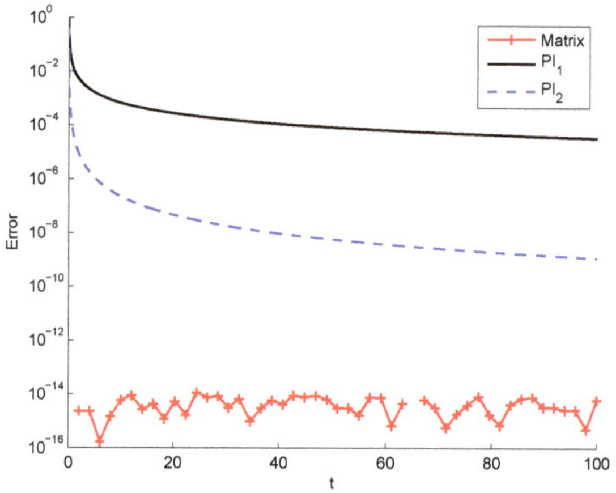

Figure 5. Relative error for the matrix approach and the product integration (PI) methods compared with the exact solution for $\alpha = 0.4$ for Equation (14).

Table 5. Comparison of execution time for the matrix approach and the product integration (PI) methods for computing the approximate solution at T for $\alpha = 0.4$.

T	Matrix	PI_1	PI_2
10	$\mathbf{5.2 \times 10^{-3}}$	8.1×10^{-2}	1.6×10^{-1}
50	$\mathbf{4.7 \times 10^{-3}}$	4.0×10^{-1}	8.0×10^{-1}
100	$\mathbf{3.2 \times 10^{-3}}$	7.9×10^{-1}	1.6×10^{0}

Acknowledgments: This work has been supported by INdAM-GNCS under the 2017 project "Analisi Numerica per modelli descritti da operatori frazionari".

Conflicts of Interest: The author declares no conflict of interest.

References

1. Gorenflo, R.; Mainardi, F. Fractional calculus: Integral and differential equations of fractional order. In *Fractals and Fractional Calculus in Continuum Mechanics (Udine, 1996)*; Carpinteri, A., Mainardi, F., Eds.; Springer: Vienna, Austria, 1997; Volume 378, pp. 223–276.
2. Mainardi, F. Fractional Calculus: Some Basic Problems in Continuum and Statistical Mechanics. In *Fractals and Fractional Calculus in Continuum Mechanics (Udine, 1996)*; Carpinteri, A., Mainardi, F., Eds.; Springer: Vienna, Austria, 1997; pp. 291–348.
3. Podlubny, I. Fractional differential equations. In *Mathematics in Science and Engineering*; Academic Press Inc.: San Diego, CA, USA, 1999; Volume 198.
4. Diethelm, K. *The Analysis of Fractional Differential Equations*; Lecture Notes in Mathematics; Springer-Verlag: Berlin, Germany, 2010; Volume 2004.
5. Mainardi, F. *Fractional Calculus and Waves in Linear Viscoelasticity*; Imperial College Press: London, UK, 2010.
6. Lubich, C. Runge-Kutta theory for Volterra and Abel integral equations of the second kind. *Math. Comput.* **1983**, *41*, 87–102.
7. Lubich, C. Fractional linear multistep methods for Abel-Volterra integral equations of the second kind. *Math. Comput.* **1985**, *45*, 463–469.
8. Lubich, C. Discretized fractional calculus. *SIAM J. Math. Anal.* **1986**, *17*, 704–719.
9. Diethelm, K.; Ford, J.M.; Ford, N.J.; Weilbeer, M. Pitfalls in fast numerical solvers for fractional differential equations. *J. Comput. Appl. Math.* **2006**, *186*, 482–503.
10. Galeone, L.; Garrappa, R. Explicit methods for fractional differential equations and their stability properties. *J. Comput. Appl. Math.* **2009**, *228*, 548–560.
11. Garrappa, R. On some explicit Adams multistep methods for fractional differential equations. *J. Comput. Appl. Math.* **2009**, *229*, 392–399.
12. Garrappa, R.; Popolizio, M. On accurate product integration rules for linear fractional differential equations. *J. Comput. Appl. Math.* **2011**, *235*, 1085–1097.
13. Garrappa, R. Trapezoidal methods for fractional differential equations: Theoretical and computational aspects. *Math. Comput. Simul.* **2015**, *110*, 96–112.
14. Baleanu, D.; Diethelm, K.; Scalas, E.; Trujillo, J. Series on Complexity, Nonlinearity and Chaos. In *Fractional Calculus: Models and Numerical Methods*; World Scientific Publ: Singapore, 2012.
15. Luchko, Y.; Gorenflo, R. An operational method for solving fractional differential equations with the Caputo derivatives. *Acta Math. Vietnam.* **1999**, *24*, 207–233.
16. Diethelm, K.; Ford, N.J. Multi-order Fractional Differential Equations and Their Numerical Solution. *Appl. Math. Comput.* **2004**, *154*, 621–640.
17. Diethelm, K.; Luchko, Y. Numerical solution of linear multi-term initial value problems of fractional order. *J. Comput. Anal. Appl.* **2004**, *6*, 243–263.
18. Luchko, Y. Initial-boundary-value problems for the generalized multi-term time-fractional diffusion equation. *J. Math. Anal. Appl.* **2011**, *374*, 538–548.

19. Garrappa, R. Stability-preserving high-order methods for multiterm fractional differential equations. *Int. J. Bifurc. Chaos Appl. Sci. Eng.* **2012**, *22*, 1250073.
20. Al-Refai, M.; Luchko, Y. Maximum principle for the multi-term time-fractional diffusion equations with the Riemann-Liouville fractional derivatives. *Appl. Math. Comput.* **2015**, *257*, 40–51.
21. Garrappa, R. Numerical Solution of Fractional Differential Equations: A Survey and a Software Tutorial. **2017**, submitted.
22. Esmaeili, S. The numerical solution of the Bagley-Torvik equation by exponential integrators. *Sci. Iran.* **2017**, 2941–2951.
23. Torvik, P.; Bagley, R. On the appearance of the fractional derivative in the behavior of real materials. *J. Appl. Mech. Trans. ASME* **1984**, *51*, 294–298.
24. Mainardi, F. Fractional relaxation-oscillation and fractional diffusion-wave phenomena. *Chaos Solitons Fractals* **1996**, *7*, 1461–1477.
25. Diethelm, K.; Ford, J. Numerical Solution of the Bagley-Torvik Equation. *BIT Numer. Math.* **2002**, *42*, 490–507.
26. Haubold, H.J.; Mathai, A.M.; Saxena, R.K. Mittag-Leffler functions and their applications. *J. Appl. Math.* **2011**, 298628.
27. Gorenflo, R.; Kilbas, A.A.; Mainardi, F.; Rogosin, S. *Mittag-Leffler functions. Theory and Applications*; Springer Monographs in Mathematics; Springer: Berlin, Germany, 2014.
28. Garrappa, R.; Popolizio, M. Evaluation of generalized Mittag–Leffler functions on the real line. *Adv. Comput. Math.* **2013**, *39*, 205–225.
29. Garrappa, R. Numerical evaluation of two and three parameter Mittag-Leffler functions. *SIAM J. Numer. Anal.* **2015**, *53*, 1350–1369.
30. Moret, I.; Novati, P. On the Convergence of Krylov Subspace Methods for Matrix Mittag–Leffler Functions. *SIAM J. Numer. Anal.* **2011**, *49*, 2144–2164.
31. Moret, I. A note on Krylov methods for fractional evolution problems. *Numer. Funct. Anal. Optim.* **2013**, *34*, 539–556.
32. Garrappa, R.; Popolizio, M. Computing the matrix Mittag-Leffler function with applications to fractional calculus. submitted.
33. Cameron, R.F.; McKee, S. Product integration methods for second-kind Abel integral equations. *J. Comput. Appl. Math.* **1984**, *11*, 1–10.
34. Diethelm, K.; Freed, A.D. The FracPECE subroutine for the numerical solution of differential equations of fractional order. Forschung und wissenschaftliches Rechnen 1998; Heinzel, S., Plesser, T., Eds.; Gessellschaft für wissenschaftliche Datenverarbeitung: Goettingen, Germany, 1999; pp. 57–71.
35. Diethelm, K.; Ford, N.J.; Freed, A.D. Detailed error analysis for a fractional Adams method. *Numer. Algorithms* **2004**, *36*, 31–52.
36. Garrappa, R.; Messina, E.; Vecchio, A. Effect of perturbation in the numerical solution of fractional differential equations. *Discret. Contin. Dyn. Syst. Ser. B* **2017**, doi:10.3934/dcdsb.2017188.
37. Higham, N.J. *Functions of matrices*; Society for Industrial and Applied Mathematics (SIAM): Philadelphia, PA, USA, 2008.
38. Garrappa, R.; Moret, I.; Popolizio, M. Solving the time-fractional Schrödinger equation by Krylov projection methods. *J. Comput. Phys.* **2015**, *293*, 115–134.
39. Garrappa, R.; Moret, I.; Popolizio, M. On the time-fractional Schrödinger equation: Theoretical analysis and numerical solution by matrix Mittag-Leffler functions. *Comput. Math. Appl.* **2017**, *5*, 977–992.
40. Mainardi, F.; Pironi, P.; Tampieri, F. A numerical approach to the generalized Basset problem for a sphere accelerating in a viscous fluid. In Proceedings of the 3rd Annual Conference of the Computational Fluid Dynamics Society of Canada, Banff, AB, Canada, 25–27 June 1995; Volume II, pp. 105–112.
41. Mainardi, F.; Pironi, P.; Tampieri, F. On a generalization of the Basset problem via fractional calculus. In Proceedings of the 15th Canadian Congress of Applied Mechanics, Victoria, BC, Canada, 28 May–2 June 1995; Volume II, pp. 836–837.
42. Garrappa, R.; Mainardi, F.; Maione, G. Models of dielectric relaxation based on completely monotone functions. *Fract. Calc. Appl. Anal.* **2016**, *19*, 1105–1160.

43. Edwards, J.T.; Ford, N.J.; Simpson, A.C. The numerical solution of linear multi-term fractional differential equations: Systems of equations. *J. Comput. Appl. Math.* **2002**, *148*, 401–418.
44. Ford, N.J.; Connolly, J.A. Systems-based decomposition schemes for the approximate solution of multi-term fractional differential equations. *J. Comput. Appl. Math.* **2009**, *229*, 382–391.

mathematics

MDPI

Article

Best Approximation of the Fractional Semi-Derivative Operator by Exponential Series

Vladimir D. Zakharchenko and Ilya G. Kovalenko *

Institute of Mathematics and Information Technologies, Volgograd State University, Volgograd 400062, Russia; zvd@volsu.ru

* Correspondence: ilya.g.kovalenko@gmail.com; Tel.: +7-8442-460812

Received: 30 November 2017; Accepted: 12 January 2018; Published: 16 January 2018

Abstract: A significant reduction in the time required to obtain an estimate of the mean frequency of the spectrum of Doppler signals when seeking to measure the instantaneous velocity of dangerous near-Earth cosmic objects (NEO) is an important task being developed to counter the threat from asteroids. Spectral analysis methods have shown that the coordinate of the centroid of the Doppler signal spectrum can be found by using operations in the time domain without spectral processing. At the same time, an increase in the speed of resolving the algorithm for estimating the mean frequency of the spectrum is achieved by using fractional differentiation without spectral processing. Thus, an accurate estimate of location of the centroid for the spectrum of Doppler signals can be obtained in the time domain as the signal arrives. This paper considers the implementation of a fractional-differentiating filter of the order of ½ by a set of automation astatic transfer elements, which greatly simplifies practical implementation. Real technical devices have the ultimate time delay, albeit small in comparison with the duration of the signal. As a result, the real filter will process the signal with some error. In accordance with this, this paper introduces and uses the concept of a "pre-derivative" of ½ of magnitude. An optimal algorithm for realizing the structure of the filter is proposed based on the criterion of minimum mean square error. Relations are obtained for the quadrature coefficients that determine the structure of the filter.

Keywords: near-earth objects; potentially hazardous asteroids; radial velocity determination; real-time measurements; differential filter; fractional derivative; approximate integration; exponential series

1. Introduction

A number of technical problems require the estimation of the parameters of the spectrum of broadband signals in real time. For example, in radar, the central frequency of the Doppler signal spectrum determines the radial component of the speed of the moving object and is necessary for prediction of its trajectory. In particular, this issue refers to Doppler radar systems that measure the spectral characteristics of beats of a probing and reflected signal. Accurate and rapid measurement of these parameters, in this case, is a necessary condition for effective operation of such systems. However, in order to estimate the parameters of the reflected signal spectrum, the use of spectral analysis does not always satisfy operational requirements. For example, the spectrum of the Doppler signal can be broadened by the motion of individual points of the object with different velocities. In the case when the various points of the object forming the reflected signal move with different velocities (for example, when the object rotates), the reflected signal can have a wide spectrum of Doppler frequencies corresponding to the spectrum of the reflecting point velocities on its surface. In such a case, the centroid of the energy spectrum of the Doppler signal is used as the Doppler frequency in the radar, which remains a stable characteristic corresponding to the motion of the center of mass of the moving object. A complex movement of an object can take place if the object maneuvers. For uncontrolled objects, in particular, asteroids, tumbling (rotation with respect to all three axes with incommensurable angular velocities) is a common phenomenon when moving along a ballistic trajectory [1]. This reduces the accuracy of estimating the instantaneous speed of the object and complicates the forecast of its trajectory. The priority of mankind to counteract the threat from near-Earth cosmic objects (NEO's) explains the importance of solving the subtask of ultra-fast and ultra-precise determination of the speed of a fast asteroid [2].

A fast and accurate forecast of the trajectory of a dangerous cosmic object is necessary for calculating the point of encounter for weapons to destroy such an object. Forecasting the trajectory is undertaken by measuring the speed of the object based on the analysis of the Doppler frequency shift of the reflected radar signal.

Increasing the accuracy and speed of measuring the frequency of the Doppler signal is a prerequisite for the effective operation of space-based radar systems. At the final stage of the weapons' approach to the object, the parameters of the asteroid's motion must be measured quickly. At velocities of cosmic objects of several tens of km/s, the error price multiplies many times, and the requirements for accuracy must be maximized.

It was shown in [3,4] that an estimate of the velocity of the "average" point of a fast moving asteroid can be reduced to a calculation of the derivative of order ½ from the so-called "Doppler signal", that is, the beat signal of the reflected and probing signals. On the basis of this, an algorithm for the operation of a fractional-differentiating filter (FDF) of order ½ has been proposed, which allows for the estimation of the centroid of the spectrum of the input signal to be obtained in real time without spectral processing. It has also been shown that the use of such a fractal-processing algorithm makes it possible to increase the rate by which an estimate can be obtained by up to six orders of magnitude [4].

Of course, modern signal-processing tools should be designed as digital, given obvious progress in the development of digital devices, and questions about the analog methods. The characteristics of digital filters are absolutely stable, while the parameters of analog filters are unstable and depend on external conditions.

Nevertheless, models of analog devices usually precede the construction of digital models. Analog algorithms are more obvious than digital ones. The analog filter in this problem is interesting in that it allows for real-time signal processing, i.e., at the rate of input of the input signal.

In the present paper, using an analog FDF as an example, we consider its possible use by simple technical means in the form of a set of operations realized by simple elements of—astatic transfer elements of the first order. We show that, mathematically, the structure of the FDF can be represented by a linear approximation of the kernel of the fractional derivative operator of the order of ½ by decaying exponents. We propose an optimal realization of the structure of the FDF by the criterion of minimum mean square error.

The problem considered in the present paper belongs to that class of problems pertaining to the recognition of signal singularities. We note the rapid growth of the development of algorithms using fractional differentiation, specifically in the field of artificial intelligence. These include, for example, image-processing algorithms [5], the identification of image features [6–8], and computer [9–11] and experimental [12,13] implementation of fractal proportional-integral-derivative (PID) controllers for industrial control systems.

2. The Possibility of Using Fractal Signal Processing

As a signal $x(t)$, we consider a class of Lebesgue measurable functions $L_p(\Omega)$ on a set $\Omega = [T_0, T]$, $-\infty \leq T_0 < T \leq \infty$. In other words, for functions $x(t)$, the inequality $\int\limits_{\Omega} |x(t)|^p dt < \infty$ must hold, where $1 \leq p < \infty$. The function $x(t)$ corresponds to the spectral density of the signal amplitude (spectrum) $S(\omega) = \mathbf{F}[x(t)]$. Here, the symbol $\mathbf{F}$ denotes the Fourier transform of the signal.

In the theory of signals, the method of moments [14] (pp. 135–136) is widely used to estimate the parameters of the spectrum, according to which the mean frequency and effective width of the signal spectrum on the positive half-axis of frequencies is defined as the centroid ω_0 and twice the radius of inertia (twice the standard deviation) $2\Delta\omega$ of the energy spectrum of the signal $E(\omega) = |S(\omega)|^2$ (see Figure 1):

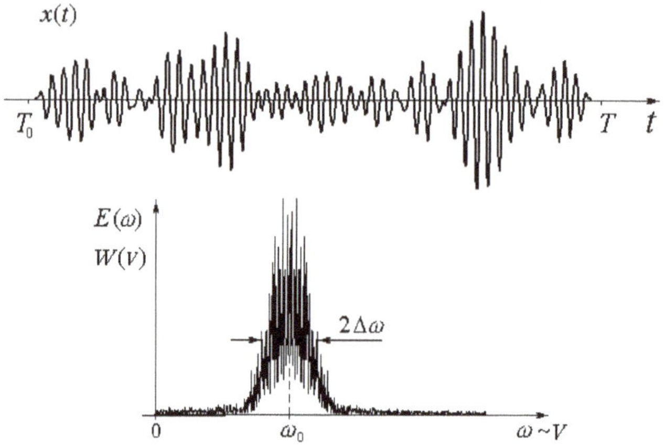

Figure 1. Determination of the parameters of the spectrum ω_0 and $\Delta\omega$ of the Doppler signal $x(t)$ by the method of moments. $W(V)$ is the velocity distribution of the reflecting elements of the object's surface.

In the case of expanding the spectrum of Doppler frequencies, it can be agreed that the "average" speed of the object is estimated from the position of the centroid of the spectrum [15] whose definition is given by the ratio:

$$\omega_0 = \frac{\int\limits_0^{\infty} \omega\,|S(\omega)|_2 d\omega}{\int\limits_0^{\infty} |S(\omega)|_2 d\omega}. \tag{1}$$

Calculations using Equation (1) require a sufficiently long time for the observation of the signal and its processing, since processing uses the procedure for calculating the spectrum $S(\omega)$, which from the computational point of view is quite laborious. With the number N of signal $x(t)$ samples, the calculation of the spectrum in the most economical case will require the number $N \log_2 N$ of multiplications by the fast Fourier transform (FFT) technique. To this, we additionally require $2N$ multiplications in order to calculate the integrals in the numerator and denominator in the fraction (1). In addition, in order to carry out these calculations it is necessary to first obtain the entire signal (all N samples), which introduces an additional delay in obtaining an estimate. Moreover, such an algorithm requires a fairly large amount of working memory for N counts of the spectrum. In this case, it is advantageous to have a flow algorithm that allows for the calculation of the numerator and denominator in Equation (1) in real time as the signal arrives. This condition is satisfied by the Duhamel integral calculation algorithm, which can be realized using a transversal filter that processes a signal in the time domain in real time. This allows us to calculate the numerator and denominator in Equation (1) at the end of the signal.

It has been shown in [3,4] that the right-hand side of Equation (1) can be represented by the ratio of the square of the fractional derivative to the square of the signal's norm (energy), which allows for us to find the centroid of the spectrum without using the Fourier transform:

$$\omega_0 = \frac{\int_{T_0}^{T} \left[D^{1/2}x(t)\right]^2 dt}{\int_{T_0}^{T} [x(t)]^2 dt}. \tag{2}$$

Here $D^{1/2}$ is the operator of the fractional Liouville derivative of half-integer order [16]

$$D^{1/2}x(t) = \frac{1}{\sqrt{\pi}} \frac{d}{dt} \int_{-\infty}^{t} \frac{x(t')\,dt'}{\sqrt{t-t'}}, \tag{3}$$

which we further refer to as the semiderivative operator in accordance with the terminology that has been accepted in the literature. The denominator of the fraction in Equation (2) is derived using Parseval's equation [14] (p. 140) in the transition from the frequency domain to the time domain:

$$\int_{0}^{\infty} |S(\omega)|^2 d\omega = \frac{1}{2}\int_{-\infty}^{\infty} |S(\omega)|^2 d\omega = \frac{1}{2}\|S(\omega)\|^2 = \pi\|x(t)\|^2 = \pi\int_{-T_0}^{T} [x(t)]^2 dt. \tag{4}$$

Similarly, on the basis of Parseval's equality, the numerator of the fraction in (2) is also transformed:

$$\int_{0}^{\infty} \omega|S(\omega)|^2 d\omega = \frac{1}{2}\int_{-\infty}^{\infty} \left|\sqrt{i\omega}S(\omega)\right|^2 d\omega = \frac{1}{2}\|\sqrt{i\omega}S(\omega)\|^2 = \pi\|D^{1/2}x(t)\|^2 = \pi\int_{-T_0}^{T} \left[D^{1/2}x(t)\right]^2 dt, \tag{5}$$

where i is the imaginary unit. From this, it follows that in order to calculate the centroid of the spectrum, it is necessary to calculate the square of the norm of the fractional derivative of the order of ½ and the signal energy.

The semi-derivative operator must satisfy the following condition:

$$D^{1/2}x(t) = \mathbf{F}^{-1}\left\{\sqrt{i\omega}\,\mathbf{F}[x(t)]\right\}. \tag{6}$$

Equation (6) shows that the fractional derivative can be calculated using linear FDF. The action of the fractional derivative operator according to Equation (6) is interpreted as the convolution of the investigated signal with the following function

$$h(t) = \mathbf{F}^{-1}\left[(i\omega)^{1/2}\right], \tag{7}$$

which is usually called the pulse response of the FDF. The pulse response (7) is identical to the kernel of the fractional differentiation operator.

It follows from Equation (2) that for a limited observation interval of signal ($T - T_0 < \infty$), the processing time $x(t)$ can be substantially reduced if we calculate the fractional derivative of the order of ½ in real time [3,4]. Estimating the location of the centroid of the spectrum (2) can be undertaken without spectral processing as the reflected signal reaches the detector and is obtained at the end of the observation interval T. For example, in the problem of the rapid estimation of the radial velocity of a potentially dangerous near-Earth asteroid, the real computational speed can be increased by a million times [4].

In fact, for a space object with a diameter of ~100 m and a rotation period of ~10 min, the width of the velocity spectrum amounts to ~1 m/s, which corresponds to the width of the Doppler spectrum in the X band ~50 Hz. In this case, the average Doppler frequency at a speed of 30 km/s is ~2 MHz; the signal sampling frequency should be appropriately selected as >4 MHz ($\Delta T \sim 0.1$ MS).

For high-speed resolution (~0.01 m/s or higher), it is necessary to measure the Doppler frequency with an accuracy of ~0.5 Hz. With a Doppler duration of ~0.1 s, this will require processing $N \sim 10^6$ of the signal samples.

A standard calculation using the FFT method will require $N_{flop} \geq 2 \cdot 10^7$ multiplication operations. For a processor with a performance of 1 Mflops, this will take about 20 s, while the signal delay in the filter for the filter of order $M = 100$ is only $T_m = M\Delta T = 10^{-5}$ s. Thus, the gain in speed will be more than six orders. Time costs will be comparable only when using the FFT method on processors with a performance of several teraflops, which is economically unprofitable and technically difficult to implement.

Thus, it is important to find the optimal algorithm for an approximate calculation of the fractional derivative of the signal (maximum accuracy with a minimum of computational operation). If fractional differentiation is performed on analog devices, then the algorithm simultaneously determines the design of real devices. The present paper is devoted to the search for the quadrature of the highest degree of accuracy for finding the approximate value of the fractional semi-derivative.

We emphasize that we do not construct an n-point quadrature of the highest degree of accuracy for calculating the semi-derivative in which the semi-derivative is approximated by the weighted sum of the values of the integrand at different nodes. Such a problem was solved long ago [17,18]. Instead, we find the best approximation precisely for the kernel $h(t)$ of the fractional derivative operator, and the calculation of the semi-derivative by the quadrature formula deduced by us reduces to the sum of integrals contained as weight-function damped exponentials.

3. The Pulse Response Characteristic of the Liouville Semi-Derivative

By virtue of the linearity property of the fractional differentiation operator and the difference kernel, expression (6) can be written as the Duhamel integral [14] (p. 142),

$$D^{1/2}x(t) = \int_{T_0}^{t} x(t')h(t-t')dt', \quad T_0 \leq t \leq T, \tag{8}$$

and this means that the fractional differentiation operator can be realized as a linear stationary FDF with specified characteristics. The general requirements imposed on the filter are expressed by the conditions (9) and (10).

Among the mandatory requirements is the condition of causality of the impulse response

$$h(t<0) \equiv 0, \tag{9}$$

which has already been taken into account in its definition in Equation (8).

The specific form of the pulse response is determined by the method of specifying a fractional derivative. In [3,4], the definition of a fractional Riemann-Liouville derivative was used. Here we confine ourselves to its particular limiting case, the Liouville derivative [16]

The Liouville derivative (3) has more habitual and evident properties, such as the identity vanishing of the derivative of a constant. The latter, as follows from Equation (8), implies another condition on the pulse response

$$\int_{0}^{\infty} h(t)\,dt = 0. \tag{10}$$

An explicit expression for the pulse response of the Liouville operator of a semi-derivative is presented in [3,4]:

$$h(t) = \lim_{\varepsilon \to 0} \frac{1}{\sqrt{\pi}} \left[\frac{\delta(t)}{\sqrt{t+\varepsilon}} - \frac{1}{2(t+\varepsilon)^{3/2}} \sigma(t) \right]. \tag{11}$$

Here $\delta(t)$ is the Dirac function, and $\sigma(t)$ is the Heaviside function. Since this expression was originally derived in an article in a hard-to-get edition [19], we reiterate its derivation below.

Definition 1. *Let $\varepsilon \geq 0$. Then the left-sided Liouville pre-derivative $D_\varepsilon^{1/2}$ of order ½ is defined by*

$$D_\varepsilon^{1/2} x(t) = \frac{1}{\sqrt{\pi}} \frac{d}{dt} \int_{-\infty}^{t} \frac{x(t')\,dt'}{\sqrt{t+\varepsilon-t'}}. \tag{12}$$

We call the real quantity $\varepsilon \geq 0$ the parameter of the pre-derivative.

Definition 2. *The Liouville fractional derivative of order ½ (see Equation (3)) is then defined as the limit of sequence of the Liouville pre-derivatives:*

$$D^{1/2} = \lim_{\varepsilon \to 0} D_\varepsilon^{1/2}. \tag{13}$$

Theorem 1. *The pulse response of the Liouville operator of a semi-derivative (13) is given by Equation (11).*

Proof. Differentiating the integral with a variable upper limit on the rhs of Equation (12) with respect to t,

$$\begin{aligned} D_\varepsilon^{1/2} x(t) &= \frac{1}{\sqrt{\pi}} \frac{d}{dt} \int_{-\infty}^{t} \frac{x(t')\,dt'}{\sqrt{t+\varepsilon-t'}} = \frac{1}{\sqrt{\pi}} \left[\frac{x(t)}{\sqrt{\varepsilon}} - \frac{1}{2} \int_{-\infty}^{t} \frac{x(t')\,dt'}{(t+\varepsilon-t')^{3/2}} \right] \\ &= \frac{1}{\sqrt{\pi}} \left[\int_{-\infty}^{t+0} \frac{x(t')\delta(t-t')\,dt'}{\sqrt{\varepsilon}} - \frac{1}{2} \int_{-\infty}^{t+0} \frac{x(t')}{(t+\varepsilon-t')^{3/2}}\,dt' \right], \end{aligned} \tag{14}$$

and using the definition (6), we find an expression for the pulse response of the pre-derivative in the form

$$h_\varepsilon(t) = \frac{1}{\sqrt{\pi}} \left[\frac{\delta(t)}{\sqrt{\varepsilon}} - \frac{\sigma(t)}{2(t+\varepsilon)^{3/2}} \right]. \tag{15}$$

Passing to the limit $\varepsilon \to 0$, we confirm that Equation (11) is satisfied.

4. Approximation of Operator Kernel of Fractional Pre-Derivative by Superposition of Damped Exponentials

For the practical implementation of analog FDF, it is desirable to use the simplest technical means. These are physical elements with pulse characteristics in the form of damped exponentials that are the first-order astatic transfer elements [20]. The time of the pulse response decay τ from the physical point of view corresponds to the memory time of the physical element.

We construct the best approximation of the impulse response of a FDF of the order of ½ by a linear combination of damped exponentials. We will take into account the fact that, in the practical implementation of differentiating or integrating filters, one always has to have finite systems in which there is a finite delay time ε. The magnitude of the time delay ε is sufficiently small and is determined by the signal delay in the input circuits of the analog device. Usually it is a fraction of a microsecond, which is commensurate with the clock interval of digital signal processing processors, and is small when compared to the time of signal processing by the filter. As digital technology develops, the value of ε will only decrease.

Our task is to approximate the component of expression (15), which has the form of a damped power function,

$$F(\xi) = \frac{1}{(\xi + 1)^{3/2}}, \xi \equiv \frac{t}{\varepsilon}, \tag{16}$$

by a set of damped exponentials on the interval $\xi = [0, \infty)$.

The direct Laplace transform of a function $f(p)$,

$$F(\xi) = \int_0^\infty e^{-\xi p} f(p)\, dp, \tag{17}$$

is the expansion of a function $F(\xi)$ over an infinite set (continuum) of exponentials. The problem is reduced to finding a quadrature formula containing a finite number of exponentials for an approximate presentation of $F(\xi)$,

$$F(\xi) = \widetilde{F}_N(\xi) + R \equiv \sum_{k=1}^{N} b_k e^{-\xi a_k} + R, \tag{18}$$

instead of the true representation (the continuum of exponentials). Here, R is the remainder term, b_k, the weight coefficients, and a_k, the nodes of the approximating quadrature formula.

Let us change a variable. We introduce the function:

$$\varphi(p) = \int_0^p f(p')\, dp'. \tag{19}$$

Then, (17) can be rewritten in the form:

$$F(\xi) = \int_0^{\varphi_{\max}} e^{-\xi p(\varphi)} d\varphi, \tag{20}$$

where $p(\varphi)$ is the inverse of (19).

From the condition for the existence of the Laplace integral,

$$\int_0^\infty \left| f(p') \right| dp' < \infty, \tag{21}$$

and from definition (19), it follows that,

$$\varphi_{\max} \equiv \varphi(\infty) < \infty, \tag{22}$$

and the variable φ itself is defined on the interval $[0,\ \varphi_{\max}]$.

The maximum accuracy when approximating the integral (20) by a quadrature of type (18), as we know, can be achieved using the Gaussian quadrature of the highest degree of accuracy [21]. To achieve maximum accuracy, when integrating the integral over a finite interval with a fixed number of nodes N, the quadrature formula follows φ_k as nodes, where $a_k = p(\varphi_k)$; we take the nodes φ_k of the Legendre polynomials, and b_k as weight coefficients and some functional combinations of them. In this case, of course, it is necessary to transform the segment $[-1,\ 1]$ on which the Legendre polynomials are defined into a segment $[0,\ \varphi_{\max}]$. The nodes and weight coefficients of the Gauss–Legendre quadrature formulas can be found in the reference books [22].

We note that the inverse Laplace transform,

$$f(p) = \frac{1}{2\pi i} \int_{c-i\infty}^{c+i\infty} e^{\xi p} F(\xi)\, d\xi, \tag{23}$$

and the relation resulting from (19),

$$f(p) = \frac{d\varphi(p)}{dp}, \tag{24}$$

allow for us to establish a connection between φ and F if (23) is integrated over p:

$$\varphi(p) = \frac{1}{2\pi i} \int_{c-i\infty}^{c+i\infty} \frac{e^{\xi p}}{\xi} F(\xi)\, d\xi. \tag{25}$$

For $F(\xi)$ in a particular form (16), the Laplace transform is known:

$$f(p) = \frac{p^{1/2}}{\Gamma(3/2)} e^{-p}. \tag{26}$$

Here, $\Gamma(x)$ is the gamma function. Integrating (24), we find:

$$\varphi(p) = \int_0^p \frac{(p')^{1/2}}{\Gamma(3/2)} e^{-p'}\, dp' = \frac{\sqrt{\pi}\operatorname{erf}(\sqrt{p}) - 2\sqrt{p}e^{-p}}{2\Gamma(3/2)} = \operatorname{erf}(\sqrt{p}) - \frac{2}{\sqrt{\pi}}\sqrt{p}e^{-p}. \tag{27}$$

The function $\varphi(p)$ that is found increases monotonically from zero to the maximum value:

$$\varphi_{\max} = 1. \tag{28}$$

Unfortunately, the function $p(\varphi)$ according to (27), although continuous on the interval $[0, 1]$, is non-smooth. When $\varphi \to 0$ it behaves like $p \sim (3/4)^{2/3}\pi^{1/3}\varphi^{2/3}$, so $p(\varphi)$ does not even have the first derivative. According to the standard estimate [22],

$$\left| \int_0^1 e^{-\xi p(\varphi)} d\varphi - \sum_{k=1}^{N} b_k e^{-\xi a_k} \right| \leq \frac{(N!)^4}{(2N+1)((2N)!)^3} M_{2N}, \tag{29}$$

where M_{2N} is the majorant estimate of the absolute value of the derivative,

$$\left| \frac{d^{2N}}{d\varphi^{2N}} e^{-\xi p(\varphi)} \right| \leq M_{2N}, \tag{30}$$

the Gauss-Legendre quadrature will not show rapid convergence with an increasing number of nodes.

5. Test Results

In numerical calculations, a harmonic signal with a constant amplitude and phase was considered as a model. The accuracy of the approximation was estimated by the mean-square error:

$$\widetilde{R}_N = \sqrt{\int_0^\infty \left[F(\xi) - \widetilde{F}_N(\xi)\right]^2 d\xi \Big/ \int_0^\infty |F(\xi)|^2 d\xi}\,. \tag{31}$$

Figure 2 demonstrates on a logarithmic scale the approximated function $F(\xi)$ (upper curve), and approximating functions $\widetilde{F}_N(\xi)$, successively approaching $F(\xi)$ at N = 6, 10, 14.

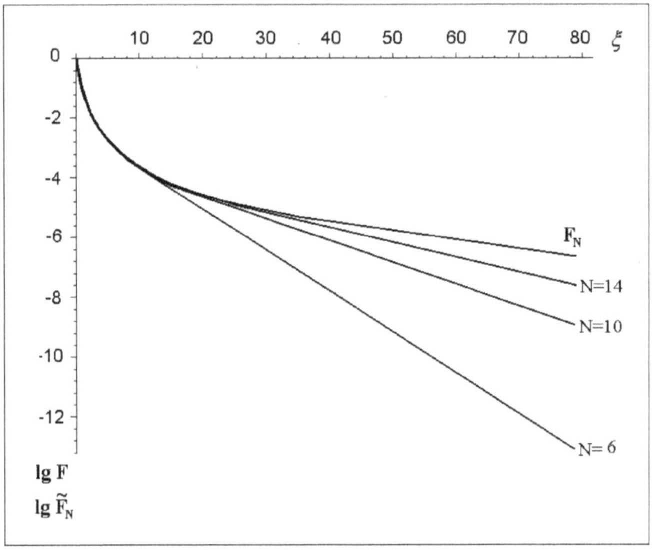

Figure 2. Dependence of the approximated function $F(\xi)$ and approximating functions $\widetilde{F}_N(\xi)$.

The calculations show that $\widetilde{R}_N$ decreases with growth N, but its tendency to zero is relatively slow because the exponents decrease faster than a polynomial of any degree. The same calculations show that to achieve acceptable accuracy in radio engineering of 1% (i.e., $\widetilde{R}_N = 0.01$), it is necessary to take N = 18 elements. It should be borne in mind that the elements making up a real filter include errors due to the imperfection of their manufacture. In this case, in practice the error introduced by each element will be accumulated separately and, therefore, it is not profitable to build a filter from such a large number of elements (N = 18).

Consider the optimization of the number of elements N in more detail. Let each element be characterized by a relative error p in the manufacture of the device. It is natural to expect the total error introduced by N elements to be defined as a statistical error proportional to the root of the number of elements. Then, the total error of the FDF approximation will be:

$$r_N = \sqrt{\int_0^\infty [F(\xi) - \widetilde{F}_N(\xi)]^2 d\xi \Big/ \int_0^\infty |F(\xi)|^2 d\xi} + p\sqrt{N}. \tag{32}$$

Figure 3 presents the graphs of the total error r_N. If the permissible error of the device is p = 0.01, then, as follows from the figure, the minimum of the total error as a function of the number of elements is reached approximately at N = 12. The minimum is thus flat, and this means that when decreasing the number of elements, the error will change insignificantly. Hence it follows that for almost the same accuracy of approximation it is economically more advantageous to use 6–8 elements to construct the FDF.

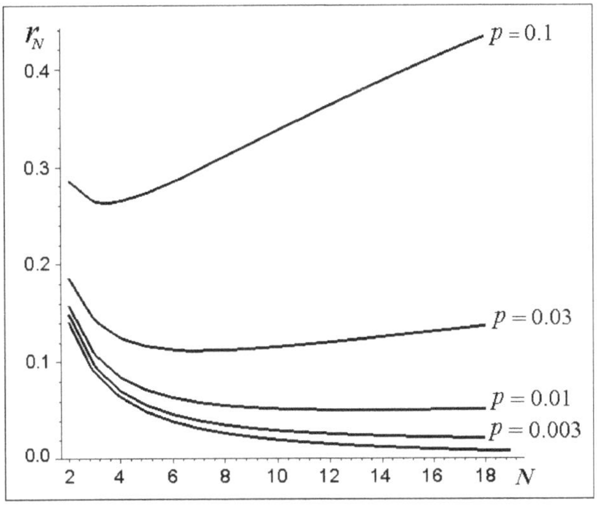

Figure 3. Graphs of functions of total errors r_N. The lower curve corresponds to the rootmean-square error $\widetilde{R}_N$. From the bottom up are the curves corresponding to the errors of the device p = 0.003, 0.01, 0.03, 0.1.

6. Conclusions

The radial velocity of fast asteroids that pose a danger to the Earth (speeds of tens of kilometers per second) can be accurately measured by means of radar by estimating the centroid of the Doppler signal spectrum.

An increase in the speed of the algorithm used for estimating the average frequency of the spectrum of the Doppler signals in the time domain as the signal is received can be achieved by using fractional differentiation without spectral processing. Thus, an accurate estimate of the centroid of the signal spectrum can be obtained almost immediately after the arrival of the signal.

The approximation of the impulse response of an analog FDF by a quadrature of the highest degree of accuracy is constructed in this paper. The kernel of the fractional pre-derivative operator is approximated by the sum of the damped exponentials. The calculation is made for FDF, as composed of the first-order astatic elements that are common in engineering, which greatly simplifies practical implementation. Relations are obtained for the quadrature coefficients that determine the structure of the FDF. The number of approximation nodes corresponds to the number of elements making up the filter. There is convergence: with the increasing number of nodes used, the accuracy of the approximation can be made arbitrarily high. The accuracy of the procedure for approximate fractional differentiation, based on the use of an approximating quadrature on the example of a function, is proved using the analytic expression of the Liouville semi-derivative for which it is known (the harmonic signal).

The problem of optimizing the analog FDF circuit, composed of first-order astatic elements, has been solved. In order to limit the number of elements used, and in order to achieve the specified accuracy, an optimal number of elements is found for a given error of each. It is shown that to achieve a precision characteristic in engineering of 99%, the optimal scheme is from $N = 8 \div 16$ elements.

7. Patents

The method for estimating the mean frequency of Doppler signals reflected by fast maneuvering targets in real time using fractional differentiation operations is protected by the Russian Federation Patent [Zakharchenko V.D. Method for estimating the mean frequency of broadband Doppler signals—RF Patent No. 2114440, 27.06.98].

Acknowledgments: This study was supported by the grant 15-47-02438-r-povolzhie_a from the Russian Foundation for Basic Research.

Author Contributions: Vladimir D. Zakharchenko suggested an idea of fractional differential analog filter composed of 1st-order lag elements; Ilya G. Kovalenko constructed an exponential series expansion; both co-wrote the paper.

Conflicts of Interest: The authors declare no conflict of interest.

References

1. Hatch, P.; Wiegert, P.A. On the rotation rates and axis ratios of the smallest known near-Earth asteroids—The archetypes of the Asteroid Redirect Mission targets. *Planet. Space Sci.* **2015**, *111*, 100–104. [CrossRef]
2. National Near-Earth Object Preparedness Strategy (December 2016). Available online: https://www.nasa.gov/sites/default/files/atoms/files/national_near-earth_object_preparedness_strategy_tagged.pdf (accessed on 29 November 2017).
3. Zakharchenko, V.D.; Kovalenko, I.G. Estimation of radial velocity of objects using fractional differentiation of Doppler radar signal. In Proceedings of the 23rd Int. Crimean Conference "Microwave & Telecommunication Technology" (CriMiCo'2013), Sevastopol, Ukraine, 9–13 September 2013; Weber Publishing: Sevastopol, Ukraine, 2013; Volume 2, pp. 1120–1121.
4. Zakharchenko, V.D.; Kovalenko, I.G. On protecting the planet against cosmic attack: ultrafast real-time estimate of the asteroid's radial velocity. *Acta Astronaut.* **2014**, *98*, 158–162. [CrossRef]
5. Khanna, S.; Chandrasekaran, V. Fractional derivative filter for image contrast enhancement with order prediction. In Proceedings of the IET Conference on Image Processing (IPR 2012), London, UK, 3–4 July 2012; Curran Associates, Inc.: Red Hook, NY, USA, 2012; pp. 73–78. [CrossRef]
6. Mathieu, B.; Melchior, P.; Oustaloup, A.; Ceyral, C. Fractional differentiation for edge detection. *Signal Process.* **2003**, *83*, 2421–2432. [CrossRef]
7. Pu, Y.; Wang, W.-X.; Zhou, J.-L.; Wang, Y.-Y.; Jia, H.-D. Fractional Differential Approach to Detecting Textural Features of Digital Image and Its Fractional Differential Filter Implementation. *Sci. China Ser. F Inf. Sci.* **2008**, *51*, 1319–1339. [CrossRef]
8. Xu, X.; Dai, F.; Long, J.; Guo, W. Fractional Differentiation-based Image Feature Extraction. *Int. J. Signal Proc. Image Proc. Pattern Recogn.* **2014**, *7*, 51–64. [CrossRef]
9. Luo, Y.; Chen, Y.Q. Fractional order [proportional derivative] controller for a class of fractional order systems. *Automatica* **2009**, *45*, 2446–2450. [CrossRef]
10. Biswas, A.; Das, S.; Abraham, A.; Dasgupta, S. Design of fractional-order $PI^{\lambda}D^{\mu}$ controllers with an improved differential evolution. *Eng. Appl. Artif. Intell.* **2009**, *22*, 343–350. [CrossRef]
11. Luo, Y.; Chen, Y.Q.; Wang, C.Y.; Pi, Y.G. Tuning fractional order proportional integral controllers for fractional order systems. *J. Process Control* **2010**, *20*, 823–831. [CrossRef]
12. Monje, C.A.; Vinagre, B.M.; Feliu, V.; Chen, Y.Q. Tuning and auto-tuning of fractional order controllers for industry applications. *Control Eng. Pract.* **2008**, *16*, 798–812. [CrossRef]
13. Luo, Y.; Chen, Y.Q.; Pi, Y. Experimental study of fractional order proportional derivative controller synthesis for fractional order systems. *Mechatronics* **2011**, *21*, 204–214. [CrossRef]
14. Franks, L.E. *Signal Theory*; Prentice-Hall: Upper Saddle River, NJ, USA, 1969.
15. Gonorovskii, I.S. *Radio Circuits and Signals*, 5th ed.; Drofa: Moscow, Russia, 2006; pp. 69–73. ISBN 5-7107-7985-7. (In Russian)
16. Samko, S.; Kilbas, A.; Marichev, O. *Fractional Integrals and Derivatives. Theory and Application*; Gordon & Breach Sci. Publishers: Philadelphia, PA, USA, 1993; p. 95. ISBN 2881248640.
17. Lether, F.G. Quadrature rules for approximating semi-integrals and semiderivatives. *J. Comput. Appl. Math.* **1987**, *17*, 115–129. [CrossRef]
18. Acharya, M.; Mohapatra, S.N.; Acharya, B.P. On Numerical Evaluation of Fractional Integrals. *Appl. Math. Sci.* **2011**, *5*, 1401–1407.
19. Zakharchenko, V.D. Mean frequency of Doppler signals estimate by the method of fractional differentiation. *Phys. Wave Process. Radiotech. Syst.* **1999**, *2*, 39–41. (In Russian)

20. Unbehauen, H. Description of Continuous Linear Time-Invariant Systems in Frequency Domain. In *Control Systems, Robotics, and Automation*; Unbehauen, H., Ed.; EOLSS Publishers Co. Ltd.: Abu Dhabi, UAE, 2009; Volume 1, pp. 164–189. ISBN 1848265905.
21. Hildebrand, F.B. *Introduction to Numerical Analysis*, 2nd ed.; McGraw-Hill: New York, NY, USA, 1974; pp. 390–394. ISBN 0-07-028761-9.
22. Kahaner, D.; Moler, C.; Nash, S. *Numerical Methods and Software*; Prentice Hall: Bergen County, NJ, USA, 1989; p. 146. ISBN 978-0-13-627258-8.

mathematics

MDPI

Article

Storage and Dissipation of Energy in Prabhakar Viscoelasticity

Ivano Colombaro [1], Andrea Giusti [2,3,4,*] and Silvia Vitali [2]

1 Department of Information and Communication Technologies, Universitat Pompeu Fabra, C/Roc Boronat 138, 08018 Barcelona, Spain; ivano.colombaro@upf.edu
2 Department of Physics & Astronomy, University of Bologna, Via Irnerio 46, 40126 Bologna, Italy; silvia.vitali4@unibo.it
3 I.N.F.N., Sezione di Bologna, via B. Pichat 6/2, I-40127 Bologna, Italy
4 Arnold Sommerfeld Center, Ludwig-Maximilians-Universität, Theresienstraße 37, 80333 München, Germany
* Correspondence: agiusti@bo.infn.it; Tel.: +49-(0)89-2180-4241

Received: 12 December 2017; Accepted: 14 January 2018; Published: 23 January 2018

Abstract: In this paper, after a brief review of the physical notion of quality factor in viscoelasticity, we present a complete discussion of the attenuation processes emerging in the Maxwell–Prabhakar model, recently developed by Giusti and Colombaro. Then, taking profit of some illuminating plots, we discuss some potential connections between the presented model and the modern mathematical modelling of seismic processes.

Keywords: Prabhakar viscoelasticity; Q-factor; fractional calculus; Mittag-Leffler functions; Prabhakar function; Integral transforms

1. Introduction

The linear theory of viscoelasticity, despite its apparent simplicity, keeps having a striking impact in geophysics, theoretical mechanics and biophysics; see, for example, [1–7]. Besides, fractional calculus [8–10] has proven itself to be one of the fundamental languages for describing processes involving memory effects, like the ones that are typically featured by viscoelastic systems. Concerning the latter, it is also worth remarking on the pivotal role of the notion of complete monotonicity, which was first (implicitly) hinted at by Gross [11] in 1953, and then brought to light by Molinari [12], in 1973. These seminal studies were then followed by many other authors; see, for example, [10,13,14].

A simple generalization of the well-known fractional Maxwell model of linear viscoelasticity was first introduced by Giusti and Colombaro in [15]. In this paper the authors provide an extension of the classical model by replacing the Caputo fractional derivative with the Prabhakar one in the constitutive equation. Concretely, if we denote with σ, ε, the stress and the strain for a given system, respectively, and we further assume that these functions are both causal such that $\sigma, \varepsilon \in AC^1(0, +\infty)$, then the constitutive equation of the Maxwell–Prabhakar model [15] reads

$$\sigma(t) + a\, {}^C\mathbf{D}^{\gamma}_{\alpha,\beta,\Omega}\, \sigma(t) = b\, {}^C\mathbf{D}^{\gamma}_{\alpha,\beta,\Omega}\, \varepsilon(t)\,, \tag{1}$$

where a and b are two suitable real constants and provided that $\alpha, \beta, \gamma, \Omega \in \mathbb{R}$, $\alpha > 0$ and $0 < \beta < 1$.

Here, ${}^C\mathbf{D}^{\gamma}_{\alpha,\beta,\Omega}$ represents the regularized Prabhakar derivative [16], which is defined by

$${}^C_a\mathbf{D}^{\gamma}_{\alpha,\beta,\Omega}\, f(t) = {}_a\mathbf{E}^{-\gamma}_{\alpha,m-\beta,\Omega}\, f^{(m)}(t), \tag{2}$$

where

$$ {}_a\mathbf{E}^{\rho}_{\mu,\nu,\lambda} f(t) = \int_a^t (t-\tau)^{\nu-1} E^{\rho}_{\mu,\nu} \left[\lambda \,(t-\tau)^{\mu}\right] f(\tau)\, d\tau $$

denotes the Prabhakar fractional integral [17] and $E^{\rho}_{\mu,\nu}(t)$ represents the Prabhakar function [18–21]. Furthermore, it is important to stress that the function $t^{\nu-1}\, E^{\rho}_{\mu,\nu}(-t^{\mu})$, for $t > 0$, is locally integrable and completely monotone provided that $0 < \mu \leq 1$ and $0 < \mu\, \rho \leq \nu \leq 1$; see, for example, [22,23].

It is important to remark that the Prabhakar fractional calculus has been attracting much attention in the mathematical community [15,17,18,23–26], particularly because of its connection with the theoretical description of the Havriliak–Negami model [18,24,27,28]. Moreover, this growing interest in Prabhakar's calculus is also reflected by the increasing literature on the recently proposed Maxwell–Prabhakar model, which was also kindly referred to as the Giusti–Colombaro model in [29].

In this paper we wish to analyze the important phenomena of storage and dissipation of energy in linear viscoelastic media, with particular regard for the class of models emerging from the constitutive equation in Equation (1). In viscoelasticity, as well as in electrical engineering, the process of dissipation of energy is usually accounted for in terms of a dimensionless parameter, called the quality factor, that is roughly defined as the ratio of the peak of energy stored in the system under a cycle of forced harmonic oscillation to the total rate of change of the energy, per cycle, by damping processes. Therefore, the aim of this paper is to compute and discuss the quality factor for the model defined in Equation (1).

2. Storage and Dissipation of Energy in Linear Viscoelasticity

In this section we wish to review the general theory, concerning the theoretical foundations, that leads to the definition of quality factor for a viscoelastic system. In order to do so, we will mimic the arguments presented in [10,30], unifying these formulations according to the notations employed in this paper.

Let us consider a quiescent viscoelastic body for $t < 0$. Then, under the hypothesis of sufficiently well-behaved causal histories, its constitutive equation in the creep representation reads

$$ \varepsilon(t) = \int_0^t J(t-\tau)\, \mathrm{d}\sigma(\tau) = \sigma(0+)\, J(t) + \int_0^t J(t-\tau)\, \dot{\sigma}(\tau)\, \mathrm{d}\tau \,, \tag{3} $$

where $\mathrm{d}\sigma(\tau)$ represents the Riemann–Stieltjes measure and $J(t)$ is the so-called creep compliance of the system, that in the Laplace domain is given by

$$ \widetilde{\varepsilon}(s) = s\, \widetilde{J}(s)\, \widetilde{\sigma}(s) \,. \tag{4} $$

In order to consider the harmonic behavior of a linear viscoelastic material, we should assume that a sufficient amount of time has elapsed since the original perturbation so that the effect of initial conditions could be considered negligible. So, let us consider some harmonic excitation of the material, which can be described in terms of the complex exponential representation, that is

$$ \sigma(t\,;\,\omega) = \chi\, \exp\left(i\,\omega\, t\right) \,, \quad \omega > 0 \,, \quad -\infty < t < \infty \,, \quad \chi \in \mathbb{C} \,. \tag{5} $$

Clearly, a similar argument can be presented in terms of the relaxation representation, however we will only focus on the creep one for sake of brevity.

If we now plug (5) into (3) we get

$$ \varepsilon(t\,;\,\omega) = i\,\omega\, \widehat{J}(\omega)\, (\chi\, \exp\left(i\,\omega\, t\right)) = i\,\omega\, \widehat{J}(\omega)\, \sigma(t\,;\,\omega) \,, \tag{6} $$

where $\widehat{J}(\omega)$ stands for the Fourier transform of $J(t)$ that, the latter being a causal function, ultimately reads

$$\widehat{J}(\omega) = \int_0^\infty \exp(-i\,\omega\,t)\,J(t)\,\mathrm{d}t\,.$$

Moreover, if we denote $J^\star(\omega) = i\,\omega\,\widehat{J}(\omega)$, as in [10], the constitutive equation, in the creep representation, for a viscoelastic body subject to a harmonic stress excitation reduces to

$$\varepsilon(t\,;\,\omega) = J^\star(\omega)\,\sigma(t\,;\,\omega)\,. \tag{7}$$

The time rate of change of energy in the system is then given by

$$\dot{\mathcal{W}} = \sigma_{\mathrm{R}}(t\,;\,\omega)\,\dot{\varepsilon}_{\mathrm{R}}(t\,;\,\omega)\,, \tag{8}$$

where the subscript R indicates that we are considering the real part of the corresponding function.

Now, one can easily solve (7) for $\sigma(t\,;\,\omega)$, then taking the real part of the resulting equation gives

$$\sigma_{\mathrm{R}}(t\,;\,\omega) = \frac{\varepsilon_{\mathrm{R}}(t\,;\,\omega)\,J^\star_{\mathrm{R}}(\omega) - \varepsilon_{\mathrm{I}}(t\,;\,\omega)\,J^\star_{\mathrm{I}}(\omega)}{|J^\star(\omega)|^2}\,, \tag{9}$$

where we denoted $\sigma \equiv \sigma_{\mathrm{R}} + i\,\sigma_{\mathrm{I}}$, $\varepsilon \equiv \varepsilon_{\mathrm{R}} + i\,\varepsilon_{\mathrm{I}}$ and $J^\star(\omega) \equiv J^\star_{\mathrm{R}}(\omega) - i\,J^\star_{\mathrm{I}}(\omega)$ for future convenience. Besides, from (5) and (7) it is also easy to see that

$$\varepsilon_{\mathrm{I}}(t\,;\,\omega) = -\frac{\dot{\varepsilon}_{\mathrm{R}}(t\,;\,\omega)}{\omega}\,. \tag{10}$$

Hence, if we plug (9) into (8) and recall the result in (10), then after some simple manipulations $\dot{\mathcal{W}}$ reduces to

$$\dot{\mathcal{W}} = \frac{\partial}{\partial t}\left[\frac{1}{2}\,\frac{J^\star_{\mathrm{R}}(\omega)}{|J^\star(\omega)|^2}\,\varepsilon^2_{\mathrm{R}}\right] + \frac{1}{\omega}\,\frac{J^\star_{\mathrm{I}}(\omega)}{|J^\star(\omega)|^2}\,\dot{\varepsilon}^2_{\mathrm{R}}\,, \tag{11}$$

where we omitted the explicit dependence on t and ω in the strain for sake of clarity.

Thus, it is easy to see that the total rate of change of energy over one cycle is accounted for by the integral over the cycle of the second term on the right-hand side of (11), namely

$$\frac{\Delta\mathcal{E}}{\text{Cycle}} = \int_t^{t+T} \dot{\mathcal{W}}(\tau)\,\mathrm{d}\tau = \pi\,\frac{J^\star_{\mathrm{I}}(\omega)}{|J^\star(\omega)|^2}\,|\chi|^2\,, \tag{12}$$

with $T = 2\pi/\omega$ the period of the cycle.

Due to the second law of thermodynamics, which requires that the total amount of energy dissipated increases with time, one can further infer that $J^\star_{\mathrm{I}}(\omega) \geq 0$.

However, despite defining a boundary term, the first piece of the right-hand side of (11) carries a very important physical meaning. Indeed, it tells us that the peak energy stored during a cycle is given by

$$\mathcal{P}_{\max} = \frac{1}{2}\,\frac{J^\star_{\mathrm{R}}(\omega)}{|J^\star(\omega)|^2}\,|\chi|^2\,. \tag{13}$$

One can now define the specific attenuation factor, or quality factor (Q-factor), as a normalized non-dimensional quantity defined by

$$Q^{-1} \equiv \frac{1}{2\pi}\,\frac{\Delta\mathcal{E}/\text{Cycle}}{\mathcal{P}_{\max}}\,. \tag{14}$$

Then, taking profit of the previous discussion it is easy to see that

$$Q^{-1} = \frac{1}{2\pi}\frac{\Delta\mathcal{E}/\text{Cycle}}{\mathcal{P}_{\max}} = \frac{J_{\mathrm{I}}^{\star}(\omega)}{J_{\mathrm{R}}^{\star}(\omega)} = -\frac{\Im\left\{i\,\omega\,\widehat{J}(\omega)\right\}}{\Re\left\{i\,\omega\,\widehat{J}(\omega)\right\}}, \tag{15}$$

recalling that $J_{\mathrm{I}}^{\star}(\omega) = -\Im\{J^{\star}(\omega)\}$.

If we combine the fact that the Fourier transform is equivalent to evaluating the bilateral Laplace transform with imaginary argument $s = i\,\omega$, together with the assumption for which $J(t)$ is a causal function, one can conclude that $\widehat{J}(\omega) = \widetilde{J}(s)\,|_{s=i\,\omega}$. This argument ultimately leads us to a very useful expression for the Q-factor, namely

$$Q^{-1}(\omega) = -\frac{\Im\left\{s\,\widetilde{J}(s)\,|_{s=i\,\omega}\right\}}{\Re\left\{s\,\widetilde{J}(s)\,|_{s=i\,\omega}\right\}}, \tag{16}$$

where we shall consider some positive real frequencies ω.

3. Quality Factor in Prabhakar-Like Viscoelasticity

Let us now compute the Q-factor for the Maxwell–Prabhakar model. Recalling that the Laplace transform of the Prabhakar integral kernel is given by

$$\mathcal{L}\left\{t^{\beta-1}\,E^{\gamma}_{\alpha,\beta}(\lambda\,t^{\alpha})\right\} = s^{-\beta}\,\left(1-\lambda\,s^{-\alpha}\right)^{-\gamma}, \tag{17}$$

where $t \in \mathbb{R}$, $\alpha, \beta, \gamma, \lambda \in \mathbb{C}$ and $\mathtt{Re}(\beta) > 0$, then it is easy to see that the creep compliance, in the Laplace domain, for a system described in terms of Equation (3) is therefore given by

$$s\,\widetilde{J}(s) = \frac{a}{b} + \frac{1}{b\,s^{\beta}\,(1-\Omega\,s^{-\alpha})^{\gamma}}. \tag{18}$$

Then, if we apply the replacement $s = i\,\omega$, the latter turns into

$$J^{\star}(\omega) = i\,\omega\,\widehat{J}(\omega) = \frac{a}{b} + \frac{1}{b\,(i\,\omega)^{\beta}\,\left[1-\Omega\,(i\,\omega)^{-\alpha}\right]^{\gamma}}. \tag{19}$$

Let us define an auxiliary variable $z(\omega) = 1 - \Omega\,(i\,\omega)^{-\alpha}$. Then, considering $\omega \in \mathbb{R}^{+}$, one can easily rewrite $s = i\,\omega = \omega\,\exp(i\,\pi/2)$, where $|s| = \omega$, that allows us to recast this new variable in the exponential representation, that is

$$z(\omega) = |z(\omega)|\,\exp\left(i\,\theta_z(\omega)\right), \tag{20}$$

with

$$|z(\omega)|^2 = 1 + \frac{\Omega^2}{\omega^{2\alpha}} - \frac{2\,\Omega}{\omega^{\alpha}}\cos\left(\frac{\alpha\,\pi}{2}\right), \tag{21}$$

$$\theta_z(\omega) = \arctan\left[\frac{\Omega\,\sin\left(\alpha\,\pi/2\right)}{\omega^{\alpha} - \Omega\,\cos\left(\alpha\,\pi/2\right)}\right]. \tag{22}$$

Then, plugging $z(\omega)$ into Equation (18), one can easily infer that

$$J^{\star}(\omega) = \frac{a}{b} + \frac{1}{b\,\omega^{\beta}\,|z(\omega)|^{\gamma}}\exp\left[-i\left(\frac{\beta\,\pi}{2} + \gamma\,\theta_z(\omega)\right)\right], \tag{23}$$

from which we can conclude that

$$\Re\{J^\star(\omega)\} = \frac{a}{b} + \frac{1}{b\,\omega^\beta\,|z(\omega)|^\gamma}\cos\left(\frac{\beta\,\pi}{2} + \gamma\,\theta_z(\omega)\right), \tag{24}$$

and

$$\Im\{J^\star(\omega)\} = -\frac{1}{b\,\omega^\beta\,|z(\omega)|^\gamma}\sin\left(\frac{\beta\,\pi}{2} + \gamma\,\theta_z(\omega)\right). \tag{25}$$

Hence, the quality factor for a Maxwell–Prabhakar viscoelastic body is given by

$$Q^{-1}(\omega) = \frac{\sin\left(\frac{\beta\pi}{2} + \gamma\,\theta_z(\omega)\right)}{a\,\omega^\beta\,|z(\omega)|^\gamma + \cos\left(\frac{\beta\pi}{2} + \gamma\,\theta_z(\omega)\right)}. \tag{26}$$

4. Quality Factor for Some Specific Realizations of the Maxwell–Prabhakar Model

In this section we discuss the quality factor for different choices of the parameters of the discussed model. Specifically, we will focus our discussion on four cases corresponding to two well-known classical viscoelastic models and the viscoelastic analogue of the Havriliak–Negami model for dielectric relaxation.

4.1. Fractional Maxwell Model

As argued in [15], it is easy to see that Equation (1) naturally reduces to the fractional Maxwell model, namely

$$\sigma(t) + a\,{}^C\mathbf{D}^\nu\,\sigma(t) = b\,{}^C\mathbf{D}^\nu\,\varepsilon(t), \tag{27}$$

where ${}^C\mathbf{D}^\nu$ represents the Caputo fractional derivative, provided that the parameters are chosen according to one of these two configurations:

(i) $\gamma = 0, a > 0, b > 0, \beta = \nu, \Omega \in \mathbb{R}$;
(ii) $\gamma \in \mathbb{R}, a > 0, b > 0, \beta = \nu, \Omega = 0$.

Here, it is trivial to infer that if $\gamma = 0$ then neither $|z(\omega)|$ nor $\theta_z(\omega)$ enter in the expression for the Q-factor, whereas if $\Omega = 0$ it is easy to see that $|z(\omega)| = 1$ and $\theta_z(\omega) = 0$. Hence, we get

$$Q^{-1}(\omega) = \frac{\sin(\nu\pi/2)}{a\,\omega^\nu + \cos(\nu\pi/2)}. \tag{28}$$

Furthermore, it is also worth remarking that if we set $\nu = 1$ (ordinary limit), we explicitly recover the Q-factor for the (ordinary) Maxwell model, that is, $Q^{-1}(\omega) = (a\,\omega)^{-1}$ (see, for example, [10]).

4.2. Fractional Voigt Model

Again, following the analysis presented in [15], one has that Equation (1) reduces to the fractional Voigt model, that is

$$\sigma(t) = M\,\varepsilon(t) + B\,{}^C\mathbf{D}^\nu\,\varepsilon(t), \tag{29}$$

by setting $\gamma = 1, a = 0, b = -B < 0, \alpha = \beta = \nu, \Omega = -M/B < 0$.

Now, if we plug this choice of parameters into Equation (26), we get

$$Q^{-1}(\omega) = \frac{\sin\left(\frac{\nu\pi}{2} + \theta_z(\omega)\right)}{\cos\left(\frac{\nu\pi}{2} + \theta_z(\omega)\right)} = \tan\left(\frac{\nu\,\pi}{2} + \theta_z(\omega)\right), \tag{30}$$

with

$$\theta_z(\omega) = -\arctan\left[\frac{M}{B}\frac{\sin(\nu\,\pi/2)}{\omega^\nu + (M/B)\,\cos(\nu\,\pi/2)}\right].$$

The latter, in the limit for $\nu = 1$, reduces to

$$Q^{-1}(\omega) = \tan\left(\frac{\pi}{2} + \theta_z(\omega)\right) = -\frac{1}{\tan\theta_z(\omega)} = \frac{B\,\omega}{M}. \tag{31}$$

which corresponds to the quality factor for the (ordinary) Voigt model (see, for example, [10]).

4.3. *Havriliak–Negami Model*

The Havriliak–Negami relaxation is an empirical model which was first introduced in order to describe the dielectric relaxation of certain types of polymers [24,25,31].

Now, it is very well known that a viscoelastic system can usually be mapped onto a class of electrical ladder networks and vice versa; see, for example, [32,33].

Following this line of thought, it is easy to see that the constitutive equation for the Havriliak–Negami viscoelastic model is given by [15]

$$\sigma(t) + a\,{}^C\mathbf{D}^{\gamma}_{\alpha,\alpha\gamma,-\lambda}\,\sigma(t) = b\,{}^C\mathbf{D}^{\gamma}_{\alpha,\alpha\gamma,-\lambda}\,\varepsilon(t), \tag{32}$$

with $\beta = \alpha\gamma$, $\Omega = -\lambda$, with $\lambda > 0$, and $0 < \alpha,\, \gamma < 1$.

Then, following the procedure presented in Section 3, one finds that

$$Q^{-1}(\omega) = \frac{\sin\left[\gamma\left(\frac{\alpha\pi}{2} + \theta_z(\omega)\right)\right]}{a\,\omega^{\alpha\gamma}\left[\omega^{2\alpha} + \lambda^2 + 2\lambda\omega^\alpha\cos\left(\frac{\alpha\pi}{2}\right)\right]^{\gamma/2} + \cos\left[\gamma\left(\frac{\alpha\pi}{2} + \theta_z(\omega)\right)\right]}, \tag{33}$$

where

$$\theta_z(\omega) = \arctan\left[-\frac{\lambda\,\sin(\alpha\,\pi/2)}{\omega^\alpha + \lambda\,\cos(\alpha\,\pi/2)}\right]. \tag{34}$$

5. Discussion and Conclusions

Attenuation effects represent one of the main fields of study in modern seismology, and consequently the specific attenuation factor (or Q-factor, for simplicity) embodies one of the key ingredients in geophysical sciences. Indeed, were it not for the damping capabilities of the soil, the energy of past earthquakes would still be resonating within the earth's interior.

In this paper, after a thorough review of the physical definition and meaning of the quality factor Q, we have provided an analysis of the storage and dissipation of energy in viscoelastic material of the Maxwell–Prabhakar class, namely, the one featured by a constitutive relation given by Equation (1). Specifically, in Section 3 we have computed the Q-factor for the general Maxwell–Prabhakar model, for which some interesting configurations of the parameters are shown in Figure 1. In Section 4 we have further provided some explicit realizations of the Maxwell–Prabhakar theory, namely the fractional Maxwell model, the fractional Voigt model and the viscoelastic equivalent of the Havriliak–Negami model for dielectric relaxation, shown in Figures 2 and 3.

Let us pay particular attention to the cases displayed in Figure 1. Indeed, as argued in [34], there exists much experimental evidence supporting the theses for which the Q-factor of homogeneous materials is substantially independent of the frequency. In this respect, it is worth noting that the Maxwell–Prabhakar class shows a very slow varying (almost constant) behaviour of the quality factor for low frequencies, for certain choices of the parameters of the model (see Figure 1). This is quite consistent with the results for the dumping of long-period teleseismic body waves and surface waves. Furthermore, for high frequencies, the model shows a power law behaviour, namely, $Q(\omega) \sim \omega^\beta$ as $\omega \to \infty$, which appears to be consistent with the experimental results concerning the attenuation of the coda of high-frequency teleseismic waves in Earth's upper mantle [35]. Furthermore, it is rather easy to prove that some very well-known constant-Q models are nothing but some specific realizations of the model defined in Equation (1). Indeed, for example, the renowned Kjartansson model [36,37] can be obtained from (18) by setting $\gamma = 0$, $\beta \equiv 2\,\eta$ with $\eta \in (0, 1/2)$ and $a = 0$, which is nothing but the Scott–Blair model [36]. In view of the last few comments, we believe that the Maxwell–Prabhakar model of viscoelasicity can potentially provide some stimulating new insights into the mathematical modelling of seismic processes and therefore is worthy of further studies.

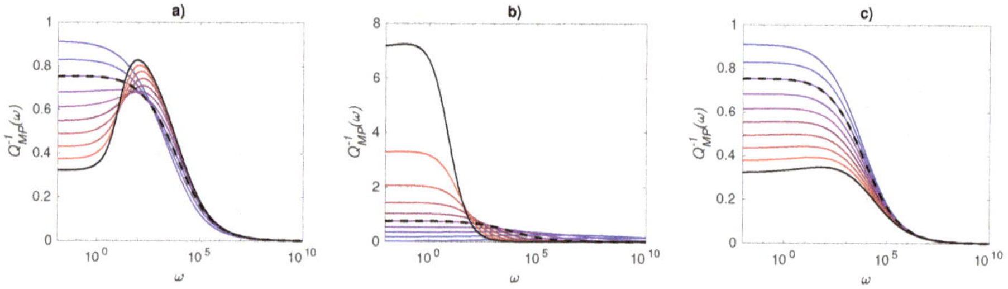

Figure 1. Q-factor of the Maxwell–Prabhakar model for $a = 0.01\,,\Omega = -10$. Model dependence on $\alpha = 0.1, 0.2, \ldots, 1$, $\beta = 0.5\,,\gamma = 0.3$ (**a**); model dependence on $\beta = 0.1, 0.2, \ldots, 1$, $\alpha = 0.3\,,\gamma = 0.3$ (**b**); model dependence on $\gamma = 0.1, 0.2, \ldots, 1$, $\beta = 0.5\,,\alpha = 0.3$ (**c**). Parameter value increasing from blue to red. In all panels the dashed line corresponds to $\alpha = 0.3\,,\beta = 0.5\,,\gamma = 0.3$.

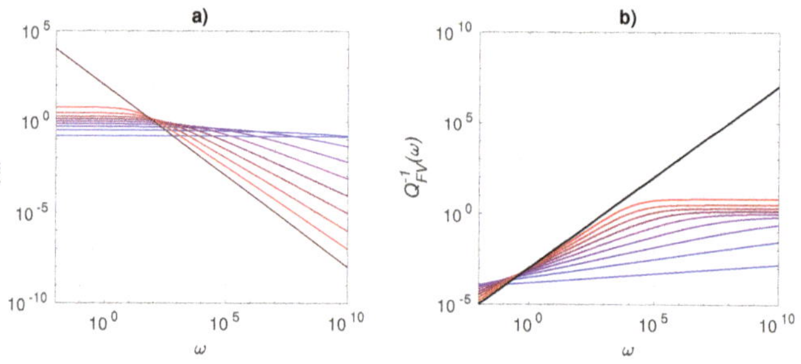

Figure 2. Q-factor of fractional Maxwell model with $a = 0.1$ (**a**), fractional Voigt model with $M = 10^3$, $B = 1$ (**b**). Parameter value increasing from blue to red.

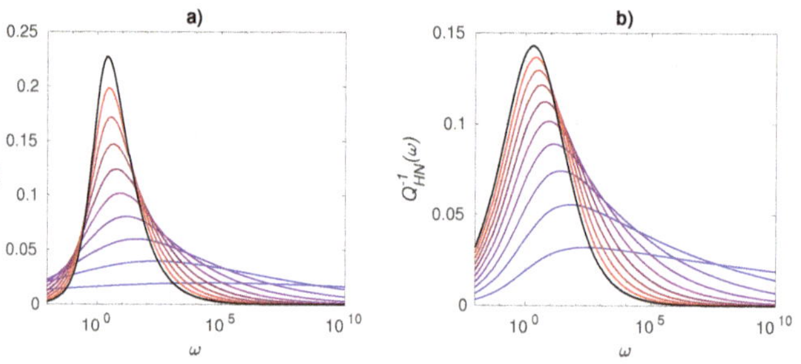

Figure 3. Q-factor of Havriliak–Negami model for dielectric relaxation with $a = 1\,,\lambda = 1$ and $\alpha = 0.1, 0.2, \ldots, 1\,,\gamma = 0.5$ (**a**), or $a = 1\,,\lambda = 1$ and $\gamma = 0.1, 0.2, \ldots, 1\,,\alpha = 0.5$ (**b**). Parameter value increasing from blue to red. Decreasing a, the widths of the curves become wider, while increasing λ, the peak results are shifted to larger ω for lower α.

Acknowledgments: The work of I.C. and A.G. has been carried out in the framework of the activities of the National Group of Mathematical Physics (GNFM, INdAM). Moreover, the work of A.G. has been partially supported by GNFM/INdAM Young Researchers Project 2017 "Analysis of Complex Biological Systems". Besides, the work of S.V. has been partially supported by the Interdepartmental Center "Luigi Galvani" for integrated studies of Bioinformatics, Biophysics and Biocomplexity of the University of Bologna.

Author Contributions: The authors contributed equally to this work.

Conflicts of Interest: The authors declare no conflict of interest.

References

1. Colombaro, I.; Giusti, A.; Mainardi, F. On transient waves in linear viscoelasticity. *Wave Motion* **2017**, *74*, 191–212, doi:10.1016/j.wavemoti.2017.07.008.
2. Colombaro, I.; Giusti, A.; Mainardi, F. On the propagation of transient waves in a viscoelastic Bessel medium. *Z. Angew. Math. Phys.* **2017**, *68*, 62–74, doi:10.1007/s00033-017-0808-6.
3. Colombaro, I.; Giusti, A.; Mainardi, F. A class of linear viscoelastic models based on Bessel functions. *Meccanica* **2017**, *52*, 825–832, doi:10.1007/s11012-016-0456-5.
4. Garra, R.; Mainardi, F.; Spada, G. A generalization of the Lomnitz logarithmic creep law via Hadamard fractional calculus. *Chaos Solitons Fractals* **2017**, *102*, 333–338, doi:10.1016/j.chaos.2017.03.032.
5. Giusti, A. On infinite order differential operators in fractional viscoelasticity. *Fract. Calc. Appl. Anal.* **2017**, *20*, 854–867, doi:10.1515/fca-2017-0045.
6. Giusti, A.; Mainardi, F. A dynamic viscoelastic analogy for fluid-filled elastic tubes. *Mecanica* **2016**, *51*, 2321–2330, doi:10.1007/s11012-016-0376-4.
7. Mainardi, F. Fractional Calculus: Some basic problems in continuum and statistical mechanics. In *Fractals and Fractional Calculus in Continuum Mechanics*; Carpinteri, A., Mainardi, F., Ed.; Springer: New York, NY, USA; Wien, Austria, 1997.
8. Giusti, A. A comment on some new definitions of fractional derivative. *arXiv* **2017**, arXiv:1710.06852.
9. Gorenflo, R.; Mainardi, F. Fractional Calculus: Integral and Differential Equations of Fractional Order. In *Fractals and Fractional Calculus in Continuum Mechanics*; Carpinteri, A., Mainardi, F., Eds.; Springer: New York, NY, USA; Wien, Austria, 1997.
10. Mainardi, F. *Fractional Calculus and Waves in Linear Viscoelasticity*; Imperial College Press: London, UK, 2010.
11. Gross, B. *Mathematical Structure of the Theories of Viscoelasticity*; Hermann & Cie: Paris, France, 1953.
12. Molinari, A. Viscoélasticité linéaire et functions complétement monotones. *Journal de Mécanique* **1973**, *12*, 541–553.
13. Mainardi, F.; Turchetti, G. Positivity constraints and approximation methods in linear viscoelasticity. *Lettere al Nuovo Cimento* **1979**, *26*, 38–40.
14. Hanyga, A. Wave propagation in linear viscoelastic media with completely monotonic relaxation moduli. *Wave Motion* **2013**, *50*, 909–928, doi:10.1016/j.wavemoti.2013.03.002.
15. Giusti, A.; Colombaro, I. Prabhakar-like fractional viscoelasticity. *Commun. Nonlinear Sci. Numer. Simul.* **2018**, *56*, 138–143, doi:10.1016/j.cnsns.2017.08.002.
16. D'Ovidio, M.; Polito, F. Fractional Diffusion-Telegraph Equations and their Associated Stochastic Solutions. *Teoriya Veroyatnostei i ee Primeneniya* **2017**, 62, 692–718, doi:10.4213/tvp5150.
17. Garra, R.; Gorenflo, R.; Polito, F.; Tomovski, Z. Hilfer-Prabhakar derivatives and some applications. *Appl. Math. Comput.* **2014**, *242*, 576–589, doi:10.1016/j.amc.2014.05.129.
18. Garra, R.; Garrappa, R. The Prabhakar or three parameter Mittag-Leffler function: Theory and application. *Commun. Nonlinear Sci. Numer. Simul.* **2018**, *56*, 314–329, doi:10.1016/j.cnsns.2017.08.018.
19. Gorenflo, R.; Kilbas, A.A.; Mainardi, F.; Rogosin, S.V. *Mittag-Leffler Functions, Related Topics and Applications*; Springer: Berlin, Germany, 2014.
20. Paneva-Konovska, J. *From Bessel to Multi-Index Mittag Leffler Functions: Enumerable Families, Series in Them and Convergence*; World Scientific Publishing: London, UK, 2016.
21. Prabhakar, T.R. A singular integral equation with a generalized Mittag Leffler function in the kernel. *Yokohama Math. J.* **1971**, *19*, 7–15.
22. Capelas de Oliveira, E.; Mainardi, F.; Vaz, J., Jr. Models based on Mittag-Leffler functions for anomalous relaxation in dielectrics. *Eur. Phys. J. Spec. Top.* **2011**, *193*, 161–171, doi:10.1140/epjst/e2011-01388-0.

23. Mainardi, F.; Garrappa, R. On complete monotonicity of the Prabhakar function and non-Debye relaxation in dielectrics. *J. Comput. Phys.* **2015**, *293*, 70–80, doi:10.1016/j.jcp.2014.08.006.
24. Garrappa, R. Grünwald-Letnikov operators for fractional relaxation in Havriliak-Negami models. *Commun. Nonlinear Sci. Numer. Simul.* **2016**, *38*, 178–191, doi:10.1016/j.cnsns.2016.02.015.
25. Garrappa, R.; Mainardi, F.; Maione, G. Models of dielectric relaxation based on completely monotone functions. *Fract. Calc. Appl. Anal.* **2016**, *19*, 1105–1160, doi:10.1515/fca-2016-0060.
26. Sandev, T. Generalized Langevin equation and the Prabhakar derivative. *Mathematics* **2017**, *5*, 66, doi:10.3390/math5040066.
27. Hanyga, A.; Seredyńska, M. On a Mathematical Framework for the Constitutive Equations of Anisotropic Dielectric Relaxation. *J. Stat. Phys.* **2008**, *131*, 269–303, doi:10.1007/s10955-008-9501-7.
28. Seredyńska, M.; Hanyga, A. Relaxation, dispersion, attenuation, and finite propagation speed in viscoelastic media. *J. Math. Phys.* **2010**, *51*, 092901, doi:10.1063/1.3478299.
29. Ding, X.; Zhang, G.; Zhao, B.; Wang, Y. Unexpected viscoelastic deformation of tight sandstone: Insights and predictions from the fractional Maxwell model. *Sci. Rep.* **2017**, *7*, 11336, doi:10.1038/s41598-017-11618-x.
30. Borcherdt, R. *Viscoelastic Waves in Layered Media*; Cambridge University Press: Cambridge, UK, 2009.
31. Havriliak, S.; Negami, S. A complex plane representation of dielectric and mechanical relaxation processes in some polymers. *Polymer* **1967**, *8*, 161–210, doi:10.1016/0032-3861(67)90021-3.
32. Giusti, A.; Mainardi, F. On infinite series concerning zeros of Bessel functions of the first kind. *Eur. Phys. J. Plus.* **2016**, *131*, 206–212, doi:10.1140/epjp/i2016-16206-4.
33. Gross, B.; Fuoss, R. Ladder structures for representation of viscoelastic systems. *J. Polym. Sci.* **1956**, *19*, 39–50.
34. Knopoff, L. Q. *Rev. Geophys.* **1964**, *2*, 625–660, doi:10.1029/RG002i004p00625.
35. Shito, A.; Karato, A.; Park, J. Frequency dependence of Q in Earth's upper mantle inferred from continuous spectra of body waves. *Geophys. Res. Lett.* **2004**, *31*, L12603, doi:10.1029/2004GL019582.
36. Carcione, J.; Cavallini, F.; Mainardi, F.; Hanyga, A. Time-domain modeling of constant-Q Seismic waves using fractional derivatives. *Pure Appl. Geophys.* **2002**, *159*, 1719–1736.
37. Kjartansson, E. Constant Q-wave propagation and attenuation. *J. Geophys. Res. Solid Earth* **1979**, *84*, 4737–4748.

mathematics

MDPI

Article

Fractional Derivatives, Memory Kernels and Solution of a Free Electron Laser Volterra Type Equation

Marcello Artioli [1], Giuseppe Dattoli [2], Silvia Licciardi [2,3,*] and Simonetta Pagnutti [1]

[1] ENEA—Bologna Research Center, Via Martiri di Monte Sole, 4, 40129 Bologna, Italy; marcello.artioli@enea.it (M.A.); simonetta.pagnutti@enea.it (S.P.)
[2] ENEA—Frascati Research Center, Via Enrico Fermi 45, 00044 Frascati, Rome, Italy; giuseppe.dattoli@enea.it
[3] Department of Mathematics and Computer Science, University of Catania, Viale A. Doria 6, 95125 Catania, Italy
* Correspondence: silviakant@gmail.com; Tel.: +39-392-509-6741

Received: 17 October 2017; Accepted: 24 November 2017; Published: 4 December 2017

Abstract: The high gain free electron laser (FEL) equation is a Volterra type integro-differential equation amenable for analytical solutions in a limited number of cases. In this note, a novel technique, based on an expansion employing a family of two variable Hermite polynomials, is shown to provide straightforward analytical solutions for cases hardly solvable with conventional means. The possibility of extending the method by the use of expansion using different polynomials (two variable Legendre like) expansion is also discussed.

Keywords: free electron laser (FEL); Volterra equations; iterative solutions; Hermite polynomials; Legendre polynomials

JEL Classification: 78A60; 45D05; 37M99; 37M99; 33C45; 33C45

1. Introduction

The free electron laser (FEL) is a device capable of transforming the kinetic energy of a releativistic electron beam into electromagnetic radiation.

The constituting elements of the device are a beam of high energy electrons, injected into an alternating magnetic field where they emit laser like radiation whose growth is ruled by the following integro differential equation [1–3]

$$\begin{aligned} &\partial_\tau\, a = i\,\pi\, g_0 \int_0^\tau \tau' e^{-i\,\nu\,\tau' - \frac{(\pi\mu_\varepsilon\tau')^2}{2}}\, a(\tau-\tau')\, d\tau', \\ &a(0) = 1 \end{aligned} \tag{1}$$

where a represents the laser field amplitude, g_0 the small signal gain coefficient, ν is linked to the laser frequency and the coefficient μ_ε is a parameter regulating the effects of the gain reduction due to the electrons' energy distribution.

The FEL equation reported in (1) belongs to the Volterra integro-differential family. It has playd an important role in the design of modern FEL devices, it is the most elementary version of the FEL linear intensity growth equation. More elaborated versions include the e-beam transverse and longitudinal distributions, the magnetic field inhomogeneities and so on. Equation (1) captures the essential features of the FEL dynamics itself and more complicated expressions can be treated with the methods outlined below [1–3]. The relevant kernel can be viewed as a memory operator, it is indeed associated with the nature of the physics it describes, namely the amplification of a laser field.

The fractional derivative operators can be ascribed to the same category, they are indeed exploited to represent phenomena implying dissipation of energy or damage [4–6].

Following such a point of view the authors of references [7,8] have proposed a further definition of fractional derivative without a singular kernel, defined as

$$D_{ex}f(t) = \frac{m(\alpha)}{1-\alpha}\int_{\beta}^{t} f(\xi)e^{-\frac{t-\xi}{1-\alpha}}d\xi \tag{2}$$

where $m(\alpha)$ is a normalization term and α constant.

Without entering into further discussion regarding the use of this derivative in applications, we note that we can define the FEL fractional (temporal) derivative as

$$D_{FEL}f(t) = \int_{0}^{\tau} \xi f(\xi)e^{-i\nu(\tau-\xi)-\frac{(\pi\mu_\varepsilon)^2}{2}(\tau-\xi)^2}d\xi \tag{3}$$

which can be considered conceptually analogous to the Caputo- Fabrizio fractional derivative [7].

The analytical/numerical method developed in this paper to deal with the solution of Equation (1) can be applied to problems involving derivatives of type (2), as discussed elsewhere.

We can write Equation (1) in a slightly different form, more suitable for our purposes

$$\begin{aligned} &\partial_\tau\, a = \hat{V}(\tau)\, a, \\ &\hat{V}(\tau) = i\,\pi\, g_0 \int_0^\tau \tau' e^{-i\,\nu\,\tau' - \tau'\partial_\tau - (\pi\mu_\varepsilon)^2\tau'^2}. \end{aligned} \tag{4}$$

The solution can accordingly be obtained by the iteration

$$\partial_\tau\, a_n(\tau) = \hat{V}^n a_0 \tag{5}$$

which can also be cast in the more familiar form [9,10]

$$\begin{aligned} &\partial_\tau\, a_n = i\,\pi\, g_0 \int_0^\tau \tau' e^{-i\,\nu\,\tau' - \frac{(\pi\mu_\varepsilon\tau')^2}{2}} a_{n-1}(\tau-\tau')\, d\tau', \\ &a_0 = 1. \end{aligned} \tag{6}$$

For later convenience we will call n the principal expansion index.

Even though the method is efficient and fast converging the integrals cannot be done analytically but requires a numerical treatment, which increases the computation times when a larger number of terms is involved.

2. Using the Hermite Expansion

A way out is the use of two variable Hermite polynomials defined as [11,12]

$$H_m(x,y) = m! \sum_{r=0}^{\lfloor \frac{m}{2} \rfloor} \frac{x^{m-2r}y^r}{(m-2r)!\,r!} \tag{7}$$

and specified by the generating function

$$\sum_{m=0}^{\infty} \frac{t^m}{m!} H_m(x,y) = e^{x\,t+y\,t^2} \tag{8}$$

$$\partial_\tau a_n = i\,\pi\, g_0 \sum_{m=0}^{\infty} \frac{1}{m!} \int_0^\tau \tau'^{(m+1)} H_m\left(-i\,\nu, -\frac{(\pi\,\mu_\varepsilon)^2}{2}\right) a_{n-1}(\tau-\tau')\, d\tau', \tag{9}$$

The expansion index m is therefore nested into the principal index n. The integration procedure is now straightforward and we get

1. *First Order Solution*

$$a_1 = i\,\pi\,g_0 \sum_{m=0}^{\infty} \frac{H_m\left(-i\,\nu, -\frac{(\pi\,\mu_\varepsilon)^2}{2}\right)}{m!} \int_0^\tau d\tau' \int_0^{\tau'} \tau''^{(m+1)} d\tau'', \tag{10}$$

which yields

$$\begin{aligned} a_1 &= i\,\pi\,g_0 \sum_{m_1=0}^{\infty} \alpha_{m_1} \tau^{m_1+3}, \\ \alpha_{m_1} &= \frac{H_{m_1}}{m_1!(m_1+2)\,(m_1+3)}. \end{aligned} \tag{11}$$

The arguments of the Hermite have been omitted to avoid cumbersome expressions.

2. *Second Order Solution*

$$a_2 = (i\,\pi\,g_0)^2 \sum_{m_2=0}^{\infty} \frac{H_{m_2}}{m_2!} \int_0^\tau d\tau' \int_0^{\tau'} \tau''^{(m_2+1)} a_1(\tau'-\tau'')d\tau'', \tag{12}$$

which after some algebra yields

$$\begin{aligned} a_2 &= (i\,\pi\,g_0)^2 \sum_{m_1,m_2=0}^{\infty} \alpha_{m_1,m_2} \tau^{m_1+m_2+6}, \\ \alpha_{m_1,m_2} &= \alpha_{m_1} \sum_{s=0}^{m_1+3} \binom{m_1+3}{s} \frac{(-1)^s H_{m_2}}{m_2!(m_2+s+2)(m_2+m_1+6)} = \\ &= \frac{H_{m_1} H_{m_2}}{m_2! m_1!(m_1+2)\,(m_1+3)(m_2+m_1+6)} \sum_{s=0}^{m_1+3} \binom{m_1+3}{s} \frac{(-1)^s}{(m_2+s+2)}. \end{aligned} \tag{13}$$

We can use the previous two contributions to check whether the results we obtain concerning e.g., the FEL gain agree with analogous conclusions already given in the literature.

It must be emphasized that further iterations are necessary, if one is interested to larger gain values. The technicalities of the expansion will be discussed in the forthcoming sections.

3. *Higher Order Solution*

The interesting aspect of the present nested procedure is that the n-th order can be computed in a modular way just looking at the symmetries of the expansion itself. The n-th order term indeed reads

$$\begin{aligned} a_n &= (i\,\pi\,g_0)^n \sum_{m_1,..m_n=0}^{\infty} \alpha_{m_1,..m_n} \tau^{(\sum_{r=1}^{n} m_r+3\,n)}, \\ \alpha_{m_1,..m_n} &= \alpha_{m_1,..m_{n-1}} \sum_{s=0}^{(\sum_{r=1}^{n-1} m_r+3\,(n-1))} \binom{\sum_{r=1}^{n-1} m_r+3\,(n-1)}{s} \frac{H_{m_n}}{m_n!(m_n+s+2)\,(\sum_{r=1}^{n} m_r+3\,n)}. \end{aligned} \tag{14}$$

A pivotal quantity in FEL physics is the gain defining the fractional intensity growth at the end of the interaction ($\tau = 1$), namely

$$G(\nu, \mu_\varepsilon) = \|a\|^2 - 1. \tag{15}$$

In the forthcoming section we will discuss the relevant numerical integration and compare the relevant results with those from a complete numerical integration.

3. Implementation and Comparison

The procedure has been implemented in Mathematica, but does not rely on its symbolic capabilities and thus can been implemented in any computer language, unless the analytical result is requested. In this case the calculation has to be done with a tool with symbolic capabilities.

The gain equation of order n is given from (15) combining solutions of (1) as

$$G(\nu, \mu_\varepsilon) = \|a_0 + a_1 + a_2 + .. + a_n\|^2 - 1. \tag{16}$$

It requires the calculation of the norm of the quadratic sum of $n+1$ terms, of which one is constant (unity for a_0) and n, according to the method outlined in previous section, are the Hermite polynomials expansion as of in Equations (11), (13) and (14). The expansion has to be truncated for computational resons and it is easy to implement, by the literal point of view, but the performance is poor. The j-th expansion, in fact, requires the evaluation of M^j terms (if all expansions are truncated at the M-th term) for every $j = 1 \dots n$, that is $\dfrac{M^{n+1}-1}{M-1}$ terms in total, which can be quite demanding. Algorithmic optimizations are thus needed.

It can be noted that some Hermite polynomials are the same in different occurrencies, and could be evaluated once and then stored for the next occurrencies.

The first optimization is to build a lookup table containing the Hermite polynomials or to define functions that remember their own values (which is a feature offered by Mathematica to implement transparently lookup tables).

The second optimization is to do index or exponent reordering with the aim to move factors in or out of the summation to avoid repetitions of the calculus, like in the r.h.s of Equation (13).

The resulting code is very compact and can be viewed in Figure A1 (Appendix A).

The comparison is reported in Figure 1 for a 2nd order gain function, where the present procedure (at different truncation levels) is confronted with those from a complete numerical integration and no differences are foreseeable. The truncation of the Hermite expansion obviously affects the speed and the approximation of the gain curve.

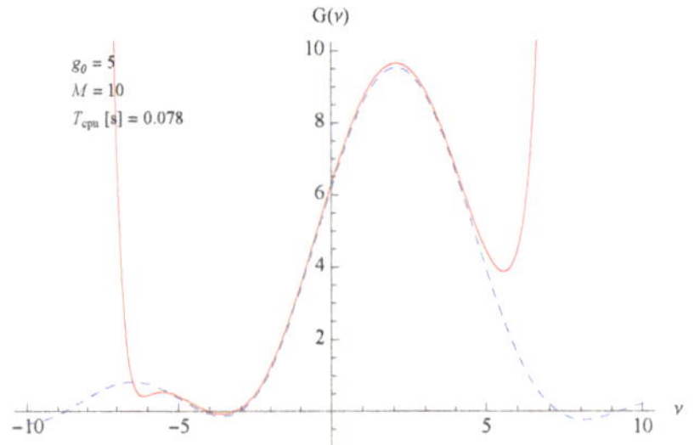

(**a**) $M = 10$

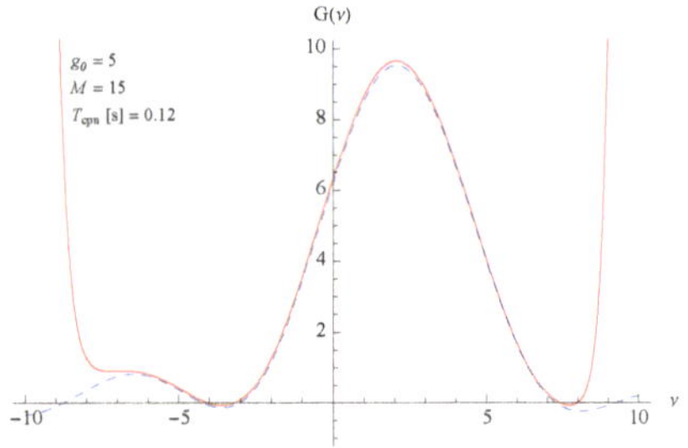

(**b**) $M = 15$

Figure 1. *Cont.*

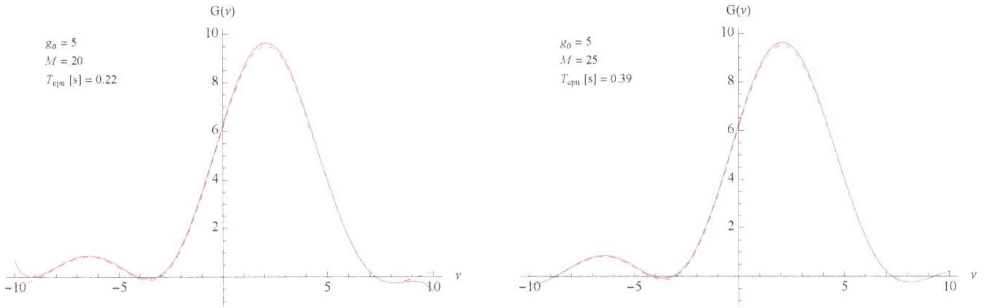

(**c**) $M = 20$ (**d**) $M = 25$

Figure 1. Comparison with the complete numerical integration (blue dashed line) at different truncation levels (M) and with no broadening effects ($\mu_\varepsilon = 0$) and $g_0 = 5$. The CPU time used refers to an Intel I7 3 GHz processor.

The method allows also taking into account the broadening effect as shown in Figure 2. It has to be noted that for higher broadening effects an higher truncation level is needed to achieve a good approximation, as shown in Figure 3. In this case, the comparison is made with the expression discussed in Appendix B.

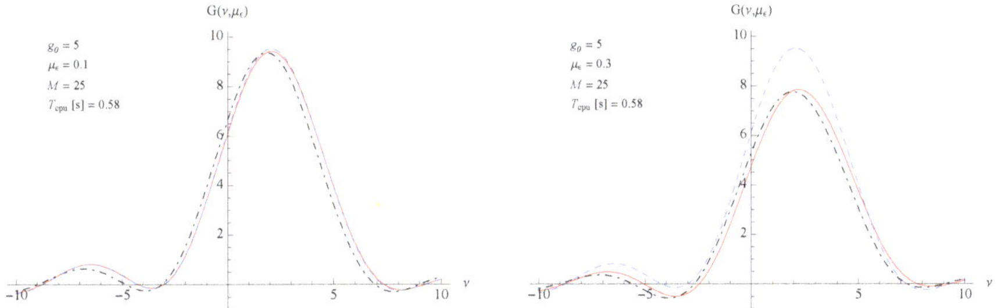

(**a**) $\mu_\varepsilon = 0.1$ (**b**) $\mu_\varepsilon = 0.3$

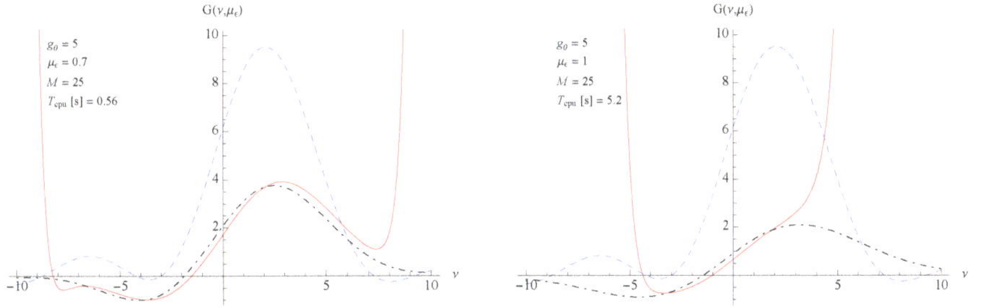

(**c**) $\mu_\varepsilon = 0.7$ (**d**) $\mu_\varepsilon = 1$

Figure 2. Comparison with analiytical expression (black dash-dotted line with different broadening effect (μ_ε)) and with the complete numerical integration (blue dashed line, which has no broadening effect) and $g_0 = 5$. The CPU time used refers an Intel I7 3 GHz processor.

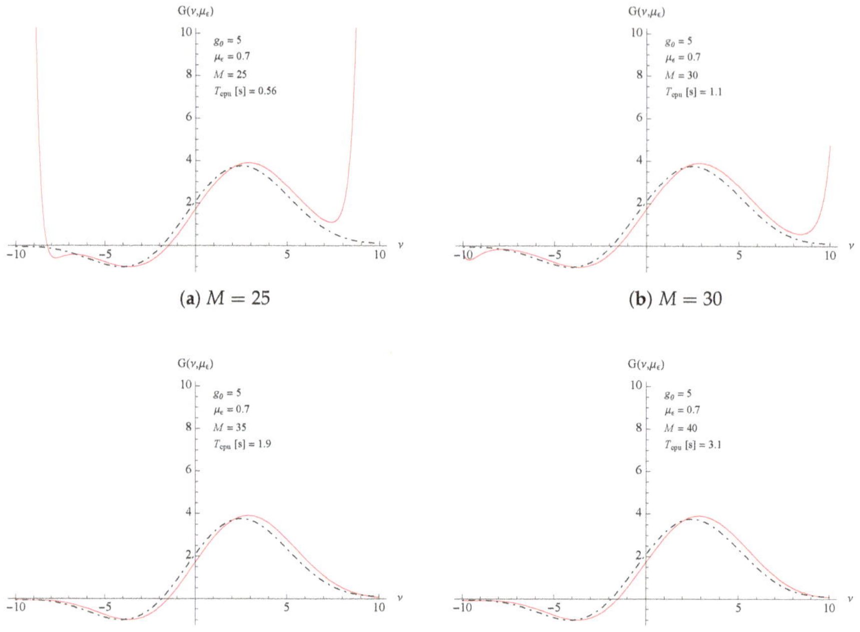

(**a**) $M = 25$ (**b**) $M = 30$

(**c**) $M = 35$ (**d**) $M = 40$

Figure 3. Comparison with analiytical expression (black dash-dotted line) in the case of higher broadening effects ($\mu_\varepsilon = 0.7$) for different truncation levels (M) and $g_0 = 5$. The CPU time used refers to an Intel I7 3 GHz processor.

The Hermite expansion approach let the integrals found in Equation (6) become polynomials and thus very easy to solve automatically. They are, of course, high in number but tractable also by a symbolic processor.

It is therefore possible to obtain, with the same degree of approximation, but with longer calculation times than the numeric case, also the full analytical expression for the curve.

The longer timings are then well payed back by the possibility to evaluate the curve on the fly with any combination of parameter values.

In Appendix A it can be found the symbolic expression of the curve for $M = 10$ (which is inaccurate, but compact enough to be checked by visual inspection and corresponding to that plotted in Figure 1).

It is clear that in our procedure there are two levels of approximation, the first associated with the principal index expansion n, depending on the values of the small signal gain coefficient g_0 which fixes how fast the laser intensity grows in time, the second is associated with the nested index m_n, depending on the values of the inhomogeneous parameter. It is worth noting that in actual FEL devices $n = 4$ is largely sufficient and since μ_ε is not larger than 0.3 the index m_n is not required to exceed 30.

4. Conclusions

The method we have foreseen is general enough to be applied to Volterra type equations with different kernels, provided that a corresponding expansion in terms of a suitable polynomial family is allowed.

To clarify the previous statement we consider the equation

$$\partial_\tau\, a(\tau) = \lambda \int_0^\tau \frac{\tau'}{1+\alpha\tau'+\beta\tau'^2} a(\tau-\tau')d\tau' \tag{17}$$

In this case the problem can be solved by the use of a two variable Legendre like polynomials

$$P_n(x,y) = \sum_{r=0}^{\lfloor \frac{n}{2} \rfloor} \frac{(n-r)!x^{n-2r}y^r}{(n-2r)!r!} \tag{18}$$

with generating function

$$\sum_{n=0}^{\infty} t^n P_n(x,y) = \frac{1}{1+x\,t+y\,t^2}. \tag{19}$$

The solution of the problem (17) looks, therefore, like the expression derived for Equation (1) with the two variable Legendre polynomials replacing the Hermite polynomials, namely

$$\partial_\tau\, a_n(\tau) = \lambda \sum_{m_n=0}^{\infty} P_{m_n}(-\alpha,-\beta) \int_0^\tau \tau'^{(m_n+1)} a_{n-1}(\tau-\tau')d\tau'. \tag{20}$$

In a forthcoming paper, dedicated to the physical aspects of the FEL, we will show how this family of polynomials allows the inclusion in the FEL high gain equation of the gain detrimental effects due to electron beam transverse and angular distributions.

Author Contributions: Each author has contributed equally to this work.

Conflicts of Interest: The authors declare no conflict of interest.

Appendix A

Wolfram Mathematica^{TM} code snippet, actually used to generate the curvesin the figures.

```
(*Hermite polynomials*)
H[n_, x_, y_] :=
  (n!) * ((x^n / (n!)) + Sum[(x^(n - 2 s) * y^s) / (((n - 2 s) !) * (s !)), {s, 1, Floor[n / 2]}]);

(*lookup table for Hermite polynomials to be used in the expansion*)
Htabular[order, v, mu] = Htabular[order_, v_, mu_] := H[order, -I * v, -1 / 2 * (Pi * mu) ^2];

(*first order solution*)
a1[v_, g0_, mu_, tau_, M_] :=
  I * Pi * g0 * Sum[Htabular[n, v, mu] * tau^(n + 3) / ((n + 3) (n + 2) (n!)), {n, 0, M}];

(*second order solution*)
a2[v_, g0_, mu_, tau_, M_] :=
  (I * Pi * g0) ^2 * tau^6 * Sum[tau^n * (n + 3) ! * (Htabular[n, v, mu]) / (n! * (n + 3) (n + 2)) *
     Sum[tau^m * (Htabular[m, v, mu]) / (m! * (n + m + 6)) *
       Sum[1 / (r! * (n + 3 - r) !) * (-1) ^r / (r + m + 2), {r, 0, n + 3}], {m, 0, M}], {n, 0, M}];

(*gain function at a second order approximation*)
gain2[v_, g0_, mu_, M_] := Simplify[Norm[1 + a1[v, g0, mu, 1, M] + a2[v, g0, mu, 1, M]] ^2 - 1];
```

Figure A1. Working Mathematica code.

If the code is run in symbolic mode, then an analytical expression can be found, as that on below

```
-1 + Norm[1 - 1/6 227 020 800
  ⅈ g0 π (-1 037 836 800 + 10 395 mu^10 π^10 + 518 918 400 ⅈ v + 155 675 520 v^2 - 34 594 560 ⅈ v^3 -
     6 177 600 v^4 + 926 640 ⅈ v^5 + 120 120 v^6 - 13 728 ⅈ v^7 - 1404 v^8 + 130 ⅈ v^9 + 11 v^10 +
     945 mu^8 π^8 (-156 + 130 ⅈ v + 55 v^2) + 630 mu^6 π^6 (2860 - 2288 ⅈ v - 936 v^2 + 260 ⅈ v^3 + 55 v^4) +
     90 mu^4 π^4 (-205 920 + 154 440 ⅈ v + 60 060 v^2 - 16 016 ⅈ v^3 - 3276 v^4 + 546 ⅈ v^5 + 77 v^6) +
     9 mu^2 π^2 (17 297 280 - 11 531 520 ⅈ v - 4 118 400 v^2 + 1 029 600 ⅈ v^3 +
        200 200 v^4 - 32 032 ⅈ v^5 - 4368 v^6 + 520 ⅈ v^7 + 55 v^8)) -
  1/403 291 461 126 605 635 584 000 000 g0^2 π^2 (560 127 029 342 507 827 200 000 +
     108 056 025 mu^20 π^20 - 320 072 588 195 718 758 400 000 ⅈ v - 100 022 683 811 162 112 000 000 v^2 +
     22 227 263 069 147 136 000 000 ⅈ v^3 + 3 889 771 037 100 748 800 000 v^4 -
     565 784 878 123 745 280 000 ⅈ v^5 - 70 723 109 765 468 160 000 v^6 + 7 771 770 303 897 600 000
      ⅈ v^7 + 763 298 869 132 800 000 v^8 - 67 848 788 367 360 000 ⅈ v^9 - 5 512 714 054 848 000 v^10 +
     385 504 479 360 000 ⅈ v^11 + 23 999 543 568 000 v^12 - 1 352 647 296 000 ⅈ v^13 -
     69 621 552 000 v^14 + 3 283 737 600 ⅈ v^15 + 141 726 000 v^16 - 5 553 600 ⅈ v^17 - 193 700 v^18 +
     5720 ⅈ v^19 + 121 v^20 + 98 232 750 mu^18 π^18 (-130 + 52 ⅈ v + 11 v^2) + 1 488 375 mu^16
      π^16 (741 520 - 370 240 ⅈ v - 116 220 v^2 + 21 736 ⅈ v^3 + 2299 v^4) + 4 422 600 mu^14 π^14
      (-19 126 800 + 9 715 200 ⅈ v + 3 284 400 v^2 - 705 600 ⅈ v^3 - 108 150 v^4 + 12 012 ⅈ v^5 + 847 v^6) +
     28 350 mu^12 π^12 (226 067 441 600 - 102 848 345 600 ⅈ v - 34 958 123 200 v^2 + 7 556 806 400 ⅈ v^3 +
        1 229 009 600 v^4 - 158 005 120 ⅈ v^5 - 16 208 920 v^6 + 1 288 144 ⅈ v^7 + 68 123 v^8) +
     3780 mu^10 π^10 (-150 989 254 416 000 + 49 510 869 408 000 ⅈ v + 16 472 127 672 000 v^2 -
        3 310 891 584 000 ⅈ v^3 - 528 515 988 000 v^4 + 69 242 659 200 ⅈ v^5 + 7 574 985 600 v^6 -
        698 006 400 ⅈ v^7 - 53 779 050 v^8 + 3 326 180 ⅈ v^9 + 140 723 v^10) + 9450 mu^8 π^8
      (1 343 758 470 912 000 - 1 075 006 776 729 600 ⅈ v - 436 721 503 046 400 v^2 + 69 995 519 193 600
         ⅈ v^3 + 10 424 998 584 000 v^4 - 1 318 168 051 200 ⅈ v^5 - 142 978 355 520 v^6 + 13 488 583 680
         ⅈ v^7 + 1 112 782 320 v^8 - 80 059 200 ⅈ v^9 - 4 945 980 v^10 + 250 536 ⅈ v^11 + 8833 v^12) +
     9000 mu^6 π^6 (-28 064 726 097 408 000 + 21 588 250 844 160 000 ⅈ v + 8 481 098 545 920 000 v^2 -
        2 261 626 278 912 000 ⅈ v^3 - 459 392 837 904 000 v^4 + 54 711 400 423 680 ⅈ v^5 +
        5 658 685 032 000 v^6 - 523 200 437 760 ⅈ v^7 - 43 336 653 360 v^8 + 3 218 483 840 ⅈ v^9 +
        214 021 808 v^10 - 12 655 552 ⅈ v^11 - 653 198 v^12 + 28 028 ⅈ v^13 + 847 v^14) + 45 mu^4 π^4
      (96 318 139 966 304 256 000 - 70 049 556 339 130 368 000 ⅈ v - 26 268 583 627 173 888 000 v^2 +
        6 735 534 263 377 920 000 ⅈ v^3 + 1 323 051 373 163 520 000 v^4 - 211 688 219 706 163 200 ⅈ v^5 -
        28 666 113 085 209 600 v^6 + 2 773 818 112 665 600 ⅈ v^7 + 230 735 769 264 000 v^8 -
        17 213 099 904 000 ⅈ v^9 - 1 165 531 086 720 v^10 + 71 812 815 360 ⅈ v^11 +
        4 015 880 960 v^12 - 202 092 800 ⅈ v^13 - 8 970 000 v^14 + 334 048 ⅈ v^15 + 8833 v^16) +
     90 mu^2 π^2 (-666 817 892 074 414 080 000 + 444 545 261 382 942 720 000 ⅈ v +
        155 590 841 484 029 952 000 v^2 - 37 718 991 874 916 352 000 ⅈ v^3 -
        7 072 310 976 546 816 000 v^4 + 1 088 047 842 545 664 000 ⅈ v^5 + 142 482 455 571 456 000 v^6 -
        16 283 709 208 166 400 ⅈ v^7 - 1 653 814 216 454 400 v^8 + 134 699 800 435 200 ⅈ v^9 +
        9 562 022 870 400 v^10 - 610 017 408 000 ⅈ v^11 - 35 429 634 240 v^12 + 1 881 834 240 ⅈ v^13 +
        91 278 720 v^14 - 4 010 240 ⅈ v^15 - 156 390 v^16 + 5148 ⅈ v^17 + 121 v^18))]^2
```

Figure A2. Mathematica code for Hermite solution.

Appendix B

The comparison of the solution method with the case including the effect of the inhomogeneous broadening has been done using as benchmark the following first order analytical solution, sufficient for the range of parameters we have quoted in Figures 2 and 3.

$$a_1 = i\,\pi\, g_0 \sqrt{\frac{\pi}{2}} \frac{1}{(\pi\mu_\varepsilon)^2} e^{-\frac{\nu^2}{2(\pi\mu_\varepsilon)^2}} \cdot A_1 \tag{A1}$$

where

$$\begin{aligned} A_1 = |\nu| \left\{ i\sqrt{\frac{2}{\pi}}(\pi\mu_\varepsilon)\left(-\nu e^{\delta^2} + \delta e^{\frac{\nu^2}{2(\pi\mu_\varepsilon)^2}}\right) + i\left[(\pi\mu_\varepsilon)^2 - \nu^2\right] Erfi\left(\frac{\nu}{\sqrt{2}(\pi\mu_\varepsilon)}\right) + \right. \\ \left. + \left[\nu\delta^* - i(\pi\mu_\varepsilon)^2\right] Erfi\left(\frac{\delta}{\sqrt{2}(\pi\mu_\varepsilon)}\right)\right\} - (\pi\mu_\varepsilon)^2\nu^2 + Erfi\left(\frac{|\nu|}{\sqrt{2}(\pi\mu_\varepsilon)}\right), \\ \delta = \nu - i(\pi\mu_\varepsilon)^2\tau. \end{aligned} \tag{A2}$$

Higher order terms and the framing within the FEL theory will be discussed elsewhere.

References

1. Colson, W.B. Classical Free Electron Laser theory. In *Laser Handbook* ; Colson, W.B., Pellegrini, C., Renieri, A., Eds.; North Holland: Amsterdam, The Netherlands, 1990; Volume VI.
2. Dattoli, G.; Renieri, A.; Torre, A. *Lectures on the Physics of Free Electron Laser and on Related Topics;* World Scientific Singapore: Singapore, 1993.
3. Saldin, E.L.; Schneidmiller, E.A.; Yurkov, M.V. *The Physics of Free Electron Laser;* Springer: Heidelberg/Berlin, Germany, 2000.
4. Mainardi, F. *Fractional Calculus and Waves in Linear Viscoelasticity: An Introduction to Mathematical Models;* World Scientific Publishing: Singapore, 2010.
5. Fabrizio, M. Fractional rheological models for thermomechanical systems. Dissipation and free energies. *Fract. Calc. Appl. Anal.* **2014**, *17*, 206–223.
6. Fabrizio, M.; Morro, A. Dissipativity and irreversibility of electromagnetic systems. *Math. Mod. Meth. Appl. Sci.* **2000**, *10*, 217–246.
7. Caputo, M.; Fabrizio, M. A new Definition of Fractional Derivative without Singular Kernel. *Progr. Fract. Differ. Appl.* **2015**, *1*, 73–85.
8. Atangana, A.; Baleanu, D. New fractional derivatives with nonlocal and non-singular kernel: Theory and application to heat transfer model. *arXiv* **2016**, arXiv:1602.03408.
9. Dattoli, G.; Torre, A.; Centioli, C.; Richetta, M. Free Electron Laser Operation in the Intermediate gain Region. *IEEE J. Quantum Electron.* **1989**, *25*, 2327.
10. Enea, for Various Material about FEL Formulae. Available online: www.fel.enea.it (accessed on 16 October 2017).
11. Appéll, P.; Kampé dé Fériét, J. *Fonctions Hypergéométriques et Hypersphériques: Polynomes d'Hermite;* Gauthier-Villars: Paris, France, 1926.
12. Artioli, M.; Dattoli, G. The Geometry of Hermite Polynomials. Available online: http://demonstrations.wolfram.com/TheGeometryOfHermitePolynomials/ (accessed on 16 October 2017).

mathematics

MDPI

Article

Application of Tempered-Stable Time Fractional-Derivative Model to Upscale Subdiffusion for Pollutant Transport in Field-Scale Discrete Fracture Networks

Bingqing Lu [1], Yong Zhang [1,*], Donald M. Reeves [2], HongGuang Sun [3] and Chunmiao Zheng [4]

[1] Department of Geological Sciences, University of Alabama, Tuscaloosa, AL 35487, USA; blu5@crimson.ua.edu
[2] Department of Geosciences, Western Michigan University, Kalamazoo, MI 49008, USA; matt.reeves@wmich.edu
[3] College of Mechanics and Materials, Hohai University, Nanjing 210098, China; shg@hhu.edu.cn
[4] School of Environmental Science & Engineering, Southern University of Science and Technology, Shenzhen 518055, Guangdong, China; zhengcm@sustc.edu.cn
* Correspondence: yzhang264@ua.edu; Tel.: +1-205-348-3317

Received: 10 December 2017; Accepted: 29 December 2017; Published: 3 January 2018

Abstract: Fractional calculus provides efficient physical models to quantify non-Fickian dynamics broadly observed within the Earth system. The potential advantages of using fractional partial differential equations (fPDEs) for real-world problems are often limited by the current lack of understanding of how earth system properties influence observed non-Fickian dynamics. This study explores non-Fickian dynamics for pollutant transport in field-scale discrete fracture networks (DFNs), by investigating how fracture and rock matrix properties influence the leading and tailing edges of pollutant breakthrough curves (BTCs). Fractured reservoirs exhibit erratic internal structures and multi-scale heterogeneity, resulting in complex non-Fickian dynamics. A Monte Carlo approach is used to simulate pollutant transport through DFNs with a systematic variation of system properties, and the resultant non-Fickian transport is upscaled using a tempered-stable fractional in time advection–dispersion equation. Numerical results serve as a basis for determining both qualitative and quantitative relationships between BTC characteristics and model parameters, in addition to the impacts of fracture density, orientation, and rock matrix permeability on non-Fickian dynamics. The observed impacts of medium heterogeneity on tracer transport at late times tend to enhance the applicability of fPDEs that may be parameterized using measurable fracture–matrix characteristics.

Keywords: fractional partial differential equations (fPDEs); discrete fracture networks (DFNs); anomalous transport; fractional advection-dispersion equations

1. Introduction

Fractional calculus, defined by non-integer order derivatives and integrals, has been applied to problems involving non-Fickian or anomalous dynamics for almost three decades [1,2]. Despite their vast potential, both theoretical development and real-world applications of fractional partial differential equations (fPDEs) have been commonly constrained by the lack of understanding of how earth system properties influence non-Fickian transport dynamics, especially for the hydrologic sciences [3]. This major challenge has historically reduced fPDEs to curve-fitting mathematical exercises, instead of routine hydrological modeling tools [4]. Intensive efforts are needed to link medium properties and the fPDE model parameters, and the commonly used Monte Carlo approach is particularly suitable for providing target dynamics for heterogeneous systems with well-controlled and definable properties.

Applications of fPDEs in the hydrologic sciences have a relatively shorter history, compared to the other stochastic hydrologic approaches, such as the single or multiple rate mobile–immobile models used extensively in chemical engineering and hydrogeology [5,6]. Anomalous transport has been observed in a variety of disordered systems in natural geological media, especially in heterogeneous porous media and fractured formations [6–10]. After Benson et al. [11] first introduced the spatial fractional advection–dispersion equation (fADE) to capture super-diffusive transport in sand tanks and a relatively homogenous aquifer, the fADEs had been applied to model anomalous transport in saturated, heterogeneous porous media [12–15] and Earth surfaces, such as natural rivers [16–18]. Applications of fADEs for flow and transport in fractured media are rather rare [19,20], although fractures are ubiquitous in geologic systems. More than 90% of natural aquifers are fractured.

Fractured media exhibit erratic heterogeneity and scale-dependent dynamics for flow and transport [21,22], challenging standard modeling tools and providing an ideal test case for fPDEs. Accurate and efficient simulation of contaminant transport within a fractured rock mass is practically important, since fractures play a large role in many natural and engineered processes, such as long-term disposal of high-level radioactive wastes in a geologic repository [21,23–25]. Fracture properties of natural rocks, such as fracture density, spatial location, length, aperture and orientation distributions, account for medium heterogeneity and result in non-Fickian contaminant transport [26–29], where the feasibility of fADEs and the potential correlation between fracture properties and fADE parameters have not been fully understood (see pioneer work in [19,20]).

This study selected the fPDE to quantify pollutant transport through field-scale fractured rock. Due to the prohibitive computational burden in mapping each individual fracture in standard grid-based models [20], different transport models have been developed to describe and predict complex transport behavior in fractured media at different scales [30–33]. For example, the fractured medium was treated as an equivalent continuum and the advection-dispersion equation (ADE) with ensemble averaged parameters was used to model contaminant transport [22,34]. However, the equivalent continuum assumption may only be valid for rocks with sufficiently high fracture densities that approximate an equivalent porous medium [35]. This condition is rare in natural fractured rock, while sparsely-fractured rock is more common, and favored for geological disposal [35,36]. Unlike the advection–dispersion equation (ADE), which characterizes Fickian diffusion at local scales, non-Fickian transport behaviors and asymmetric plumes, due to heterogeneity in fractures, have been observed in both the laboratory experiments and field tests [7,31,37]. This provides our motivation to select a non-local fADE in this study, which can account for both spatial and temporal nonlocal processes [38–41] in describing non-Fickian transport in fractured media.

The rest of the paper is organized as follows. In Section 2, we present the Monte Carlo approach used to simulate multiple discrete fracture network (DFN) flow and transport scenarios. The DFN is an efficient and conceptually robust numerical approach to simulate the dynamics of flow and transport in fractured media with definable fracture properties. It assumes that the rock mass is dissected by a network of discrete fractures, and that fluid flow and contaminant transport in a low-permeability rock mass are controlled by geometric and hydraulic properties of interconnected fractures of a network [42,43]. Results of the Monte Carlo simulations and applications of the fADE are then shown in Section 3. In Section 4, we explore the emergence of anomalous transport and its characterization, to gain insight into the effective solute dynamics and major mechanisms behind the observed anomalous behavior. We focus on the impact of fracture density and orientation and rock matrix permeability on non-Fickian dynamics and the fADE parameters. Section 5 draws the main conclusions.

2. Monte Carlo Simulations and the Fractional Advection–Dispersion Equation

The Monte Carlo approach utilized in this study contains three major steps. First, we generated multiple scenarios of stochastic fracture networks, with varying properties, using HydroGeoSphere (HGS) software (v.111, Aquanty Inc., Waterloo, ON, Canada) [44], which is a multi-dimensional, control–volume, finite element simulator designed to quantify the hydrologic cycle, including simulating flow and transport in fracture networks embedded in a porous medium with the discrete facture (DF) approach. Second, groundwater flow through the generated DFNs was modeled, which was expected to lead to strongly non-Fickian dynamics for conservative tracer transport (see further discussion below), providing the synthetic data to evaluate the influence of the fractured media properties on non-Fickian dynamics. Third, pollutant transport, characterized by breakthrough curves (BTCs), was then accounted for by a truncated power–law distribution memory function embodied in the fADE model. FracFit [41], a robust parameter estimation tool using a weighted nonlinear least squares (WNLS) algorithm, was used to parameterize our fractional calculus model, which could capture salient features of anomalous transport, such as skewness and power-law tails. Weights assigned by WNLS [45] were proportional to the reciprocal of the measured concentrations. Therefore, areas of lower concentration (representing low probability regions) received greater weight, which was essential for capturing early arrivals and late time-tail characteristics.

2.1. Generating the Random Discrete Fracture Networks

The model geometry, shown in Figure 1, is 50 × 25 m^2, and represents a horizontal, two-dimensional (2D) map view of the fractures, with unit thickness and uniform blocks (100 × 50) along the x and z axes. One hundred 2D DFN realizations (for each scenario) were generated by superimposing two different sets of fractures onto the grid, with links between the fracture elements and matrix block elements, leading to realistic DFNs [46,47]. Each scenario represented the ensemble average of 100 total individual DFN flow and transport realizations. Fracture locations, orientations, lengths, and hydraulic conductivities were generated from predefined distributions obtained in literature [32,48]:

(1) Fracture location: distributed randomly over the entire domain, with a random seed selected for each realization;
(2) Fracture orientation: two fracture sets followed mixed Gaussian distributions (each Gaussian distribution is equally weighted and has its own mean and variance), with a mean and standard deviation of $0° \pm 10°$ for the first set, and $90° \pm 10°$ for the second set;
(3) Fracture length: log-normal distribution, with 10 length bins—the mean of the smallest length bin is 2.5 m (which is 1/10 of the minimum length of the x and z domain sizes), and the mean of the largest length bin is 25 m (which is the minimum length of the x and z domain sizes);
(4) Fracture aperture: exponential distribution, with 10 aperture bins—the mean of the smallest aperture bin is 0.001 m, and the mean of the largest aperture bin is 0.0015 m.

Fracture permeability was calculated using the parallel plate law: $k_f = \omega_f^2/12$ [7,49–51], which relates aperture ω_f to permeability k_f. In our DFNs, fractures were open and act as flow conduits. Fracture permeability was dependent on aperture, and varied from 8.3×10^{-8} m^2 to 1.9×10^{-7} m^2 given an exponential distribution. We tested the matrix hydraulic conductivity of 5×10^{-8}, 1×10^{-7} and 1×10^{-6} m/s, with the corresponding permeability of 5.5×10^{-15}, 1.1×10^{-14} and 1.1×10^{-13} m^2, which are six to seven orders of magnitude smaller than the fracture permeability. The matrix permeability was assumed to be isotropic and in the range of an un-fractured, low-permeability rock, like a real silt medium ([1.15 × 10−16, 2.3 × 10−12] m^2). One realization of the DFN, with 100 fractures, is shown in Figure 1. A horizontal fracture with the length of 10 m and aperture of 0.001 m was inserted into the middle of the left boundary, in order to inject a conservative tracer. This zone provided a mobile source of pollutant, representing real-world contamination, where a pollutant usually enters firstly the mobile region in aquifers before spreading downstream.

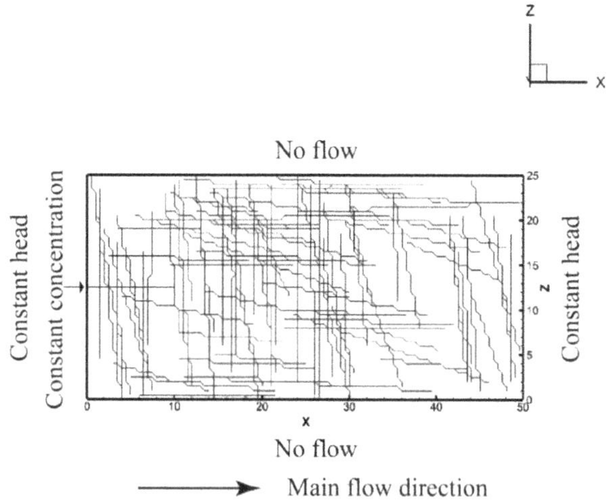

Figure 1. Map of the fractured domain, showing flow geometry and boundary conditions. Note that the discrete fractures are curved in the figure, after they are incorporated into the grid by HydroGeoSphere (HGS).

2.2. *Modeling Groundwater Flow and Pollutant Transport through the Discrete Fracture Networks*

Steady-state groundwater flow through the confined aquifer generated above was then solved by HGS. Parameters for the flow model, including the specific storage coefficient and porosity, are defined in Table 1. Boundary conditions were chosen so that the main flow direction was from left to right, with a general hydraulic gradient of approximately 0.2. The high hydraulic gradient was selected for faster convergence, and decreased simulation times in this work (so that we could observe the late-time behavior of solute transport, which is critical to identify the dynamics of sub-diffusive non-Fickian transport). The left and right boundaries were set to be constant head boundary conditions, and had values as 50 m and 40 m, respectively.

Table 1. Model parameters and initial and boundary conditions used in the simulations.

Parameter	Symbol	Value	Unit
Matrix	-	-	-
Hydraulic conductivity	$K_{xx} = K_{yy} = K_{zz}$	$5 \times 10^{-8}, 10^{-7}, 10^{-6}$	$m{\cdot}s^{-1}$
Specific storage	S_s	10^{-3}	m^{-1}
Porosity	θ_s	0.3	-
Longitudinal dispersivity	α_L	0.01	m
Transverse dispersivity	α_T	0.001	m
Vertical_transverse dispersivity	α_V	0.001	m
Tortuosity	τ	1	-
Fractures	-	-	-
Conductivity (computed)	K_f	$(\rho g \omega_f^2)/(12\mu)$ [a]	$m{\cdot}s^{-1}$
Specific storage (computed)	S_{sf}	$\rho g \alpha_\omega$ [b] $= 4.3 \times 10^{-6}$	m^{-1}
Longitudinal dispersivity	α_{Lf}	0.01	m
Transverse dispersivity	α_{Tf}	0.001	m
Vertical_transverse dispersivity	α_{Vf}	0.001	m
Solute	-	-	-
Free-solution diffusion coefficient	D_{free}	10^{-9}	m^2/s
Initial conditions	-	-	-
Concentration	$C_{t=0}$	0	kg/m^3
Boundary conditions	-	-	-
Inflow concentration	$C_{x=0}$	1	kg/m^3
Simulation Settings	-	-	-
Tracer pulse	T_{tracer}	10	s
[a] ω_f stands for aperture; μ stands for fluid viscosity		-	-
[b] α_ω stands for fluid compressibility		-	-

Pollutant transport through the saturated DFN was then solved by HGS, using the same ILU (Incomplete lower upper factorization)-preconditioned ORTHOMIN (an iterative method using orthogonalization and minimization to achieve fast convergence) solver as is used for the flow problem. Transport parameters are shown in Table 1. The longitudinal, transverse, and vertical transverse dispersivities assigned to both the un-fractured rock matrix and rock fractures were 0.01, 0.001, and 0.001 m, respectively, where the longitudinal dispersivity was one order of magnitude smaller than the grid size, and the horizontal dispersivity was one order of magnitude smaller than that along the longitudinal direction. Tortuosity was assumed to be 1, so that we could focus on the impact of other measurable parameters (such as matrix permeability and porosity) on transport behaviors. A 10 s pulse of one conservative solute, representing an example of sodium chloride, with a nominal concentration of 1 kg/m^3, was injected into the fracture, explicitly inserted at the middle of the left boundary. The free solution diffusion coefficient was 1×10^{-9} m^2/s (which is on the same order of the diffusion coefficient for hydrogen in water). The matrix porosity (0.30) was selected after referring to literature [52,53] where a value of 0.25–0.27 was assumed for the matrix porosity. We also tested a much smaller matrix porosity (0.05), and results (not shown here) revealed a tracer BTC (shifted toward younger times, likely due to a higher diffusion rate) with the overall trend similar to that with a porosity of 0.30.

The HGS transport time weighting factor was set to a fully implicit numerical scheme. Fully implicit time weighting is less prone to exhibit oscillations than central and explicit time weighting.

The governing equation for subsurface flow in porous media used in the HGS was a modified form of Richards' equation [54]:

$$-\nabla \bullet (\omega_m q) + \sum \Gamma_{ex} \pm Q = \omega_m \frac{\partial \theta_s}{\partial t} \tag{1}$$

where ω_m [dimensionless] is the volumetric fraction of the total porosity occupied by the porous medium, Γ_{ex} $[L^3 L^{-3} T^{-1}]$ represents the volumetric fluid exchange rate between the subsurface domain and discrete fractures, Q $[L^3 L^{-3} T^{-1}]$ represents a source/sink term to the porous medium system, and θ_s [dimensionless] is the saturated water content which is assumed equal to the porosity. The fluid flux q $[LT^{-1}]$ in (1) is given by Darcy's law:

$$q = -K \bullet k_r \nabla(\psi + z) \tag{2}$$

where K $[LT^{-1}]$ is hydraulic conductivity tensor, k_r [dimensionless] represents the relative permeability of the medium, ψ $[L]$ is the pressure head, and z $[L]$ is the elevation head.

A fracture is idealized as the space between two-dimensional parallel surfaces, with the tacit assumption that the total head is uniform across the fracture width. The equation for saturated flow in a fracture of width/aperture ω_f $[L]$ can be written by using the analogy of Richards Equation (1) for the porous matrix. The governing two-dimensional flow equation in a fracture has the form [44]:

$$-\bar{\bar{\nabla}} \bullet (\omega_f q_f) - \omega_f \Gamma_f = \omega_f S_{sf} \frac{\partial \psi_f}{\partial t} \tag{3}$$

where the fluid flux q_f $[LT^{-1}]$ is given by:

$$q_f = -K_f \bullet k_{rf} \bar{\bar{\nabla}}(\psi_f + z_f) \tag{4}$$

where $\bar{\bar{\nabla}}$ is the two-dimensional gradient operator defined in the fracture plane, k_{rf} [dimensionless] is the permeability of the fracture, ψ_f and z_f $[L]$ are the pressure and the elevation heads within the fracture, and S_{sf} $[L^{-1}]$ is the specific storage coefficient for the fractures.

The saturated hydraulic conductivity of a fracture K_f $[LT^{-1}]$ with a uniform aperture ω_f is given by: $K_f = \rho g \omega_f^2 / 12\mu$. Because it was assumed that the fractures were non-deformable and fluid-filled, there was no contribution to the storage term from fracture compressibility. Thus, the specific storage coefficient for a fracture under saturated conditions was only related to the compressibility of water α_w $[LT^2M^{-1}]$, according to: $S_{sf} = \rho g \alpha_w$.

Three-dimensional transport of solutes in a saturated porous matrix is described by the following advection-dispersion equation [44]:

$$-\nabla \bullet \omega_m(qC - \theta_s D\nabla C) + [\omega_m \theta_s R\lambda C] + \sum \Omega_{ex} \pm Q_c = \omega_m[\frac{\partial(\theta_s RC)}{\partial t} + \theta_s R\lambda C] \quad (5)$$

where C is the solute concentration $[ML^{-3}]$, R [dimensionless] is the retardation factor, Ω_{ex} $[ML^{-3}T^{-1}]$ represents the mass exchange rate of solutes per unit volume between the subsurface domain and discrete fractures, and λ is a first-order decay constant $[L^{-1}]$. Solute exchange with the outside of the simulation domain is represented by Q_c $[ML^{-3}T^{-1}]$ which represents a source or a sink to the porous medium system.

The equation for solute transport in a saturated fracture is:

$$-\nabla \bullet (\omega_f q_f C_f - \omega_f \theta_f D_f \nabla C_f) + \omega_f [R_f \lambda_f C_f] - \omega_f \Omega_f = \omega_f [\frac{\partial(\theta_f R_f C_f)}{\partial t} + \theta_f R_f \lambda_f C_f] \quad (6)$$

where C_f is the concentration in a fracture $[ML^{-3}]$, λ_f is a first-order decay constant $[L^{-1}]$, D_f is the hydrodynamic dispersion tensor of the fracture $[L^2T^{-1}]$, and R_f [dimensionless] represents retardation factor.

2.3. *Applying the Fractional Advection–Dispersion Equations to Quantify Transport*

If the solute is initially placed in the mobile zone (i.e., a fracture along the main flow direction), the corresponding fADE takes the form [55]:

$$\frac{\partial C_m}{\partial t} + \beta \frac{\partial^\gamma C_m}{\partial t^\gamma} = -\nabla \cdot [v\, C_m - D\, \nabla C_m] - \beta\, C_m(x, t=0)\, \frac{t^{-\gamma}}{\Gamma(1-\gamma)} \quad (7)$$

$$\frac{\partial C_{im}}{\partial t} + \beta \frac{\partial^\gamma C_{im}}{\partial t^\gamma} = -\nabla \cdot [v\, C_{im} - D\, \nabla C_{im}] + C_m(x, t=0)\, \frac{t^{-\gamma}}{\Gamma(1-\gamma)} \quad (8)$$

where C_m and C_{im} $[ML^{-3}]$ denote the solute concentration in the mobile and immobile phases, respectively; γ [dimensionless] $(0 < \gamma \leq 1)$ is the time index; β $[T^{\gamma-1}]$ is the fractional-order capacity coefficient; v $[LT^{-1}]$ and D $[L^2T^{-1}]$ are the effective velocity and dispersion coefficient, respectively; and $\Gamma(\cdot)$ is the Gamma function.

Meerschaert et al. [56] generalized the fADE (7) and (8) by introducing an exponentially-truncated power-law function, which is an incomplete gamma function, as the memory function:

$$g(t) = \int_t^\infty e^{-\lambda\, s} \frac{\gamma\, s^{-\gamma-1}}{\Gamma(1-\gamma)}\, ds \quad (9)$$

where $\lambda > 0$ $[T^{-1}]$ is the truncation parameter in time. This modification leads to the tempered time fADE (tt-fADE) [56,57]:

$$\frac{\partial C_m}{\partial t} + \beta\, e^{-\lambda t} \frac{\partial^\gamma}{\partial t^\gamma} \left[e^{\lambda t}\, C_m\right] - \beta\, \lambda^\gamma\, C_m = -\nabla \cdot [vC_m - D\, \nabla C_m] - \beta\, C_m^0 \int_t^\infty e^{-\lambda \tau} \frac{\tau^{-\gamma-1}}{\Gamma(1-\gamma)}\, d\tau \quad (10)$$

$$\frac{\partial C_{im}}{\partial t} + \beta\, e^{-\lambda t} \frac{\partial^\gamma}{\partial t^\gamma} \left[e^{\lambda t}\, C_{im}\right] - \beta\, \lambda^\gamma\, C_{im} = -\nabla \cdot [v\, C_{im} - D\, \nabla C_{im}] + C_m^0 \int_t^\infty e^{-\lambda \tau} \frac{\tau^{-\gamma-1}}{\Gamma(1-\gamma)}\, d\tau \quad (11)$$

where $C_m^0 = C_m(x, t=0)$ denotes the initial source, located in the mobile phase or fractures. A few parameters in the tt-fADE (10) and (11) including the effective velocity v and the dispersion coefficient

D, may be determined by field experiments (i.e., monitoring wells) and laboratory experiments (transport through fractured media), while the other parameters should be approximated given the plumes observed in the field. The poor predictability of the fADE model motivated us to explore the relationship between DFN properties and model parameters in Section 4.

At the late time

$$t << 1/\lambda \tag{12}$$

the tail of BTC for solutes in the mobile declines as a power-law function:

$$C_m(x,t) \propto t^{-1-\gamma} \tag{13}$$

The tail of the total-phase BTC also declines as a straight line in a log–log plot with a rate:

$$C(x,t) \propto t^{-\gamma} \tag{14}$$

while at a much later time $t >> 1/\lambda$, the slope of the mobile/total-phase BTC reaches infinity (i.e., the late-time BTC tail declines exponentially). Therefore, the value of λ controls the transition of the BTC late-time tail from a power-law function to exponential function [57].

In the following sections, we try to link the time index γ to characteristics of the fractured porous media.

3. Results of Monte Carlo Simulations

The ensemble average of BTCs for all 100 realizations for each DFN scenario are shown in Figure 2, along with the best-fit solutions using the tt-fADE model (10) and (11). Three scenarios of DFN were considered, by setting the total number of fractures to 20, 60, and 80, and holding mean fracture set orientations constant to 0° and 90°. An additional fourth scenario explored the impact of fracture orientation on transport dynamics, by changing mean fracture set orientation to −45° and 45° for a 60 fracture DFN.

The best-fit parameters of the tt-fADE model (10) and (11) for all DFNs are listed in Table 2. The effective velocity v, dispersion coefficient D, and fractional-order capacity coefficient β increase with an increasing fracture density. It is also noteworthy that β has the units of fractional-order in time, which can be converted to a dimensionless capacity coefficient $\beta_0 = \beta\ \gamma\ \lambda^{\gamma-1}$ (where β_0 represents the long-term ratio of immobile versus mobile mass). The other two parameters, including the time index γ and the truncation parameter λ, first decrease and then increase as the number of fractures increases. This trend implies that the DFN domain containing 60 fractures might exhibit higher degree of heterogeneity, and therefore translate from non-Fickian to Fickian transport at a much later time, compared to the DFN with 20 or 80 fractures. When the matrix hydraulic conductivity is set to be 10^{-7} m/s, the BTCs are more sensitive to fracture density. For example, when the number of fractures is 20, the peak concentration appears at day 365, while the DFN with 60 (or 80) fractures has the peak BTC at day 50 (or day 100). The truncation parameter λ can be calculated by the reciprocal of the cutoff time, representing the transition time from non-Fickian to Fickian transport; however, in this study, the power-law portion of BTCs persisted and the transition time could not be identified directly.

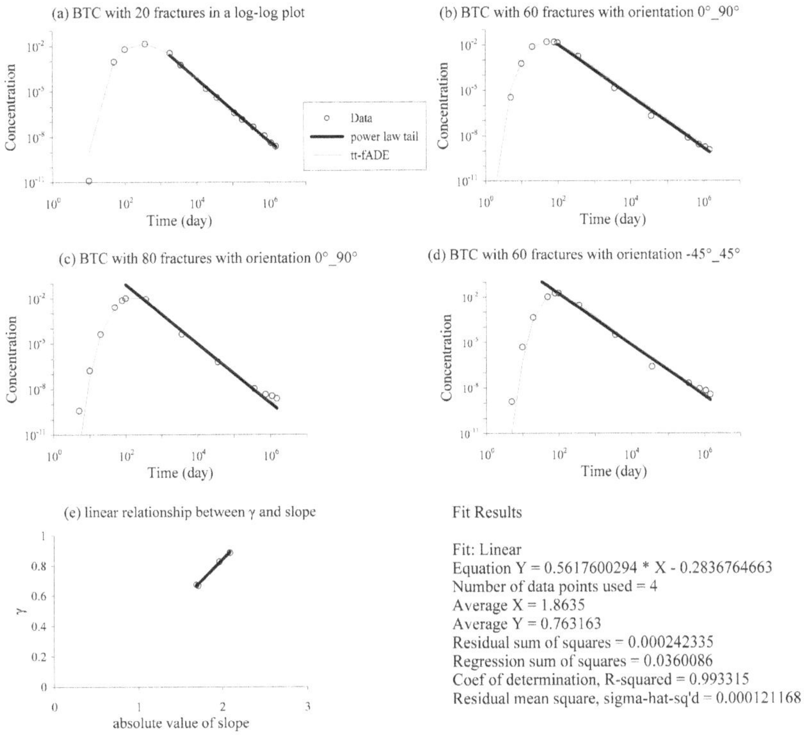

Figure 2. Breakthrough curves (BTCs) at the control plane $x = 50$ m for three different fracture densities (**a–c**); (**b**,**d**) have the same fracture density with different orientations. The BTC data (i.e., symbols) are simulated by HGS. The red line is the solution of the tempered time fADE (tt-fADE) model (10) and (11). The black line is a power law approximation to the late time decay in concentration. (**e**) shows the empirical linear relationship between the slope of the power-law decay in concentration and the time index γ in the tt-fADE model (10) and (11).

Table 2. Best-fit parameters for the tt-fADE model (10) and (11) for discrete fracture networks (DFNs) with different fracture densities and orientation scenarios. In the legend, v denotes the effective velocity, D is the dispersion coefficient, γ is the time/scale index, β is the fractional-order capacity coefficient, λ is the truncation parameter, and S_{BTC} stands for the late-time BTC slope in a log–log plot.

Fracture Numbers	v [m/day]	D [m^2/day]	β [day$^{\gamma-1}$]	γ [Dimensionless]	λ [day^{-1}]	S_{BTC} [Dimensionless]
20 (0°~90°)	0.46	9.40	2.17	0.885	3.29×10^{-10}	−2.088
60 (0°~90°)	10.53	40.36	5.46	0.667	3.76×10^{-14}	−1.712
80 (0°~90°)	12.11	93.84	27.66	0.827	1.75×10^{-8}	−1.965
60 (−45°~45°)	20.72	29.37	14.75	0.673	3.24×10^{-8}	−1.689

4. Discussion

4.1. Impact of Fracture Density on Non-Fickian Transport

The simulations indicated that fracture density dominates non-Fickian transport dynamics, and therefore significantly affects tt-fADE model parameters. For example, a denser DFN contains more connected flow paths, leading to faster plume movement (i.e., a larger effective velocity v), greater plume spreading in space (corresponding to a larger dispersion coefficient D), and more fracture/matrix interaction (i.e., tracer particles have a higher probability of entering the surrounding matrix, which correlates to a larger value of the capacity coefficient β). In addition, the complex relationship between DFN density and other tt-fADE model parameters (time index γ and truncation parameter λ) can be explained by the hypothesis that moderately dense DFNs (i.e., fracture scenario with 60 fractures) maximize variations in matrix–fracture exchange rates, which leads to the heaviest late-time BTC tail (i.e., the smallest time index γ) and the persistent power-law BTC (i.e., the smallest truncation parameter λ). DFNs with a broader range of densities are needed in a future study to further investigate the above hypothesis.

The dominant impact of fracture density on pollutant transport may be due to two reasons. First, the number of fractures significantly affects the architecture of DFNs, including both the interconnectivity of fractures and the thickness of surrounding matrix. It is well-known that the interconnected fracture network forms the preferred flow paths in the formation [22,58,59]. These pathways are "paths of least resistance", where most of the flow and solute is concentrated [7,26,36,60], resulting in early arrivals. Matrix block size, which is typically characterized using fracture spacing, affects trapping time and influences late-time tailing in BTCs. Second, networks containing higher fracture densities may yield a broader distribution of apertures [61], resulting in a larger variation of advective velocities in the system. Flow in the fracture is constrained to the two-dimensional domain of the fracture aperture, where average flux at any location within a fracture is proportional to the cube of the aperture [35,50,62]. Small variations in the aperture (which is 1×10^3~1.5×10^3 μm in this study), therefore, can cause large, non-linear variations in advective velocity.

4.2. Impact of Fracture Orientation on Non-Fickian Transport

Fracture orientation has a complex impact on pollutant transport behavior. The DFN with mean fracture set orientations of $-45°$ and $45°$ (where the overall flow direction is along $0°$) provides two major directions for pollutant particles to move, while the DFN with mean fracture set orientations of $0°$ and $90°$ offers horizontally dominant motion for pollutants. Hence, the former has a relatively larger effective velocity and less spatial spreading (i.e., a smaller dispersion coefficient). Mean fracture set orientations of $-45°$ and $45°$ also enhance the movement of solute particles into matrix blocks, resulting in a greater immobile mass (and hence a larger capacity coefficient β) than that for the cases with mean fracture set orientations of $0°$ and $90°$. It is noteworthy, however, that the BTC peak occurs at a time determined by both the effective velocity v and the capacity coefficient β. The larger for v and/or the smaller for β, the earlier for the BTC peak. Although the DFNs with mean fracture orientations of $-45°$ and $45°$ have a relatively larger v, the ensemble BTC peak appears later (at day 100) than that for the DFNs with mean fracture orientations of $0°$ and $90°$ (BTC peak occurs at day 50), because the DFNs with mean fracture orientations of $-45°$ and $45°$ have a much larger β, which acts as part of the retardation coefficient $(1 + \beta)$ when $\gamma = 1$. The DFNs with mean fracture orientations of $-45°$ and $45°$ might also produce more tortuous pathways, which lead to greater transverse dispersion than the DFNs with mean fracture orientations of $0°$ and $90°$ [19]. Finally, regardless of fracture orientation, late-time BTCs exhibit similar slopes, and therefore the similar time index γ, if the DFNs have the same density and a constant distribution of fracture length. The latter impact (caused by the fracture length distribution), however, cannot be reliably investigated in this study, due to the field-scale DFNs (with a relatively small domain size). In a future study, we will increase the model domain size (to regional scale) and systematically explore the impact of fracture length distributions on non-Fickian dynamics.

The above analysis implies that the fracture density is more important than the fracture orientation in affecting the power-law late-time tailing of the BTC, while both properties can affect the plume mean displacement (captured mainly by the velocity in the tt-fADE model (10) and (11)), plume spreading (captured by the dispersion coefficient in model (10) and (11)), and fracture-matrix mass exchange ratio (captured by the capacity coefficient in (10) and (11)).

4.3. Impact of Matrix Permeability on Late-Time BTC

Rock matrix permeability may affect the late-time behavior of chemical transport. To explore this potential impact, here we conducted additional Monte Carlo simulations using different matrix permeability: (1) increasing matrix permeability by 10 times (from 1.1×10^{-14} m^2 to 1.1×10^{-13} m^2) and (2) decreasing matrix permeability by 2 times (from 1.1×10^{-14} m^2 to 5.5×10^{-15} m^2). Results are shown in Figure 3. We did not select the smaller value for matrix permeability, because preliminary numerical tests showed that if matrix permeability was smaller than 1.1×10^{-15} m^2, the numerical iteration (in the flow model) could not converge.

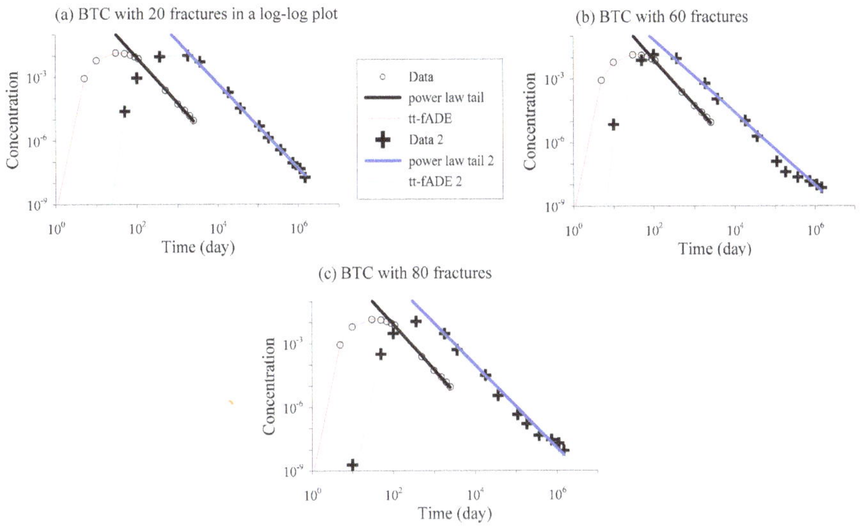

Figure 3. BTCs at the control plane $x = 50$ m for three different fracture densities (**a**–**c**) with matrix permeability of 1.1×10^{-13} m^2 (red lines and circles) and 5.5×10^{-15} m^2 (green lines and crosses). The BTC data (i.e., symbols) are simulated by HGS. The red/green line is the solution of the tt-fADE model (10) and (11). The black/blue line is a power-law approximation to the late time decay in concentration.

The BTC declines faster at the late time with an increasing matrix permeability. This is because the matrix domain with higher permeability leads to shorter trapping times for chemical particles. Contrarily, the average velocity is smaller for a rock matrix with smaller permeability, resulting in heavier late-time tailing in the BTC.

4.4. How to Approximate the Time Index γ Given the Limited Late-Time BTC

Although the transport time considered by our Monte Carlo simulation is very long (about 5000 years), we could not observe the transition in the BTC from persistent non-Fickian to Fickian transport behaviors (in the Monte Carlo simulations performed in this study, the total modeling time could not be longer than 2×10^6 days, since the concentration at the outlet after 2×10^6 days is so small that HGS generates negative BTCs). It is common that the measured BTC cannot last long enough to cover the whole power-law portion and its transition in the BTC at late times, due to the usually limited sampling period, as well as the low concentration at the late time, which might be below the detection limit. The time index γ cannot be directly estimated when the power-law portion of the BTC at the late time cannot be identified.

One possible way to solve the above challenge is to approximate γ using the BTC's recession limb. In this study, the recession limb of each measured BTC exhibits a straight line in a log–log plot (Figure 2), which indicates prolonged non-Fickian or anomalous transport. There is a linear relationship between the slope of the BTC recession limb (whose absolute value is denoted as $|S_{BTC}|$) and the time index γ in the tt-fADE model (10) and (11):

$$\gamma = 0.5618 \times |S_{BTC}| - 0.2837 \tag{15}$$

where the corresponding coefficient of determination is 0.9933 (i.e., strong correlation). Since $|S_{BTC}|$ can be measured easily (here "easily" means that one can use the observed BTC, shown for example in Figure 2, to directly obtain the slope of the BTC, which is the negative value of $|S_{BTC}|$), the above linear relationship might be practically useful. However, this approximation will be ineffective if the power-law portion of the BTC is too short to observe (which represents a fast transfer from non-Fickian to Fickian diffusion), or the data contains apparent noises (which can be common for the measurement of low concentrations at the BTC tails in the field).

4.5. Subdiffusive Transport Shown by Plume Snapshots

We further explored the spatial distribution of a chemical in the DFNs, by depicting plume snapshots at different times for a single realization, and the ensemble average of the Monte Carlo results (Figures 4 and 5).

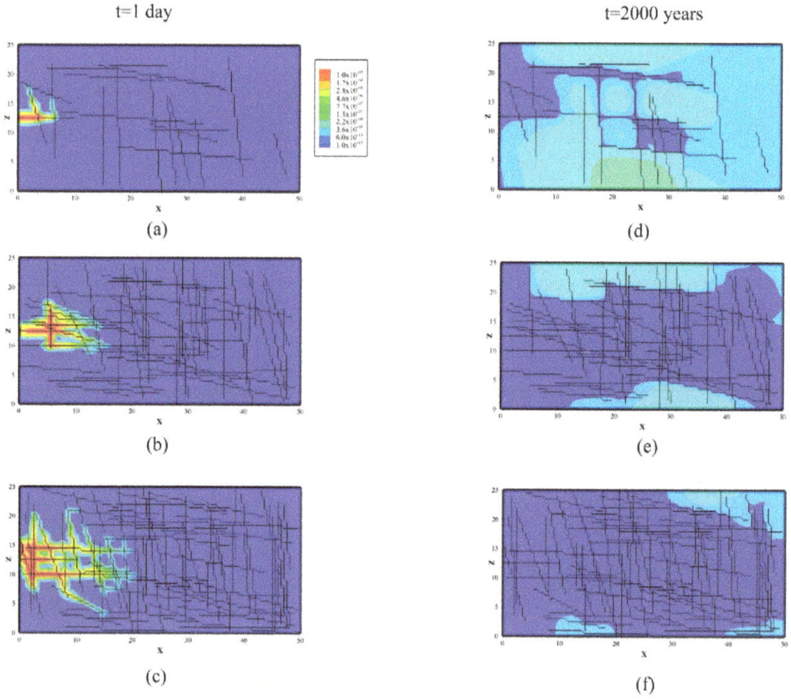

Figure 4. One realization of plume snapshots for the first time step (t = 1 day) for the three different fracture densities: (**a**) the DFNs with 20 fractures; (**b**) 60 fractures; and (**c**) 80 fractures. The 13th time step (t = 2000 years) for the three different fracture densities: (**d**) 20 fractures; (**e**) 60 fractures; and (**f**) 80 fractures.

We chose the first time step (t = 1 day) to show the early arrivals for the three different fracture densities with matrix permeability of 1.1×10^{-14} m^2. All the three cases show that chemical particles move along the preferential flow paths. With increasing fractures in the DFNs, the plume spreads to a wider area in the vertical direction with a more uniform moving front. For the 13th time step (t = 2000 years), the snapshots exhibited apparent retention and sequestration (meaning that the surrounding low-permeable matrix trapped chemical particles, especially near the source [39,63]) in the least-dense fractured network. Competition between channeling and sequestration leads to persistent non-Fickian dynamics. Fast- and slow-moving chemical particles are counterbalanced, and the effective velocity becomes larger for the DFNs with more fractures.

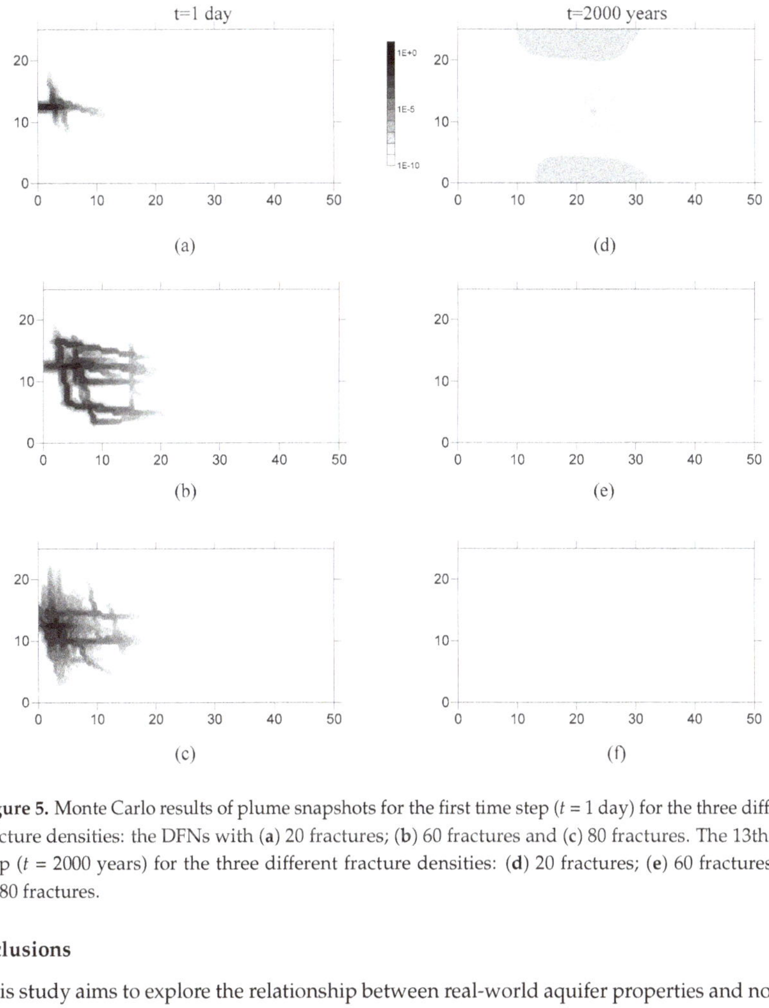

Figure 5. Monte Carlo results of plume snapshots for the first time step (t = 1 day) for the three different fracture densities: the DFNs with (**a**) 20 fractures; (**b**) 60 fractures and (**c**) 80 fractures. The 13th time step (t = 2000 years) for the three different fracture densities: (**d**) 20 fractures; (**e**) 60 fractures and (**f**) 80 fractures.

5. Conclusions

This study aims to explore the relationship between real-world aquifer properties and non-Fickian transport dynamics, so that the fractional partial differential equations built upon fractional calculus can be reliably applied with appropriate hydrogeologic interpretations. We use the Monte Carlo approach to generate field-scale DFNs where the fracture properties change systematically, and then to simulate groundwater flow and pollutant transport through the complex DFNs. For a point source located initially in the mobile phase or fracture, the late-time behavior for the BTCs simulated by the Monte Carlo approach is then explained by the tempered–stable time fractional advection–dispersion equation. We build the relationship between medium heterogeneity and transport dynamics through the combination of numerical experiments and stochastic analysis. The following five primary conclusions were found.

First, the DFN density is the major factor affecting non-Fickian transport at the late time, since (1) the number of fractures significantly affects the internal structure of DFNs; and (2) DFNs containing a higher fracture density may yield broader distribution of apertures, resulting in greater variation of advective velocities. A moderate dense DFN (filled with 60 fractures) results in the most heterogenous domain and the heaviest late-time tail in the BTCs, implying that there might exist a potential threshold of fracture density for non-Fickian dynamics when both the domain size and the fracture length distribution remain unchanged.

Second, for the DFNs with the same density and different mean orientations, the late time tailing behavior is affected by the ratio of fractures along different directions. More longitudinal fractures aligned with the overall flow direction will lead to an increase in earlier arrivals, while more transverse fractures, with flow deviating from the general flow direction, will enhance chemical particles to interact with the rock matrix, resulting in a greater immobile mass (a larger β in the tt-fADE) and a heavier late-time tailing. The BTC peak time is determined by both the effective velocity and capacity coefficient β (representing the immobile mass ratio). The time index and the slope of the late-time BTC, however, are similar for the DFNs with the similar density and different mean orientations, since the fracture density dominates the distribution of the mass exchange rate between fracture and matrix (and therefore defines the index of the time fractional derivative in the tt-fADE model).

Third, a rock matrix with higher permeability will lead to a relatively more homogeneous domain with a larger time index and a steeper slope in the late-time BTC, since the trapping times in rock matrix becomes relatively shorter.

Fourth, the DFN properties can be quantitatively linked to the slope of the recession limb of the BTC. Particularly, the time index γ in the tt-fADE model has a simple, linear empirical expression with the power-law slope of the BTC recession limb, which can help estimate the index γ in practical applications where the chemical transport time is usually not long enough to cover the full range of late-time transport behavior.

Fifth, plume snapshots show that all the three fracture densities built in this study exhibit early arrivals and sequestration effects. The denser DFNs can produce more early arrivals, due to more preferential flow paths, while the sparse DFNs tend to show heavier sequestration near the source. Therefore, competition between the channeling effect (in relative high-permeable fractures or the mobile phase) and the trapping effect (in the surrounding rock matrix or immobile domains) results in complex prolonged non-Fickian characteristics in chemical transport. With predictable parameters, the fractional partial differential equations can be used as an efficient upscaling tool to address real-world contamination problems.

Acknowledgments: This work was funded partially by the National Natural Science Foundation of China under grants 41628202 and 41330632, and the University of Alabama (RGC). This paper does not necessarily reflect the views of the funding agencies.

Author Contributions: The results are part of Bingqing Lu's dissertation under Yong Zhang's supervision. Donald M. Reeves, HongGuang Sun and Chunmiao Zheng contributed to manuscript preparation and revision.

Conflicts of Interest: The authors declare no conflict of interest.

References

1. Gorenflo, R.; Mainardi, F. Fractional Calculus. In *Fractals and Fractional Calculus in Continuum Mechanics*; Mainardi, F., Ed.; Springer: Vienna, Austria, 1997; pp. 223–276.
2. Metzler, R.; Klafter, J. The random walk's guide to anomalous diffusion: A fractional dynamics approach. *Phys. Rep.* **2000**, *339*, 1–77. [CrossRef]
3. Zhang, Y.; Sun, H.G.; Stowell, H.H.; Zayernouri, M.; Hansen, S.E. A review of applications of fractional calculus in Earth system dynamics. *Chaos Solitons Fractals* **2017**, *102*, 29–46. [CrossRef]

4. Fogg, G.E.; Zhang, Y. Debates—Stochastic subsurface hydrology from theory to practice: A geologic perspective. *Water Resour. Res.* **2016**, *52*, 9235–9245. [CrossRef]
5. Coats, K.H.; Smith, B.D.; van Genuchten, M.T.; Wierenga, P.J. Dead-End Pore Volume and Dispersion in Porous Media. *Soil Sci. Soc. Am. J.* **1964**, *4*, 73–84. [CrossRef]
6. Haggerty, R.; McKenna, S.; Meigs, L.C. On the late-time behaviour of tracer breakthrough curves. *Water Resour. Res.* **2000**, *36*, 3467–3479. [CrossRef]
7. Klepikova, M.V.; Le Borgne, T.; Bour, O.; Dentz, M.; Hochreutener, R.; Lavenant, N. Heat as a tracer for understanding transport processes in fractured media: Theory and field assessment from multiscale thermal push-pull tracer tests. *Water Resour. Res.* **2016**, *52*, 5442–5457. [CrossRef]
8. Becker, M.W.; Shapiro, A.M. Tracer transport in fractured crystalline rock: Evidence of nondiffusive breakthrough tailing. *Water Resour. Res.* **2000**, *36*, 1677–1686. [CrossRef]
9. Kang, P.K.; Le Borgne, T.; Dentz, M.; Bour, O.; Juanes, R. Impact of velocity correlation and distribution on transport in fractured media: Field evidence and theoretical model. *Water Resour. Res.* **2015**, *51*, 940–959. [CrossRef]
10. Becker, M.W.; Shapiro, A.M. Interpreting tracer breakthrough tailing from different forced-gradient tracer experiment configurations in fractured bedrock. *Water Resour. Res.* **2003**, *39*. [CrossRef]
11. Benson, D.A.; Wheatcraft, S.W.; Meerschaert, M.M. Application of a fractional advection-dispersion equation. *Water Resour. Res.* **2000**, *36*, 1403–1412. [CrossRef]
12. Chang, F.X.; Chen, J.; Huang, W. Anomalous diffusion and fractional advection-diffusion equation. *Acta Phys. Sin.* **2005**, *54*, 1113–1117.
13. Huang, G.; Huang, Q.; Zhan, H. Evidence of one-dimensional scale-dependent fractional advection-dispersion. *J. Contam. Hydrol.* **2006**, *85*, 53–71. [CrossRef] [PubMed]
14. Green, C.T.; Zhang, Y.; Jurgens, B.C.; Starn, J.J.; Landon, M.K. Accuracy of travel time distribution (TTD) models as affected by TTD complexity, observation errors, and model and tracer selection. *Water Resour. Res.* **2014**, *50*, 6191–6213. [CrossRef]
15. Garrard, R.M.; Zhang, Y.; Wei, S.; Sun, H.; Qian, J. Can a Time Fractional-Derivative Model Capture Scale-Dependent Dispersion in Saturated Soils? *Groundwater* **2017**, *55*, 857–870. [CrossRef] [PubMed]
16. Schumer, R.; Meerschaert, M.M.; Baeumer, B. Fractional advection-dispersion equations for modeling transport at the Earth surface. *J. Geophys. Res. Earth Surf.* **2009**, *114*, 1–15. [CrossRef]
17. Drummond, J.D.; Aubeneau, A.F.; Packman, A.I. Stochastic modeling of fine particulate organic carbon dynamics in rivers. *Water Resour. Res.* **2014**, *50*, 4341–4356. [CrossRef]
18. Sun, H.G.; Chen, D.; Zhang, Y.; Chen, L. Understanding partial bed-load transport: Experiments and stochastic model analysis. *J. Hydrol.* **2015**, *521*, 196–204. [CrossRef]
19. Reeves, D.M.; Benson, D.A.; Meerschaert, M.M.; Scheffler, H.P. Transport of conservative solutes in simulated fracture networks: 2. Ensemble solute transport and the correspondence to operator-stable limit distributions. *Water Resour. Res.* **2008**, *44*, 1–20. [CrossRef]
20. Zhang, Y.; Baeumer, B.; Reeves, D.M. A tempered multiscaling stable model to simulate transport in regional-scale fractured media. *Geophys. Res. Lett.* **2010**, *37*, 1–5. [CrossRef]
21. Berkowitz, B. Characterizing flow and transport in fractured geological media: A review. *Adv. Water Resour.* **2002**, *25*, 861–884. [CrossRef]
22. Neuman, S.P. Trends, prospects and challenges in quantifying flow and transport through fractured rocks. *Hydrogeol. J.* **2005**, *13*, 124–147. [CrossRef]
23. Zhao, Z.; Rutqvist, J.; Leung, C.; Hokr, M.; Liu, Q.; Neretnieks, I.; Hoch, A.; Havlíček, J.; Wang, Y.; Wang, Z.; et al. Impact of stress on solute transport in a fracture network: A comparison study. *J. Rock Mech. Geotech. Eng.* **2013**, *5*, 110–123. [CrossRef]
24. Zhao, Z.; Jing, L.; Neretnieks, I.; Moreno, L. Numerical modeling of stress effects on solute transport in fractured rocks. *Comput. Geotech.* **2011**, *38*, 113–126. [CrossRef]
25. Selroos, J.; Walker, D.D.; Ström, A.; Gylling, B.; Follin, S. Comparison of alternative modelling approaches for groundwater flow in fractured rock. *J. Hydrol.* **2002**, *257*, 174–188. [CrossRef]
26. Mukhopadhyay, S.; Cushman, J.H. Monte Carlo Simulation of Contaminant Transport: II. Morphological Disorder in Fracture Connectivity. *Transp. Porous Media* **1998**, *31*, 183–211. [CrossRef]
27. Outters, N. *A Generic Study of Discrete Fracture Network Transport Properties Using FracMan/MAFIC*; SKB-R-03-13; International Atomic Energy Agency (IAEA): Vienna, Austria, 2003.

28. Gustafson, G.; Fransson, Å. The use of the Pareto distribution for fracture transmissivity assessment. *Hydrogeol. J.* **2006**, *14*, 15–20. [CrossRef]
29. Liu, R.; Li, B.; Jiang, Y. A fractal model based on a new governing equation of fluid flow in fractures for characterizing hydraulic properties of rock fracture networks. *Comput. Geotech.* **2016**, *75*, 57–68. [CrossRef]
30. Zhang, Y.; Benson, D.A.; Meerschaert, M.M.; Labolle, E.M.; Scheffler, H.P. Random walk approximation of fractional-order multiscaling anomalous diffusion. *Phys. Rev. E Stat. Nonlinear Soft Matter Phys.* **2006**, *74*, 1–10. [CrossRef] [PubMed]
31. Cortis, A.; Birkholzer, J. Continuous time random walk analysis of solute transport in fractured porous media. *Water Resour. Res.* **2008**, *44*, 1–11. [CrossRef]
32. McKenna, S.A.; Meigs, L.C.; Haggerty, R. Tracer tests in a fractured dolomite: 3. Double porosity, multiple-rate mass transfer processes in convergent flow tracer tests. *Water Resour. Res.* **2001**, *37*, 1143–1154. [CrossRef]
33. Cook, P.G. *A Guide to Regional Groundwater Flow in Fractured Rock Aquifers*; CSIRO: Canberra, Australia, 2003; p. 107.
34. Bear, J. *Dynamics of Fluids in Porous Media*; Dover Publications: New York, NY, USA, 1972; ISBN 0-486-65675-6.
35. Painter, S.; Cvetkovic, V. Upscaling discrete fracture network simulations: An alternative to continuum transport models. *Water Resour. Res.* **2005**, *41*, 1–10. [CrossRef]
36. Lei, Q. Characterisation and Modelling of Natural Fracture Networks: Geometry, Geomechanics and Fluid Flow. Ph.D. Thesis, Imperial College London, London, UK, 2016.
37. Bakshevskaia, V.A.; Pozdniakov, S.P. Simulation of Hydraulic Heterogeneity and Upscaling Permeability and Dispersivity in Sandy-Clay Formations. *Math. Geosci.* **2016**, *48*, 45–64. [CrossRef]
38. Kang, P.K.; Dentz, M.; Le Borgne, T.; Juanes, R. Anomalous transport on regular fracture networks: Impact of conductivity heterogeneity and mixing at fracture intersections. *Phys. Rev. E Stat. Nonlinear Soft Matter Phys.* **2015**, *92*, 1–15. [CrossRef] [PubMed]
39. Zhang, Y.; Benson, D.A.; Baeumer, B. Predicting the tails of breakthrough curves in regional-scale alluvial systems. *Ground Water* **2007**, *45*, 473–484. [CrossRef] [PubMed]
40. Zhang, Y.; Benson, D.A.; LaBolle, E.M.; Reeves, D.M. *Spatiotemporal Memory and Conditioning on Local Aquifer Properties*; DOE/NV/0000939-01, Publication #45244; Desert Research Institute: Las Vegas, NV, USA, 2009; p. 64.
41. Kelly, J.F.; Bolster, D.; Meerschaert, M.M.; Drummond, J.D.; Packman, A.I. FracFit: A robust parameter estimation tool for fractional calculus models. *Water Resour. Res.* **2017**, *53*, 2559–2567. [CrossRef]
42. Cacas, M.C.; Ledoux, E.; de Marsily, G.; Tillie, B.; Barbreau, A.; Durand, E.; Feuga, B.; Peaudecerf, P. Modeling fracture flow with a stochastic discrete fracture network: Calibration and validation: 1. The flow model. *Water Resour. Res.* **1990**, *26*, 479–489. [CrossRef]
43. McClure, M.W.; Mark, W.; Horne, R.N. *Discrete Fracture Network Modeling of Hydraulic Stimulation: Coupling Flow and Geomechanics*; Springer: Berlin, Germany, 2013; ISBN 3319003836.
44. Therrien, R.; McLaren, R.G.; Sudicky, E.; Panday, S.M. *HydroGeoSphere a Three-dimensional Numerical Model Describing Fully-integrated Subsurface and Surface Flow and Solute Transport*; Groundwater Simulations Group, University of Waterloo: Waterloo, ON, Canada, 2010; p. 429.
45. Chakraborty, P.; Meerschaert, M.M.; Lim, C.Y. Parameter estimation for fractional transport: A particle-tracking approach. *Water Resour. Res.* **2009**, *45*. [CrossRef]
46. Geiger, S.; Cortis, A.; Birkholzer, J.T. Upscaling solute transport in naturally fractured porous media with the continuous time random walk method. *Water Resour. Res.* **2010**, *46*. [CrossRef]
47. Long, J.C.S.; Remer, J.S.; Wilson, C.R.; Witherspoon, P.A. Porous media equivalents for network of discontinuous fractures. *Water Resour. Res.* **1982**, *18*, 645–658. [CrossRef]
48. Fiori, A.; Becker, M.W. Power law breakthrough curve tailing in a fracture: The role of advection. *J. Hydrol.* **2015**, *525*, 706–710. [CrossRef]
49. Klimczak, C.; Schultz, R.A.; Parashar, R.; Reeves, D.M. Cubic law with aperture-length correlation: Implications for network scale fluid flow. *Hydrogeol. J.* **2010**, *18*, 851–862. [CrossRef]
50. Keller, A.A.; Roberts, P.V.; Blunt, M.J. Effect of fracture aperture variations on the dispersion of contaminants. *Water Resour. Res.* **1999**, *35*, 55–63. [CrossRef]
51. Zhao, Z.; Li, B.; Jiang, Y. Effects of fracture surface roughness on macroscopic fluid flow and solute transport in fracture networks. *Rock Mech. Rock Eng.* **2014**, *47*, 2279. [CrossRef]

52. Edery, Y.; Geiger, S.; Berkowitz, B. Structural controls on anomalous transport in fractured porous rock. *Water Resour. Res.* **2016**, *52*, 5634–5643. [CrossRef]
53. Callahan, T.J.; Reimus, P.W.; Bowman, R.S.; Haga, M.J. Using multiple experimental methods to determine fracture/matrix interactions and dispersion of nonreactive solutes in saturated volcanic tuff. *Water Resour. Res.* **2000**, *36*, 3547–3558. [CrossRef]
54. Therrien, R.; Sudicky, E. Three-dimensional analysis of variably-saturated flow and solute transport in discretely-fractured porous media. *J. Contam. Hydrol.* **1996**, *23*, 1–44. [CrossRef]
55. Schumer, R.; Benson, D.A.; Meerschaert, M.; Baeumer, B. Fractal mobile/immobile solute transport. *Water Resour. Res.* **2003**, *39*. [CrossRef]
56. Meerschaert, M.M.; Zhang, Y.; Baeumer, B. Tempered anomalous diffusion in heterogeneous systems. *Geophys. Res. Lett.* **2008**, *35*, 1–5. [CrossRef]
57. Zhang, Y.; Green, C.T.; Baeumer, B. Linking aquifer spatial properties and non-Fickian transport in mobile-immobile like alluvial settings. *J. Hydrol.* **2014**, *512*, 315–331. [CrossRef]
58. Moreno, L.; Tsang, C.F. Multiple-Peak Response to Tracer Injection Tests in Single Fractures: A Numerical Study. *Water Resour. Res.* **1991**, *27*, 2143–2150. [CrossRef]
59. Gerke, H.H. Preferential flow descriptions for structured soils. *J. Plant Nutr. Soil Sci.* **2006**, *169*, 382–400. [CrossRef]
60. Reeves, D.M.; Parashar, R.; Pohll, G.; Carroll, R.; Badger, T.; Willoughby, K. Practical guidelines for horizontal hillslope drainage networks in fractured rock. *Eng. Geol.* **2013**, *163*, 132–143. [CrossRef]
61. Neuman, S.P. Multiscale relationships between fracture length, aperture, density and permeability. *Geophys. Res. Lett.* **2008**, *35*, L22402. [CrossRef]
62. Hirthe, E.M.; Graf, T. Fracture network optimization for simulating 2D variable-density flow and transport. *Adv. Water Resour.* **2015**, *83*, 364–375. [CrossRef]
63. LaBolle, E.M.; Fogg, G.E. Role of Molecular Diffusion in Contaminant Migration and Recovery in an Alluvial Aquifer System. *Transp. Porous Media* **2001**, *42*, 155–179. [CrossRef]

mathematics

MDPI

Article

Analysis of PFG Anomalous Diffusion via Real-Space and Phase-Space Approaches

Guoxing Lin

Carlson School of Chemistry and Biochemistry, Clark University, Worcester, MA 01610, USA; glin@clarku.edu; Tel.: +1-50-8793-7594

Received: 12 December 2017; Accepted: 22 January 2018; Published: 29 January 2018

Abstract: Pulsed-field gradient (PFG) diffusion experiments can be used to measure anomalous diffusion in many polymer or biological systems. However, it is still complicated to analyze PFG anomalous diffusion, particularly the finite gradient pulse width (FGPW) effect. In practical applications, the FGPW effect may not be neglected, such as in clinical diffusion magnetic resonance imaging (MRI). Here, two significantly different methods are proposed to analyze PFG anomalous diffusion: the effective phase-shift diffusion equation (EPSDE) method and a method based on observing the signal intensity at the origin. The EPSDE method describes the phase evolution in virtual phase space, while the method to observe the signal intensity at the origin describes the magnetization evolution in real space. However, these two approaches give the same general PFG signal attenuation including the FGPW effect, which can be numerically evaluated by a direct integration method. The direct integration method is fast and without overflow. It is a convenient numerical evaluation method for Mittag-Leffler function-type PFG signal attenuation. The methods here provide a clear view of spin evolution under a field gradient, and their results will help the analysis of PFG anomalous diffusion.

Keywords: pulsed-field gradient (PFG) anomalous diffusion; fractional derivative; nuclear magnetic resonance (NMR); magnetic resonance imaging (MRI)

1. Introduction

Anomalous dynamic behavior exists in many polymer or biological systems [1–6]. These systems often consist of various molecules such as macromolecules and small penetrant molecules like water. These molecules often demonstrate anomalous dynamics behavior: their rotational motion time constants usually obey a stretched exponential distribution, namely, the Kohlrausch–Williams–Watts (KWW) function distribution [7,8], and their translational diffusion jump time and jump length could follow power law distributions [2]. Nuclear magnetic resonance (NMR) is one of the important techniques used to detect these anomalous dynamic behaviors. For instance, anomalous diffusion can be detected by pulsed-field gradient (PFG) NMR experiments.

PFG diffusion NMR [9–15] has been a powerful tool for measuring normal diffusion. The history of using a field gradient to measure diffusion can be tracked back to Hahn, who first observed the influence of the molecule diffusion under a gradient magnetic field upon echo amplitudes in 1950 [9]. PFG diffusion measurements have broad applications in NMR and magnetic resonance imaging (MRI) [13–16]. For instance, the water diffusion difference in different parts of the brain can be used as an important contrast factor to build imaging of acute strokes [16]. The PFG technique, being an ultra-valuable tool for normal disunion, could play an increasingly important role in monitoring anomalous diffusion occurring in many polymer and biological systems.

PFG anomalous diffusion [17–23] is different from PFG normal diffusion. Unlike normal diffusion, anomalous diffusion has a non-Gaussian probability distribution [2,24], and its mean square displacement is not linearly proportional to diffusion time [2]. These non-Gaussian characteristics make it complicated to analyze PFG anomalous diffusion. Although PFG anomalous diffusion can be approximately analyzed by traditional methods such as the apparent diffusion coefficient method, PFG theories based on the fractional derivative could not only improve the analysis accuracy on the diffusion domain size, diffusion constant, and other variables [17,18,25–28], but also yield additional information such as the time derivative order α and space derivative order β that are related with diffusion jump time and jump length distributions determined by material properties.

Much effort has been devoted to studying PFG anomalous diffusion based on fractional calculus, which includes the propagator method [29], the modified Bloch equation method [17,25,30,31], the effective phase-shift diffusion equation (EPSDE)method [18], the instantaneous signal attenuation method [26], the modified-Gaussian or non-Gaussian distribution method [27], etc. Additionally, PFG anomalous diffusions in restricted geometries such as plate, sphere, and cylinder have been investigated [28]. These theoretical methods analyze PFG anomalous diffusion from different angles. Therefore, each of them can have its own advantages in handling certain types of PFG anomalous diffusion. To better apply the PFG technique to studying anomalous diffusion, it is still valuable to develop new theoretical treatments for PFG anomalous diffusion.

In this paper, two methods based on the fractional derivative [1,2,6,32–34] are proposed to give general analytical PFG signal attenuation expressions for anomalous diffusion. The first method is the recently developed EPSDE method [18]. This method describes the spin phase evolution by an effective phase diffusion process in virtual phase space [18]. Solving the EPSDE gives valuable information about the phase evolution process such as the phase probability distribution function and the moment of mean phase displacement. Meanwhile, other conventional methods render it difficult to get this phase information, and usually assume an approximate phase distribution such as Gaussian phase distribution [13]. In this paper, it will be shown that a solvable PFG signal attenuation equation can be derived by applying a Fourier transform to the effective phase-shift equation. The second method is to observe the signal intensity at the origin, and is an ultra-simple new method. For a homogeneous diffusion spin system, although the magnetization amplitude attenuates because of the gradient magnetic field effect, the phase of magnetization keeps constant at the origin of the gradient field. Such a specific phase property is employed to derive a PFG signal attenuation equation in this paper. The above two methods give the same signal attenuation equation, from which the general PFG signal attenuation expression can be derived by the Adomian decomposition method [35–39]. Besides the Adomian decomposition method, a direct integration method was proposed for the numerical evaluation of the PFG signal attenuation, which is a fast and simple method. The results include the finite gradient pulse width (FGPW) effect [13–15], namely, the signal attenuation during each gradient pulse application period. Theoretically, during a short gradient pulse, the PFG signal attenuation can be neglected; nevertheless, the gradient pulse used in a clinical MRI is usually long [16,40]. Additionally, a longer gradient pulse allows the measuring of slower diffusion under the same gradient maximum intensity, which matters in the study of polymer and biological systems where the molecule diffusion is often slow. Therefore, we need to consider the FGPW effect in many real applications. The two methods here give the same results in terms of the FGPW effect. These results agree with reported results from some other methods [18,26,27] and the continuous time random walk simulation [26,29]. Furthermore, PFG anomalous diffusion of intramolecular multiple quantum coherence (MQC) is also discussed [27]. The MQC effect has the benefit of enhancing the gradient effect on PFG signal attenuation [41]. The two methods provide complementary views of PFG anomalous diffusion from both the real space and phase space.

2. Theory

2.1. The Phase-Space Method: The Effective Phase-Shift Diffusion Method

For the sake of simplicity, only one-dimensional anomalous diffusion is considered in this paper. In PFG experiments, the field gradient pulse results in a time- and space-dependent magnetic field $B(z,t) = B_0 + g(t) \cdot z$, where B_0 is the exterior magnetic field, z is the position, and $g(t)$ is the time-dependent gradient [13–15]. The magnetic field exerts a torque on each spin moment. The torque changes the spin angular momentum direction, which leads the spin to process about the magnetic field with Larmor frequency $\omega(z,t) = -\gamma B(z,t)$. In a rotating frame with angular frequency $\omega_0 = -\gamma B_0$, a diffusing spin at position $z(t')$ has a time-dependent angular frequency $\gamma g(t') \cdot z(t')$, and its phase accumulated along the diffusion path is [13–15,19]

$$\phi(t) = -\int_0^t \gamma g(t') \cdot z(t') dt', \tag{1}$$

where $\phi(t)$ is the net-accumulating phase. The range of $\phi(t)$ is $-\infty < \phi(t) < \infty$ rather than $-\pi \leq \phi(t) \leq \pi$, and $\cos(\phi(t))$ is the projection factor of the spin magnetization to the observing coordinate axis. The PFG signal comes from the ensemble contribution from all spins by averaging over all possible phases [13–15]:

$$S(t) = S(0) \int_{-\infty}^{\infty} P(\phi, t) \exp(+i\phi) d\phi, \tag{2}$$

where $S(0)$ is the signal intensity at the beginning of the first gradient pulse, $S(t)$ is the signal intensity at time t, and $P(\phi, t)$ is the accumulating phase probability distribution function. In a symmetric diffusion system, Equation (2) can be further written as

$$S(t) = S(0) \int_{-\infty}^{\infty} P(\phi, t) \cos(\phi) d\phi. \tag{3}$$

Because the NMR signal comes from numerous spins, the percentage of spins with potentially infinite $\phi(t)$ is very small, and can therefore be neglected. Additionally, $-1 \leq \cos(\phi(t)) \leq 1$; therefore, $0 \leq |S(t)| \leq S(0)$, and $S(t)$ is a finite quantity that can be measured in NMR experiments. A possibly infinite "mean square phase displacement" $\left\langle |\phi(t)|^\beta \right\rangle$ is not necessarily an obstacle in PFG NMR measurement, although it may have a significant effect in other academic fields. The PFG signal attenuation can be obtained based on Equation (2) or (3), but the phase evolution process based on the phase path integral Equation (1) is complicated. Nevertheless, the recently developed EPSDE method provides a simple way to describe the phase evolution based on an effective phase diffusion process [18]. In the following, the effective phase diffusion equation method [18] will be briefly reintroduced. Additionally, a general solution including the FGPW effect will be given that has not been reported in [18].

For simplicity, only diffusion with a symmetric probability distribution is studied here. The self-diffusion process can be described by a random walk, which consists of a sequence of independent random jumps with waiting times $\Delta t_1, \Delta t_2, \Delta t_3, \ldots, \Delta t_n$, and corresponding displacement lengths $\Delta z_1, \Delta z_2, \Delta z_3, \ldots, \Delta z_n$. Based on the random walk, Equation (1) can be rewritten as [18]

$$\begin{aligned} \phi(t_{tot}) &= -\sum_{i=1}^{n} \gamma \Delta t_i g(t_i) \cdot \left(\sum_{m=1}^{i} \Delta z_m + z_0 \right) \\ &= -\gamma \sum_{m=1}^{n} \left[\sum_{i}^{n} \Delta t_i g(t_i) - \sum_{i}^{m-1} \Delta t_i g(t_i) \right] \cdot \Delta z_m - \gamma \sum_{i=1}^{n} \Delta t_i g(t_i) \cdot z_0 \\ &= -\sum_{m=1}^{n} \left[K(t) - K(t_{m-1}) \right] \cdot \Delta z_m - K(t) \cdot z_0 \\ &= -\sum_{m=1}^{n} \left[K(t) - K(t_m) \right] \cdot \Delta z_m - K(t) \cdot z_0 \end{aligned} \tag{4}$$

where

$$K(t) = \int_0^t \gamma g(t')dt' \tag{5}$$

is the wavenumber [13–15]. In Equation (4), $K(t_m) \approx K(t_{m-1})$ is used because $[K(t_m) - K(t_{m-1})] \cdot \Delta z_m$ is negligible, and the temporal and spatial summation orders are interchanged because $\Delta t_i g(t_i)$ and Δz_m are often noncorrelated. In most PFG experiments, $K(t_{tot}) = 0$ where t_{tot} is the time at the end of the rephasing gradient pulse, so Equation (4) can be further simplified to [18]

$$\phi(t_{tot}) = -\sum_m K(t_m) \cdot \Delta z_m. \tag{6}$$

The $-\sum\limits_m K(t_m) \cdot \Delta z_m$ term in Equation (6) is an effective phase random walk with a jump length $-K(t_m) \cdot \Delta z_m$ and a jump waiting time Δt_m The phase random walk can be treated as an effective phase diffusion, and can be obtained from the corresponding spin diffusion in the real space by scaling the jump length $\Delta z(t)$ by a factor $-K(t)$. Because $\left\langle |\Delta z|^\beta \right\rangle / \langle \Delta t^\alpha \rangle \propto D_f$ and $\left\langle |K(t)\Delta z|^\beta \right\rangle / \langle \Delta t^\alpha \rangle \propto D_{\phi eff}(t)$, the effective phase diffusion constant is $D_{\phi eff}(t) = K^\beta(t) D_f$ (oftentimes $K(t) \geq 0$) [18], where D_f and $D_{\phi eff}(t)$ are the fractional diffusion coefficients for the real-space diffusion and the effective phase diffusion, respectively, and their units are m^β / s^α and rad^β / s^α. As both the effective phase diffusion and the real spin particle diffusion belong to the same type of diffusion, they should obey the same type of diffusion equation. The one-dimensional real-space diffusion equation based on the fractional derivative is [6,31–34]

$${}_tD_*^\alpha M_{xy}(z,t) = D_f \frac{\partial^\beta}{\partial |z|^\beta} M_{xy}(z,t), \tag{7}$$

where $M_{xy}(z,t)$ is the spin magnetization in PFG experiments, and ${}_tD_*^\alpha$ is the Caputo fractional derivative defined as [6,31–34]

$${}_tD_*^\alpha f(t) := \begin{cases} \frac{1}{\Gamma(m-\alpha)} \int_0^t \frac{f^m(\tau)d\tau}{(t-\tau)^{\alpha+1-m}}, & m-1 < \alpha < m \\ \frac{d^m}{dt^m} f(t), & \alpha = m \end{cases} \tag{8}$$

and $\frac{\partial^\beta}{\partial |z|^\beta}$ is the space fractional derivative [6,31–34] defined in Appendix A. By replacing coordinate z with ϕ, and D_f with $D_{\phi eff}(t)$, respectively, from real-space diffusion Equation (9), the effective phase diffusion equations can be obtained [18] as

$${}_tD_*^\alpha P(\phi,t) = K^\beta(t) D_f \frac{\partial^\beta}{\partial |\phi|^\beta} P(\phi,t). \tag{9}$$

As $\hat{F}\{f(x,t)\} = f(q,t)$ and $\hat{F}\left\{ \frac{\partial^\beta}{\partial |x|^\beta} f(x,t); q \right\} = -q^\beta f(q)$ [31–34] where q is the wavenumber, applying a Fourier transform $\hat{F}\{f(x)\} = \int_{-\infty}^{\infty} f(x) \exp(+iqx)dx$ to both sides of Equation (9) gives

$${}_tD_*^\alpha P(q,t) = -K^\beta(t) q^\beta D_f P(q,t), \tag{10}$$

which is a q-space phase diffusion equation. Here, q is the wavenumber dedicated to the above Fourier transform, which differs from the field gradient pulse-induced wavenumber $K(t)$ defined by Equation (5). When q = 1, $P(q,t)$ in Equation (10) equals the attenuated signal amplitude because, in PFG experiments, the signal attenuation can be calculated based on Equation (2) as [18]

$$S(t) = S(0) \int_{-\infty}^{\infty} P(\phi,t) \exp(+i\phi)d\phi = P(1,t). \tag{11}$$

The normalized value of $S(0)$ equaling 1 will be used and dropped out throughout this paper. Substituting Equation (11) and $q = 1$ into Equation (10), we get

$$ {}_tD_*^{\alpha} S(t) = -D_f K^{\beta}(t) S(t), \tag{12} $$

which is the PFG signal attenuation equation.

The solution of Equation (12) will be given in Section 2.3 as the same equation will be obtained by observing the signal intensity at the origin in the subsequent section.

2.2. Observing the Signal Intensity at the Origin in Real Space

For a nondiffusing spin in a rotating frame with angular frequency $\omega_0 = -\gamma B_0$, its precession phase can be described as [13–15]

$$ \psi(z,t) = (\omega - \omega_0)t = -K(t) \cdot z, \tag{13} $$

where $\psi(z,t)$ is the precession phase, which is different from the accumulated phase $\phi(t)$ described by (1). The nondiffusing spin system has a time- and space-modulated magnetization, which can be described as

$$ M_{xy}(z,t) = S(t) \exp(-iK(t) \cdot z), \tag{14} $$

where $S(t)$ is the magnetization amplitude, which can be referred to as signal intensity, and $\exp(-iK(t) \cdot z)$ is the phase term. For a pulsed gradient spin echo (PGSE) or pulsed gradient stimulated echo (PGSTE) experiment with a constant gradient, as shown in Figure 1, the wavenumber is

$$ K(t) = \begin{cases} \gamma g t, \ 0 < t \leq \delta, \text{ dephasing gradient} \\ \gamma g \delta, \ \delta < t \leq \Delta \\ \gamma g(\Delta + \delta - t), \ \Delta < t \leq \Delta + \delta, \text{ rephasing gradient} \end{cases} \tag{15} $$

where δ is the gradient pulse length, and Δ is the diffusion delay starting from the beginning of the first dephasing gradient pulse to the beginning of the last rephrasing gradient pulse. Because of the two counteracting gradient pulses—a dephasing pulse and a rephasing pulse—$K(t)$ returns to 0 at the end of the rephasing pulse, and the phase of nondiffusing spins will be refocused. Therefore, the signal of the nondiffusing spin system does not attenuate when the T_2 relaxation is neglected.

For the spin diffusion in a homogeneous sample, its magnetization still can be described by Equation (14), $M_{xy}(z,t) = S(t) \exp(-iK(t) \cdot z)$. At a random position z, the possibilities of spins diffusing to opposite directions $z \pm \Delta z$ are equal [26]. The opposite movements yield $\exp(-iK(t) \cdot (z + \Delta z)) + \exp(-iK(t) \cdot (z - \Delta z)) = \cos(K(t) \cdot \Delta z) \exp(-iK(t) \cdot z)$, and have no effect on the phase, $\exp(-iK(t) \cdot z)$, but do make the signal intensity decay by a factor of $\cos(K(t) \cdot \Delta z)$ for these spins. By substituting $M_{xy}(z,t) = S(t) \exp(-iK(t) \cdot z)$ into diffusion Equation (7), and applying $\frac{\partial^{\beta}}{\partial |z|^{\beta}} \exp(-iK(t) \cdot z) = -K^{\beta}(t) \exp(-iK(t) \cdot z)$, we have

$$ {}_tD_*^{\alpha} [S(t) \exp(-iK(t) \cdot z)] = -D_f K^{\beta}(t) [S(t) \exp(-iK(t) \cdot z)]. \tag{16} $$

At the origin, $z = 0$, $i\gamma g z = 0$, $\exp(-iK(t) \cdot z) = 1$, and $S(t) \exp(-iK(t) \cdot z) = S(t)$; we thus have

$$ {}_tD_*^{\alpha} S(t) = -D_f K^{\beta}(t) S(t). \tag{17} $$

Equation (17) is identical to Equation (12) obtained previously by the EPSDE method in Section 2.1.

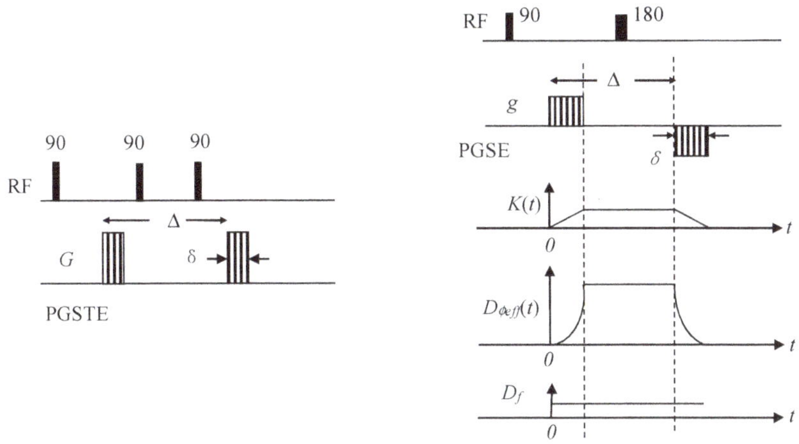

(**a**) (**b**)

Figure 1. (**a**) Pulsed gradient stimulated echo (PGSTE) pulse sequences; (**b**) pulsed gradient spin echo (PGSE) pulse sequence. The gradient pulse width is δ, and the diffusion delay is Δ. The time-dependent behavior of wavenumber $K(t)$, diffusion constant D_f, and the effective phase diffusion constant $D_{\phi eff}(t) = K^{\beta}(t)D_f$ are also demonstrated below the PGSE pulse sequence.

This method of observing the signal intensity at the origin could be understood in another way: if only the signal intensity or amplitude is observed from a selected position at different diffusion times t_i, $i = 1, 2, 3 \cdots$ in a PFG anomalous diffusion experiment, the observed signal intensity or amplitude shall still obey Equation (17) regardless of whether the phase is observed or not. It is reasonable that both ways can give the same signal attenuation expression, because the signal amplitude is homogeneous throughout the sample at each instant, and selecting the origin at a random position in the sample does not affect the result.

2.3. Analytical Solution by the Adomian Decomposition Method

The same equation as Equation (12) or (17) has been obtained by the fractional integral modified Bloch equation method. The three different methods get the same PFG signal attenuation equation. Reference [30] employed the Adomian decomposition method to give a general analytical solution to Equation (12). In the following, the general solution from the Adomian decomposition will be given in more detail for the cases of normal diffusion and general anomalous diffusion.

2.3.1. Normal Diffusion

First, the signal attenuation equation—Equation (12) or (17)—can be used to obtain the PFG signal attenuation expression for normal diffusion, which is a specific case of anomalous diffusion with $\alpha = 1$ and $\beta = 2$. When $\alpha = 1$ and $\beta = 2$, Equation (17) reduces to

$$\frac{d}{dt}S(t) = -K^2(t)DS(t), \tag{18}$$

where D is the diffusion constant of normal diffusion. The solution of Equation (18) is

$$S(t) = \exp\left(-\int_0^t DK^2(t')dt'\right), \tag{19}$$

which is the PFG signal attenuation expression for normal diffusion. The same result as expression (19) was obtained in [42]. For PGSE or PGSTE experiments, the PFG signal attenuation calculated based on Equation (19) is $\exp\left(-D_f\gamma^2 g^2\delta^2(\Delta-\delta/3)\right)$, which is a routinely used expression for PFG normal diffusion [13–15].

2.3.2. General Anomalous Diffusion

Second, the PFG signal attenuation for general anomalous diffusion can be obtained. The same equation as Equation (17) and its solution have been obtained by the modified Bloch equation method in [30]. According to [30], Equations (12) or (17) can be written equivalently as

$$S(t)=\sum_{k=0}^{m-1} S^{(k)}(0^+)\frac{t^k}{k!}+J^{\alpha}(a(t)S(t)),\ m-1<\alpha<m \tag{20}$$

where

$$a(t)=-K^{\beta}(t)D_f=\begin{cases}-(\gamma g t)^{\beta}D_f,\ 0<t\le\delta\\ -(\gamma g\delta)^{\beta}D_f,\ \delta<t\le\Delta\\ -(\gamma g)^{\beta}(\Delta+\delta-t)^{\beta}D_f,\ \Delta<t\le\Delta+\delta.\end{cases} \tag{21}$$

Based on the Adomian decomposition method [35–39], the solution of Equation (20) is [30,31]

$$S(t)=\sum_{n=0}^{\infty} S_n(t)\,, \tag{22a}$$

where

$$S_0(t)=\sum_{k=0}^{m-1} S^{(k)}(0^+)\frac{t^k}{k!},\ m-1<\alpha<m \tag{22b}$$

and

$$\begin{aligned}S_n(t)&=J^{\alpha}(a(t)S_{n-1}(t))\\ &=-\int_0^t \frac{(t-\tau)^{\alpha-1}}{\Gamma(\alpha)}D_fK^{\beta}(\tau)S_{n-1}(\tau)d\tau\\ &=-\int_0^t \frac{D_fK^{\beta}(\tau)S_{n-1}(\tau)d(t-\tau)^{\alpha}}{\alpha\Gamma(\alpha)}.\end{aligned} \tag{22c}$$

In PGSE or PGSTE experiments, the time t can be separated into three periods: $0<t\le\delta$, $\delta<t\le\Delta$, and $\Delta<t\le\Delta+\delta$. If the initial condition, $S^{(1)}(0^+)=0$, is used [31–34] for free diffusion in a homogeneous sample, we can get the following:

(a) *PFG signal attenuation under short gradient pulse (SGP) approximation*: δ is short enough and the diffusion inside each gradient pulse can be neglected. We get

$$S(\Delta)=\sum_{n=0}^{\infty} S_n(\Delta)=E_{\alpha,1}\left(-D_f{K_{SGP}}^{\beta}\Delta^{\alpha}\right),\ K_{SGP}=\gamma g\delta, \tag{23}$$

where $S_0(\Delta)=1$, and $S_n(\Delta)=\frac{\left(-D_f{K_{SGP}}^{\beta}\Delta^{\alpha}\right)^n}{\Gamma(1+n\alpha)}$. Equation (23) replicates the SGP approximation result obtained in references [18,26,27].

(b) *Single pulse attenuation*: This is an ideal situation—the first gradient pulse is regular, but the second gradient pulse is infinitely narrow and has the purpose of counteracting the first gradient pulse. We get

$$\begin{aligned}S(t)&=\sum_{n=0}^{\infty} S_n(t)=1+\sum_{n=1}^{\infty}\left(-D_f(\gamma g)^{\beta}t^{\alpha+\beta}\right)^n\prod_{k=1}^{n}\frac{\Gamma(1+(k-1)\alpha+n\beta)}{\Gamma(1+k(\alpha+\beta))}\\ &=E_{\alpha,1+\beta/\alpha,\beta/\alpha}\left(-D_f(\gamma g)^{\beta}t^{\alpha+\beta}\right),\end{aligned} \tag{24}$$

where $S_0(t) = 1$, $S_n(t) = \left(-D_f(\gamma g)^\beta t^{\alpha+\beta}\right)^n \prod_{k=1}^{n} \frac{\Gamma(1+(k-1)\alpha+n\beta)}{\Gamma(1+k(\alpha+\beta))}$, and $E_{\alpha,\eta,\gamma}(x) = \sum_{n=0}^{\infty} c_n x^n, c_0 = 1$, $c_n = \prod_{k=0}^{n-1} \frac{\Gamma((k\eta+\gamma)\alpha+1)}{\Gamma((k\eta+\gamma+1)\alpha+1)}$ is a Mittag-Leffler-type function [43]. Equation (24) is consistent with the results obtained by the modified Bloch equation proposed in [25].

(c) *General PFG signal attenuation*: The PGSE or PGSTE experiment includes three periods: $0 < t_1 \le \delta$, $\delta < t_2 \le \Delta$, and $\Delta < t_3 \le \Delta + \delta$. The integration in Equation (22c) during these three periods is tedious, but can be calculated with computer assistance. Nevertheless, we can get the first and second terms of Equation (22a) as the following:

$$S_0(t) = 1 \tag{25a}$$

$$S_1(t) = J^\alpha(a(t)S_0(t)) = -\int_0^t \frac{(t-\tau)^{\alpha-1}}{\Gamma(\alpha)} D_f K^\beta(\tau)d\tau = -\frac{D_f(\gamma g)^\beta}{\Gamma(1+\alpha)} \times \begin{cases} \alpha t^{\alpha+\beta} B(\beta+1,\alpha),\ 0 < t \le \delta \\ \alpha t^{\alpha+\beta}\left[B(\beta+1,\alpha) - B(\frac{\delta}{t};\beta+1,\alpha)\right] - \delta^\beta (t-\delta)^\alpha,\ \delta < t \le \Delta \\ \left\{\alpha t^{\alpha+\beta}\left[B(\beta+1,\alpha) - B(\frac{\delta}{t};\beta+1,\alpha)\right] - \delta^\beta\left[(t-\delta)^\alpha - (t-\Delta)^\alpha\right]\right\} -,\ \Delta < t \le \Delta+\delta \\ \int_\Delta^t \alpha(t-\tau)^{\alpha-1}(\Delta+\delta-t)^\beta d\tau \end{cases} \tag{25b}$$

where $B(x,y)$ and $B(a;x,y)$ are the Beta function and incomplete Beta function, respectively. When $t = \Delta + \delta$, Equation (25b) gives

$$\begin{aligned} S_1(\Delta+\delta) &= J^\alpha(a(t)S_0(t)) \\ &= -\int_0^{\Delta+\delta} \frac{(\Delta+\delta-\tau)^{\alpha-1}}{\Gamma(\alpha)} D_f K^\beta(\tau)d\tau \\ &= -\frac{D_f(\gamma g)^\beta}{\Gamma(1+\alpha)}\left\{\alpha(\Delta+\delta)^{\alpha+\beta}\left[B(\beta+1,\alpha) - B(\tfrac{\delta}{\Delta+\delta};\beta+1,\alpha)\right] + \delta^\beta(\Delta^\alpha - \delta^\alpha) + \tfrac{\alpha\delta^{\alpha+\beta}}{\alpha+\beta}\right\}. \end{aligned} \tag{26}$$

Equation (26) agrees with the signal attenuation expression $S(t) = E_{\alpha,1}\left(-\int_0^t D_f K^\beta(t')dt'^\alpha\right)$ obtained by the instantaneous signal attenuation method [26]. At small signal attenuation, $S(t) = E_{\alpha,1}\left(-\int_0^t D_f K^\beta(t')dt'^\alpha\right)$ can be approximately expanded as

$$\begin{aligned} S(t) &= E_{\alpha,1}\left(-\int_0^t D_f K^\beta(t')dt'^\alpha\right) \\ &\approx 1 - \frac{D_f(\gamma g)^\beta}{\Gamma(1+\alpha)}\left\{\alpha(\Delta+\delta)^{\alpha+\beta}\left[B(\alpha,\beta+1) - B(\tfrac{\Delta}{\Delta+\delta};\alpha,\beta+1)\right] + \delta^\beta(\Delta^\alpha - \delta^\alpha) + \tfrac{\alpha\delta^{\alpha+\beta}}{\alpha+\beta}\right\}. \end{aligned} \tag{27}$$

Equation (27) is the same as that given by the combination of Equations (25a) and (26) (note that $B(x,y) = B(y,x)$ and $B(a;x,y) = B(1-a;y,x)$); however, at large signal attenuation, $E_{\alpha,1}\left(-\int_0^t D_f K^\beta(t')dt'^\alpha\right)$ may deviate from the combination of Equations (25a) and (26) as the higher-order terms in the expansion cannot be omitted. This agreement can be explained by the following: in typical PGSE or PGSTE experiments, both the gradient intensity and pulse length of the dephasing pulse and the rephrasing pulse are identical, $|g_1| = |g_2|$ and $\delta_1 = \delta_2$; thus, $K(\Delta+\delta-\tau) = K(\tau)$, which results in

$$\begin{aligned} -\int_0^{\Delta+\delta} \frac{(\Delta+\delta-\tau)^{\alpha-1}}{\Gamma(\alpha)} D_f K^\beta(\tau)d\tau &\overset{\tau'=\Delta+\delta-\tau}{\rightarrow} = -\int_0^{\Delta+\delta} \frac{\tau'^{\alpha-1}}{\Gamma(\alpha)} D_f K^\beta(\Delta+\delta-\tau')d\tau' \\ &\overset{K(\Delta+\delta-\tau')=K(\tau')}{\rightarrow} -\int_0^{\Delta+\delta} \frac{\tau'^{\alpha-1}}{\Gamma(\alpha)} D_f K^\beta(\tau')d\tau'. \end{aligned} \tag{28}$$

When $0 < \alpha \leq 2$, $\beta = 2$, the general anomalous diffusion reduces to time-fractional diffusion and Equation (25b) can be further calculated as

$$S_1(\Delta+\delta) = -\frac{D_f(\gamma g)^2}{\Gamma(1+\alpha)} \times \left\{ \frac{\alpha\delta^{\alpha+2}}{\alpha+2} + \delta^2(\Delta^\alpha - \delta^\alpha) + \frac{2}{(\alpha+1)(\alpha+2)}\left[(\Delta+\delta)^{2+\alpha} - \Delta^{2+\alpha}\right] - \frac{2}{\alpha+1}\Delta^{1+\alpha}\delta - \Delta^\alpha\delta^2 \right\}. \tag{29}$$

When $\alpha = 1$, $0 < \beta \leq 2$, the general anomalous diffusion reduces to space-fractional diffusion. Equation (22) reduces to $S(t) = \exp\left(-\int_0^t D_f K^\beta(t')dt'\right)$, which is the same as that obtained by [18,26,30,31].

2.4. Numerical Evaluation of Mittag-Leffler Function-Based PFG Signal Attenuation by the Direct Integration Method

Although the Adomian decomposition method can be used to numerically evaluate the PFG signal attenuation, its calculation speed is slow and it could cause overflow at large signal attenuation. References [30,31,44] proposed an alternative method: a direct integration method that can give the same numerical results, but with a much faster speed and without overflow. From Equation (20), if we set $\sum_{k=0}^{m-1} S^{(k)}(0^+)\frac{t^k}{k!} = 1$, we have [44]

$$S(t) = 1 - \int_0^t \frac{(t-\tau)^{\alpha-1}}{\Gamma(\alpha)} G(\tau)S(\tau)d\tau, \tag{30}$$

where $G(t) = K^\beta(t)D_f$. If we denote $E_{\alpha,1,G(t)}(-t) = S(t)$, Equation (30) can be further written as [44]

$$E_{\alpha,1,G(t)}(-t) = 1 - \int_0^t \frac{(t-\tau)^{\alpha-1}}{\Gamma(\alpha)} G(\tau)E_{\alpha,1,G(t)}(-\tau)d\tau. \tag{31a}$$

When $G(t) = 1$, $E_{\alpha,1,G(t)}(-t) = E_{\alpha,1}(-t^\alpha)$, which is a Mittag-Leffler function (MLF); when $G(t) = c$, $c > 0$, $E_{\alpha,1,G(t)}(-t) = E_{\alpha,1}(-ct^\alpha)$. Hence, we have [44]

$$E_{\alpha,1}(-t^\alpha) = 1 - \int_0^t \frac{(t-\tau)^{\alpha-1}}{\Gamma(\alpha)} E_{\alpha,1}(-\tau^\alpha)d\tau, \tag{31b}$$

$$E_{\alpha,1}(-ct^\alpha) = 1 - \int_0^t \frac{(t-\tau)^{\alpha-1}}{\Gamma(\alpha)} cE_{\alpha,1}(-c\tau^\alpha)d\tau. \tag{31c}$$

Equation (30) can be rewritten in a discrete form as [44]

$$S(t_j) = 1 - \sum_{k=1}^{j} a(t_k)S(t_{k-1})\left[(t_j - t_{k-1})^\alpha - (t_j - t_k)^\alpha\right]/\Gamma(1+\alpha), \tag{32}$$

where $t_j = \sum_{k=1}^{j} \Delta t_k$. Based on Equation (32), the PFG signal attenuations $S(t_1), S(t_2), \ldots, S(t_n)$ can be calculated step by step starting from $S(t_1)$ through to $S(t_n)$. Similarly, the Mittag-Leffler-type function and its derivative can be numerically evaluated by [44]

$$E_{\alpha,1,G(t)}(-t_j) = 1 - \sum_{k=1}^{j} G(t_k)E_{\alpha,1,G(t)}(-t_{k-1})\left[(t_j - t_{k-1})^\alpha - (t_j - t_k)^\alpha\right]/\Gamma(1+\alpha), \tag{33}$$

$$E'_{\alpha,1,G(t)}(-t_j) = \frac{E_{\alpha,1,G(t)}(-t_j) - E_{\alpha,1,G(t)}(-t_{j-1})}{\Delta t_j}. \quad (34)$$

This method is simple, and there is no overflow issue in the calculation.

3. Continuous-Time Random Walk (CTRW) in a Lattice Model

Random walk simulation is a powerful numerical method that employs a stochastic jump process to model normal and anomalous diffusion in physics, chemistry, biology, and many other disciplines. It can be used to simulate the PFG signal attenuation in mathematically tractable or intractable systems. In this paper, the continuous-time random walk (CTRW) simulation method developed in [26] is used to verify the theoretical results; it is based on two models: the CTRW model [45] and the Lattice model [46,47]. In the simulation, a sequence of independent random waiting time and jump length combinations $(\Delta t_1, \Delta\xi_1)$, $(\Delta t_2, \Delta\xi_2)$, $(\Delta t_3, \Delta\xi_3)$, ... , $(\Delta t_n, \Delta\xi_n)$ is produced by the computer program. The individual waiting time Δt and jump length $\Delta\xi$ are given according to [26] by the following expressions:

$$\Delta t = -\eta_t \log U \left(\frac{\sin(\alpha\pi)}{\tan(\alpha\pi V)} - \cos(\alpha\pi) \right)^{\frac{1}{\alpha}} \quad (35)$$

and

$$\Delta\xi = \eta_z \left(\frac{-\log U \cos(\Phi)}{\cos((1-\beta)\Phi)} \right)^{1-\frac{1}{\beta}} \frac{\sin(\beta\Phi)}{\cos(\Phi)} \quad (36)$$

where η_t and η_z are scale constants, $\Phi = \pi(V - 1/2)$, and $U, V \in (0,1)$ are two independent random numbers. The CTRW simulation based on the above waiting time and jump length satisfies the time–space fractional diffusion equation under the diffusive limit, providing a simple way to simulate anomalous diffusion in various research areas, such as physics and economics [45].

Although continuous waiting time and jump length are used in the simulation, the accumulating spin phase associated with the diffusion path is recorded in a discrete manner. Such a discrete phase recording manner is convenient and reasonable. The simulation can be viewed as a numerical PFG experiment, whose observables such as phase can be observed in a discrete time selected by an experiment observer; the observing or recording manner will not affect the fundamental numerical experiment process, namely the production of the continuous random walk sequence $(\Delta t_1, \Delta\xi_1)$, $(\Delta t_2, \Delta\xi_2)$, $(\Delta t_3, \Delta\xi_3)$, ... , $(\Delta t_n, \Delta\xi_n)$. The spin phase was recorded by the lattice model developed in [46,47], which has been applied to simulate PFG diffusion in polymer systems [48]. The spin phase in the simulation is

$$\phi(t'_{j'}) = \sum_{j=1}^{l} \gamma g(t_j)\xi(t_j)\Delta t_j + \gamma g(t'_{j'})\xi(t_{l+1})(t'_{j'} - t_l),\ t_l \le t'_{j'} \le t_{l+1} \quad (37)$$

where $t_l = \sum_{j=1}^{l} \Delta t_j$, $\xi(t_l) = \sum_{j=1}^{l} \Delta\xi_j$, and $t'_{j'}$ is the discrete record time, which takes place between the lth and $(l+1)$th steps of the random walk. The second term $\gamma g(t'_{j'})\xi(t_{l+1})(t'_{j'} - t_l)$ on the right-hand side of Equation (37) corresponds to the partial phase evolution of the $(l+1)$th jump step that needs to be recorded at the time $t'_{j'}$. The PFG signal attenuation in the simulation can be obtained by averaging over all the walkers in the simulation [26,47,48]:

$$S(t) = \langle \cos[\phi_i(t)] \rangle = \frac{1}{N_{walks}} \sum_{i=1}^{N_{walks}} \cos[\phi_i(t)], \quad (38)$$

where N_{walks} is the total number of walks. The total number of walks used in each simulation is 1,000,000.

Because the CTRW model proposed in [25] is only for subdiffusion, the simulation here is limited to the subdiffusion. Please refer to references [26,30,45–48] for more detailed information.

4. Results and Discussion

This paper uses two different methods to describe PFG anomalous diffusion: the EPSDE method [18] and the method of observing the signal intensity at the origin. The major results in this paper are summarized in Table 1. The EPSDE method [18] uses an effective phase diffusion process to describe the phase evolution, while observing the signal intensity at the origin only monitors the real-space signal amplitude change at the origin where the phase of the magnetization remains constant. Each of these two approaches has its unique advantages. The EPSDE method can directly obtain the phase distribution, which greatly simplifies PFG diffusion analysis [18]; in contrast, the traditional methods often need to get the real-space spin particle probability distribution first, which is then used to obtain the phase distribution. The ultra-simple method of observing the signal intensity at the origin obtains the signal attenuation equation by substituting the magnetization $M_{xy}(z,t) = S(t)\exp(-iK(t)\cdot z)$ directly into the diffusion equation and using the phase property $\exp(-iK(t)\cdot z) = 1$ and $i\gamma gz = 0$ at the origin. These two approaches view the spin evolution process from two different spaces—the real space and the phase space—providing a clear picture of spin dynamics in PFG diffusion experiments.

Table 1. Comparison of pulsed-field gradient (PFG) anomalous diffusion results by the effective phase-shift diffusion method, the observing the signal intensity at the origin method and other methods.

Virtual Phase-Space Diffusion	**Real-Space Spin Diffusion**
Effective phase-shift diffusion equation [18]	Observing the signal intensity at the origin
Diffusion equations:	
${}_tD_*^{\alpha}P(\phi,t) = K^{\beta}(t)D_f\frac{\partial^{\beta}}{\partial\|\phi\|^{\beta}}P(\phi,t)$ [18]	${}_tD_*^{\alpha}M_{xy}(z,t) = D_f\frac{\partial^{\beta}}{\partial\|z\|^{\beta}}M_{xy}(z,t)$ [32]
Solving methods:	
Fourier transform	Substitute $M_{xy}(z,t) = S(t)\exp(-iK(t)\cdot z)$ into equation
PFG signal attenuation equation*:	
${}_tD_*^{\alpha}S(t) = -D_fK^{\beta}(t)S(t)$,	
or equivalently $S(t) = \sum\limits_{k=0}^{m-1} S^{(k)}(0^+)\frac{t^k}{k!} + J^{\alpha}(a(t)S(t))$, $m-1<\alpha<m$	
PFG signal attenuation expression by the Adomian decomposition method [30,31]*:	
$S(t) = \sum\limits_{n=0}^{\infty} S_n(t)$,	
where $S_0(t) = \sum\limits_{k=0}^{m-1} S^{(k)}(0^+)\frac{t^k}{k!}$, $m-1<\alpha<m$, and $S_n(t) = -\int\limits_0^t \frac{(t-\tau)^{\alpha-1}D_fK^{\beta}(\tau)S_{n-1}(\tau)d\tau}{\Gamma(\alpha)}$	
Under short gradient pulse (SGP) approximation: $S(\Delta) = E_{\alpha,1}\left(-D_fK_{SGP}{}^{\beta}\Delta^{\alpha}\right)$, $K_{SGP} = \gamma g\delta$	
At small attenuation:	$S(t) \approx E_{\alpha,1}\left(-\int_0^t D_fK^{\beta}(t')dt'^{\alpha}\right)$
Other methods:	
(1) $S(t) = E_{\alpha,1+\beta/\alpha,\beta/\alpha}\left(-\gamma^{\beta}t^{\alpha+\beta}\int\limits_g \|\mathbf{g}\cdot\mathbf{y}\|^{\beta}m(d\mathbf{y})\right)$ [25]	
(2) $E_{\alpha,1}\left[-i\gamma g\tau(t/\tau)^{\alpha}\right]\exp\left[-B(t/\tau)^{\alpha}\right]$, $B = \frac{2\Gamma(2-\alpha)D_f\gamma^2g^2\tau^3}{3\alpha^2\Gamma(2-\alpha)}$ [17]	
(3) $S(t) \approx E_{\alpha,1}\left(-\int_0^t D_fK^{\beta}(t')dt'^{\alpha}\right)$ [26]	
(4) $S(\Delta) = E_{\alpha,1}\left(-D_fK_{SGP}{}^{\beta}\Delta^{\alpha}\right)$, $K_{SGP} = \gamma g\delta$ [18,26,27]	

* The PFG signal attenuation equation and expression obtained in this paper are the same as that obtained by the integral modified Bloch equation method [30].

These two methods agree with each other. They get the same PFG signal attenuation Equation (11) or (16), which can be solved by the Adomian decomposition method. The solution gives PFG signal attenuation expressions (21a)–(21c), which include the FGPW effect. Understanding the FGPW effect is important as clinical MRI applications often use a long gradient pulse. Additionally, using long gradient pulses allows researchers to monitor slower diffusion in polymer or biological systems where the molecules or ions often diffuse slowly. The consistent results from two different methods help us better understand the FGPW effect.

Additionally, the two methods agree with other methods. First, the PFG signal attenuation Equation (17) obtained here is the same as that obtained by the integral modified Bloch equation method [30,31]. Second, the results agree with the CTRW simulation as shown in Figure 2. In Figure 2a, the fractional diffusion constant D_f = 6.6 × 10^{-11} m^β/s^α was obtained from fitting the curve of mean square displacement versus t by equation $\left\langle |z|^\beta \right\rangle = 2D_f t^\alpha/\Gamma(1+\alpha)$. Figure 2b shows the PFG signal attenuation of anomalous diffusion obtained by Equations (22a)–(22c) based on the Adomian decomposition method, Equation (32) based on the direct integration method, and the CTRW simulation. Other parameters used in Figure 2 are $\alpha = 0.5$, $\beta = 2$, $g = 0.1$ T/m, $\Delta - \delta$ equaling 0 ms, 20 ms. Interested readers can refer to [30] for additional comparisons between the theoretical prediction and the CTRW simulation. Third, the PFG signal attenuation expressions (22a)–(22c) are also consistent with those obtained by various approximation methods such as the non-Gaussian phase distribution (nGPD) method [27] and the instantaneous signal attenuation method [26]. The same SGP approximation results, Equation (23), can be obtained by all these five different methods: the EPSDE method [18], the instantaneous signal attenuation method [26], the nGPD method [27], the method to observe the signal intensity at the origin, and the modified Bloch equation method [30]. Moreover, Equation (27), $S(t) \approx E_{\alpha,1}\left(-\int_0^t D_f K^\beta(t')dt'^\alpha\right)$, can be approximately obtained in this paper; this agrees with the result obtained from the instantaneous signal attenuation method [26].

The numerical evaluation of PFG signal attenuation by Equation (20) or (30) can be conveniently performed by the direct integration method. Figure 2b shows that the results of the direct integration method agree with those obtained from the Adomian decomposition method. Compared to the Adomian decomposition method, the computing speed of the direct integration method improves by orders of magnitudes. The speed improvement is because in the direct integration method, each $S(t_j)$ in Equation (32) is only a single time integration, while in the Adomian decomposition method, $S(t_j) = \sum_{n=0}^{\infty} S_n(t_j)$ is a superposition of many terms of time integration. Additionally, from Equations (33) and (34), the direct integration method can be used to calculate Mittag-Leffler-type functions and their derivatives. Figure 3 shows that the MLF calculated from the direct integration method agrees with that calculated by other methods in [49,50]. The Pade approximation method in [50] is only for subdiffusion. Because the direct integration method does not cause overflow, it can be a useful method for calculating Mittag-Leffler-type functions. The Fortran code for MLF calculation by the direct integration method can be obtained from the following link: https://github.com/GLin2017/Mittag-Leffler-function-calculated-by-Direct-Integration. Matlab code for calculating the MLF function by other methods to a desired accuracy can be found at www.mathworks.com/matlabcentral/fileexchange/8738-mittag-leffer-function, 2005. This direct integration method will be a great help to the application of current theoretical results to PFG anomalous diffusion in NMR and MRI.

In the current results, only PFG anomalous diffusion of single quantum coherence is considered. The results here can be easily extended to handle PFG anomalous diffusion of intramolecular MQC [27,40]. As the gradient field-induced phase evolution of an n-order intramolecular MQC is n times faster than the corresponding single quantum coherence, the PFG signal attenuation of intramolecular MQC is the same as that of single quantum coherence with an effective gradient intensity ng [42]. Therefore, for intramolecular MQC in PGSE or PGSTE experiments, the PFG signal attenuation equation is

$$S(t) = \sum_{k=0}^{m-1} S^{(k)}(0^+)\frac{t^k}{k!} + J^\alpha\left(a_{MQC}(t)S(t)\right),\ m-1 < \alpha < m, \tag{39}$$

where

$$a_{MQC}(t) = -K_{MQC}{}^{\beta}(t)D_f = \begin{cases} -n^{\beta}(\gamma g t)^{\beta} D_f,\ 0 < t \le \delta \\ -n^{\beta}(\gamma g \delta)^{\beta} D_f,\ \delta < t \le \Delta \\ -n^{\beta}(\gamma g)^{\beta}(\Delta + \delta - t)^{\beta} D_f,\ \Delta < t \le \Delta + \delta. \end{cases} \tag{40}$$

From Equation (40), it is obvious that the $a_{MQC}(t)$ in the MQC equals $n^{\beta}a(t)$.

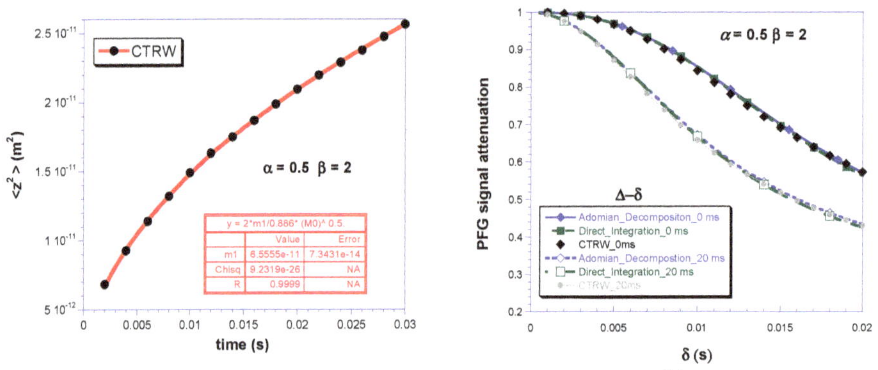

(**a**) (**b**)

Figure 2. Comparison of the PFG signal attenuation from theoretical predictions with continuous-time random walk (CTRW) simulation: (**a**) mean square displacement from CTRW simulation; (**b**) PFG signal attenuation from different methods: the Adomian decomposition method with Equations (22a)–(22c), the direct integration method with Equation (32), and the CTRW simulation. The parameters used are $\alpha = 0.5$, $\beta = 2$, $D_f = 6.6 \times 10^{-11}$ $\mathrm{m}^{\beta}/\mathrm{s}^{\alpha}$, and $g = 0.1$ T/m.

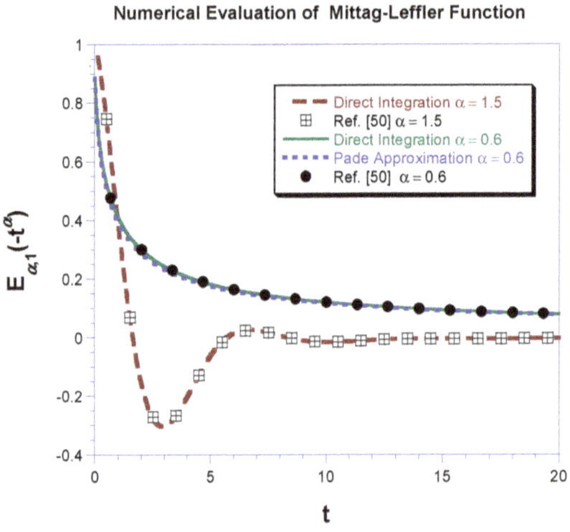

Figure 3. Comparing the Mittag-Leffler function (MLF) calculated by different methods: Equation (33) from the direct integration method, the Pade approximation method [50], and the method used in [49].

This paper only considers the spin self-diffusion that can be described by the time–space fractional diffusion equation based on the fractional derivative. In general, the two methods used in this paper can be applied to other types of anomalous diffusions such as that described by the time–space diffusion equation based on the fractal derivative [51], etc. Additionally, only the symmetric anomalous diffusion in homogeneous spin systems is studied here. In real applications, the anomalous diffusion can take place in complicated systems such as inhomogeneous systems, restricted geometries [28], anisotropic systems [44], nonsymmetric systems, etc. Additionally, the gradient field may be nonlinear [52]. The current methods may face challenges in these complicated situations, which reminds us that much effort is required for the study of PFG anomalous diffusion.

Conflicts of Interest: The author declares no conflicts of interest.

Appendix A. Definition of the Fractional Derivative

The definition of the space fractional derivative [6,32–34] is given by

$$\frac{\partial^\beta}{\partial|z|^\beta} = -\frac{1}{2\cos\left(\frac{\pi\alpha}{2}\right)}\left[{}_{-\infty}D_z^\beta + {}_zD_\infty^\beta\right] \quad \text{(A1)}$$

where

$${}_{-\infty}D_z^\beta f(z) = \frac{1}{\Gamma(m-\beta)}\frac{d^m}{dz^m}\int_{-\infty}^{z}\frac{f(y)dy}{(z-y)^{\beta+1-m}},\ \beta > 0,\ m-1 < \beta < m \quad \text{(A2)}$$

and

$${}_zD_\infty^\beta f(z) = \frac{(-1)^m}{\Gamma(m-\beta)}\frac{d^m}{dz^m}\int_{z}^{\infty}\frac{f(y)dy}{(z-y)^{\beta+1-m}},\ \beta > 0,\ m-1 < \beta < m. \quad \text{(A3)}$$

References

1. Wyss, W. The fractional diffusion equation. *J. Math. Phys.* **1986**, *27*, 2782–2785. [CrossRef]
2. Metzler, R.; Klafter, J. The random walk's guide to anomalous diffusion: A fractional dynamics approach. *Phys. Rep.* **2000**, *339*, 1–77. [CrossRef]
3. Sokolov, I.M. Models of anomalous diffusion in crowded environments. *Soft Matter* **2012**, *8*, 9043–9052. [CrossRef]
4. Povstenko, Y. *Linear Fractional Diffusion-Wave Equation for Scientists and Engineers*; Birkhäuser: New York, NY, USA, 2015.
5. Köpf, M.; Corinth, C.; Haferkamp, O.; Nonnenmacher, T.F. Anomalous diffusion of water in biological tissues. *Biophys. J.* **1996**, *70*, 2950–2958. [CrossRef]
6. Saichev, A.I.; Zaslavsky, G.M. Fractional kinetic equations: Solutions and applications. *Chaos* **1997**, *7*, 753–764. [CrossRef] [PubMed]
7. Lindsey, C.P.; Patterson, G.D. Detailed comparison of the Williams–Watts and Cole-Davidson functions. *J. Chem. Phys.* **1980**, *73*, 3348–3357. [CrossRef]
8. Kaplan, J.I.; Garroway, A.N. Homogeneous and inhomogeneous distributions of correlation times. Lineshapes for chemical exchange. *J. Magn. Reson.* **1982**, *49*, 464–475. [CrossRef]
9. Hahn, E.L. Spin echoes. *Phys. Rev.* **1950**, *80*, 580–594. [CrossRef]
10. Torrey, H.C. Bloch Equations with Diffusion Terms. *Phys. Rev.* **1956**, *104*, 563–565. [CrossRef]
11. McCall, D.W.; Douglass, D.C.; Anderson, E.W. Self-diffusion studies by means of nuclear magnetic resonance spin-echo techniques. *Ber. Bunsenges. Phys. Chem.* **1963**, *67*, 336–340. [CrossRef]
12. Stejskal, E.O.; Tanner, J.E. Spin diffusion measurements: Spin echoes in the presence of a time-dependent field gradient. *J. Chem. Phys.* **1965**, *42*, 288–292. [CrossRef]
13. Price, W.S. Pulsed-field gradient nuclear magnetic resonance as a tool for studying translational diffusion: Part 1. Basic theory. *Concepts Magn. Reson.* **1997**, *9*, 299–336. [CrossRef]
14. Price, W.S. *NMR Studies of Translational Motion: Principles and Applications*; Cambridge University Press: Cambridge, UK, 2009.
15. Callaghan, P. *Translational Dynamics and Magnetic Resonance: Principles of Pulsed Gradient Spin Echo NMR*; Oxford University Press: Oxford, UK, 2011.
16. McRobbie, D.W.; Moore, E.A.; Graves, M.J.; Prince, M.R. *MRI from Picture to Proton*, 2nd ed.; Cambridge University Press: Cambridge, UK, 2007; p. 338.

17. Magin, R.L.; Abdullah, O.; Baleanu, D.; Zhou, X.J. Anomalous diffusion expressed through fractional order differential operators in the Bloch–Torrey equation. *J. Magn. Reson.* **2008**, *190*, 255–270. [CrossRef] [PubMed]
18. Lin, G. An effective phase shift diffusion equation method for analysis of PFG normal and fractional diffusions. *J. Magn. Reson.* **2015**, *259*, 232–240. [CrossRef] [PubMed]
19. Kärger, J.; Pfeifer, H.; Vojta, G. Time correlation during anomalous diffusion in fractal systems and signal attenuation in NMR field-gradient spectroscopy. *Phys. Rev. A* **1988**, *37*, 4514–4517. [CrossRef]
20. Kimmich, R. *NMR: Tomography, Diffusometry, Relaxometry*; Springer: Heidelberg, Germany, 1997.
21. Fatkullin, N.; Kimmich, R. Theory of field-gradient NMR diffusometry of polymer segment displacements in the tube-reptation model. *Phys. Rev. E* **1995**, *52*, 3273–3276. [CrossRef]
22. Bennett, K.M.; Schmainda, K.M.; Bennett, R.T.; Rowe, D.B.; Lu, H.; Hyde, J.S. Characterization of continuously distributed cortical water diffusion rates with a stretched-exponential model. *Magn. Reson. Med.* **2003**, *50*, 727–734. [CrossRef] [PubMed]
23. Bennett, K.M.; Hyde, J.S.; Schmainda, K.M. Water diffusion heterogeneity index in the human brain is insensitive to the orientation of applied magnetic field gradients. *Magn. Reson. Med.* **2006**, *56*, 235–239. [CrossRef] [PubMed]
24. Klafter, J.; Sokolov, I.M. *First Step in Random Walks. From Tools to Applications*; Oxford University Press: New York, NY, USA, 2011.
25. Hanyga, A.; Seredyńska, M. Anisotropy in high-resolution diffusion-weighted MRI and anomalous diffusion. *J. Magn. Reson.* **2012**, *220*, 85–93. [CrossRef] [PubMed]
26. Lin, G. Instantaneous signal attenuation method for analysis of PFG fractional diffusions. *J. Magn. Reson.* **2016**, *269*, 36–49. [CrossRef] [PubMed]
27. Lin, G. Analyzing signal attenuation in PFG anomalous diffusion via a non-gaussian phase distribution approximation approach by fractional derivatives. *J. Chem. Phys.* **2016**, *145*, 194202. [CrossRef] [PubMed]
28. Lin, G.; Zheng, S.; Liao, X. Signal attenuation of PFG restricted anomalous diffusions in plate, sphere, and cylinder. *J. Magn. Reson.* **2016**, *272*, 25–36. [CrossRef] [PubMed]
29. Damion, R.A.; Packer, K.J. Predictions for pulsed-field-gradient NMR experiments of diffusion in fractal spaces. *Proc. Math. Phys. Eng. Sci.* **1997**, *453*, 205–211. [CrossRef]
30. Lin, G. The exact PFG signal attenuation expression based on a fractional integral modified-Bloch equation. *arXiv*, 2017.
31. Lin, G. Fractional differential and fractional integral modified-Bloch equations for PFG anomalous diffusion and their general solutions. *arXiv*, 2017.
32. Mainardi, F.; Luchko, Y.; Pagnini, G. The fundamental solution of the space-time-fractional diffusion equation. *Fract. Calc. Appl. Anal.* **2001**, *4*, 153–192.
33. Gorenflo, R.; Mainardi, F. Fractional Diffusion Processes: Probability Distributions and Continuous Time Random Walk. In *Processes with Long-Range Correlations. Theory and Applications*; Rangarajan, G., Ding, M., Eds.; Lecture Notes in Physics, No. 621; Springer: Berlin, Germany, 2003; pp. 148–166.
34. Balescu, R. V-Langevin equations, continuous time random walks and fractional diffusion. *Chaos Solitons Fract.* **2007**, *34*, 62–80. [CrossRef]
35. Mittal, R.C.; Nigam, R. Solution of fractional integro-differential equations by Adomian decomposition method. *Int. J. Appl. Math. Mech.* **2008**, *4*, 87–94.
36. Adomian, G. *Solving Frontier Problems of Physics: The Decomposition Method*; Kluwer Academic: Dordrecht, The Netherlands, 1994.
37. Adomian, G.; Rach, R. Inversion of nonlinear stochastic operators. *J. Math. Anal. Appl.* **1983**, *91*, 39–46. [CrossRef]
38. Adomian, G.; Rach, R. On the solution of algebraic equations by the decomposition method. *J. Math. Anal. Appl.* **1985**, *105*, 141–166. [CrossRef]
39. Duan, J.-S.; Rach, R.; Baleanu, D.; Wazwaz, A.-M. A review of the Adomian decomposition method and its applications to fractional differential equations. *Commun. Fract. Calc.* **2012**, *3*, 73–99.
40. Grinberg, F.; Farrher, E.; Ciobanu, L.; Geffroy, F.; Le Bihan, D.; Shah, N.J. Non-Gaussian diffusion imaging for enhanced contrast of brain tissue affected by ischemic stroke. *PLoS ONE* **2014**, *9*, e89225. [CrossRef] [PubMed]
41. Zax, D.; Pines, A. Study of anisotropic diffusion of oriented molecules by multiple quantum spin echoes. *J. Chem. Phys.* **1983**, *78*, 6333–6334. [CrossRef]

42. Karlicek, R.F., Jr.; Lowe, I.J. A modified pulsed gradient technique for measuring diffusion in the presence of large background gradients. *J. Magn. Reson.* **1980**, *37*, 75–91. [CrossRef]
43. Kilbas, A.A.; Saigo, M. Solutions of integral equation of Abel–Volterra type. *Differ. Integral Equ.* **1995**, *8*, 993–1011.
44. Lin, G. General PFG signal attenuation expressions for anisotropic anomalous diffusion by modified-Bloch equations. *Physica A* **2018**. [CrossRef]
45. Germano, G.; Politi, M.; Scalas, E.; Schilling, R.L. Stochastic calculus for uncoupled continuous-time random walks. *Phys. Rev. E* **2009**, *79*, 066102. [CrossRef] [PubMed]
46. Cicerone, M.T.; Wagner, P.A.; Ediger, M.D. Translational diffusion on heterogeneous lattices: A model for dynamics in glass forming materials. *J. Phys. Chem. B* **1997**, *101*, 8727–8734. [CrossRef]
47. Lin, G.; Zhang, J.; Cao, H.; Jones, A.A. A lattice model for the simulation of diffusion in heterogeneous polymer systems. Simulation of apparent diffusion constants as determined by pulse-field-gradient nuclear magnetic resonance. *J. Phys. Chem. B* **2003**, *107*, 6179–6186. [CrossRef]
48. Lin, G.; Aucoin, D.; Giotto, M.; Canfield, A.; Wen, W.; Jones, A.A. Lattice model simulation of penetrant diffusion along hexagonally packed rods in a barrier matrix as determined by pulsed-field-gradient nuclear magnetic resonance. *Macromolecules* **2007**, *40*, 1521–1528. [CrossRef]
49. Gorenflo, R.; Loutchko, J.; Luchko, Y. Computation of the Mittag-Leffler function $E_{\alpha,\beta}(z)$ and its derivative. *Fract. Calc. Appl. Anal.* **2002**, *5*, 491–518.
50. Zeng, C.; Chen, Y. Global Pade approximations of the generalized Mittag-Leffler function and its inverse. *Fract. Calc. Appl. Anal.* **2015**, *18*, 1492–1506. [CrossRef]
51. Chen, W.; Sun, H.; Zhang, X.; Korošak, D. Anomalous diffusion modeling by fractal and fractional derivatives. *Comput. Math. Appl.* **2010**, *59*, 1754–1758. [CrossRef]
52. Le Doussal, P.; Sen, P.N. Decay of nuclear magnetization by diffusion in a parabolic magnetic field: An exactly solvable model. *Phys. Rev. B* **1992**, *46*, 3465–3485. [CrossRef]

MDPI
St. Alban-Anlage 66
4052 Basel
Switzerland
Tel. +41 61 683 77 34
Fax +41 61 302 89 18
www.mdpi.com

Mathematics Editorial Office
E-mail: mathematics@mdpi.com
www.mdpi.com/journal/mathematics

www.ingramcontent.com/pod-product-compliance
Lightning Source LLC
Chambersburg PA
CBHW051848210326
41597CB00033B/5822
* 9 7 8 3 0 3 8 9 7 2 0 6 8 *